# Student Solutions Manual and Study Guide

for
Serway and Jewett's

# Principles of Physics
## A Calculus-Based Text

Fourth Edition
Volume 2

**John R. Gordon**
*James Madison University*

**Ralph McGrew**
*State University of New York*

**Raymond A. Serway**
*Emeritus, James Madison University*

**John W. Jewett, Jr.**
*California State Polytechnic University – Pomona*

Australia • Canada • Mexico • Singapore • Spain • United Kingdom • United States

Printed in the United States of America
1 2 3 4 5 6 7 09 08 07 06 05

Printer: Thomson/West

0-534-49147-2



**Thomson Higher Education**
**10 Davis Drive**
**Belmont, CA 94002-3098**
**USA**

**Asia (including India)**
Thomson Learning
5 Shenton Way
#01-01 UIC Building
Singapore 068808

**Australia/New Zealand**
Thomson Learning Australia
102 Dodds Street
Southbank, Victoria 3006
Australia

**Canada**
Thomson Nelson
1120 Birchmount Road
Toronto, Ontario M1K 5G4
Canada

**UK/Europe/Middle East/Africa**
Thomson Learning
High Holborn House
50–51 Bedford Road
London WC1R 4LR
United Kingdom

**Latin America**
Thomson Learning
Seneca, 53
Colonia Polanco
11560 Mexico
D.F. Mexico

**Spain (including Portugal)**
Thomson Paraninfo
Calle Magallanes, 25
28015 Madrid, Spain

# CONTENTS

# PREFACE

This *Student Solutions Manual and Study Guide* has been written to accompany the textbook **Principles of Physics**, Fourth Edition, Volume II, by Raymond A. Serway and John W. Jewett, Jr. The purpose of this Student Solutions Manual and Study Guide is to provide students with a convenient review of the basic concepts and applications presented in the textbook, together with solutions to selected end-of-chapter problems. This is not an attempt to rewrite the textbook in a condensed fashion. Rather, emphasis is placed upon clarifying typical troublesome points, and providing further practice in methods of problem solving.

Every textbook chapter has matching chapter in this book and each chapter is divided into several parts. Very often, reference is made to specific equations or figures in the textbook. Each feature of this Study Guide has been included to ensure that it serves as a useful supplement to the textbook. Most chapters contain the following components:

- **Notes From Selected Chapter Sections:** This is a summary of important concepts, newly defined physical quantities, and rules governing their behavior.

- **Equations and Concepts:** This is a review of the chapter, with emphasis on highlighting important concepts and describing important equations and formalisms.

- **Suggestions, Skills, and Strategies:** This offers hints and strategies for solving typical problems that the student will often encounter in the course. In some sections, suggestions are made concerning mathematical skills that are necessary in the analysis of problems.

- **Review Checklist:** This is a list of topics and techniques the student should master after reading the chapter and working the assigned problems.

- **Answers to Selected Conceptual Questions:** Suggested responses are provided for twenty percent of the Conceptual Questions.

- **Solutions to Selected End-of-Chapter Problems:** Solutions are given for about twelve of the odd-numbered problems in each textbook chapter. Problems were selected to illustrate important concepts in each chapter.

- **Tables:** A list of selected Physical Constants is printed on the inside front cover; and a table of some Conversion Factors is provided on the inside back cover.

An important note concerning significant figures: The answers to all end-of-chapter problems are stated to three significant figures, though calculations are carried out with as many digits as possible. We sincerely hope that this *Student Solutions Manual and Study Guide* will be useful to you in reviewing the material presented in the text, and in improving your ability to solve problems and score well on exams. We welcome any comments or suggestions that could help improve the content of this study guide in future editions; and we wish you success in your study.

John R. Gordon
Harrisonburg, VA

Ralph McGrew
Binghamton, NY

Raymond A. Serway
Leesburg, VA

John W. Jewett, Jr
Pomona, CA

# ACKNOWLEDGMENTS

We take this opportunity to thank everyone who contributed to this Fourth Edition of Student Solutions Manual and Study Guide to Accompany Principles of Physics.

Special thanks for managing and directing this project go to Assistant Editor for the Physical Sciences Sarah Lowe, Development Editor for the Physical Sciences Jay Campbell, Acquisitions Editor for Physics Chris Hall, Senior Developmental Editor for Physics Susan Pashos, and Publisher for the Physical Sciences David Harris.

Our appreciation goes to our reviewer, Richard Miers of Indiana Purdue Fort Wayne. His careful reading of the manuscript and checking the accuracy of the problem solutions contributed in an important way to the quality of the final product. Any errors remaining in the manual are the responsibility of the authors.

It is a pleasure to acknowledge the excellent work of the staff of M and N Toscano who prepared the final page layout and camera-ready copy. Their technical skills and attention to detail added much to the appearance and usefulness of this volume.

Finally, we express our appreciation to our families for their inspiration, patience, and encouragement.

# SUGGESTIONS FOR STUDY

We have seen a lot of successful physics students. The question, "How should I study this subject?" has no single answer, but we offer some suggestions that may be useful to you.

1. Work to understand the basic concepts and principles before attempting to solve assigned problems. Carefully read the textbook before attending your lecture on that material. Jot down points that are not clear to you, take careful notes in class, and ask questions. Reduce memorization to a minimum. Memorizing sections of a text or derivations does not necessarily mean you understand the material.

2. After reading a chapter, you should be able to define any new quantities that were introduced and discuss the first principles that were used to derive fundamental equations. A review is provided in each chapter of the Study Guide for this purpose, and the marginal notes in the textbook (or the index) will help you locate these topics. You should be able to correctly associate with each *physical quantity* the *symbol* used to represent that quantity (including vector notation, if appropriate) and the SI *unit* in which the quantity is specified. Furthermore, you should be able to express each important principle or equation in a concise and accurate prose statement. Perhaps the best test of your understanding of the material will be your ability to answer questions and solve problems in the text, or those given on exams.

3. Try to solve plenty of the problems at the end of the chapter. The worked examples in the text will serve as a basis for your study. This Study Guide contains detailed solutions to about twelve of the problems at the end of each chapter. You will be able to check the accuracy of your calculations for any odd-numbered problem, since the answers to these are given at the back of the text.

4. Besides what you might expect to learn about physics concepts, a very valuable skill you can take away from your physics course is the ability to solve complicated problems. The way physicists approach complex situations and break them down into manageable pieces is widely useful. Starting in Section 1.10, the textbook develops a general problem-solving strategy that guides you through the steps. To help you remember the steps of the strategy, they are called *Conceptualize, Categorize, Analyze,* and *Finalize.*

## GENERAL PROBLEM-SOLVING STRATEGY

### Conceptualize

- The first thing to do when approaching a problem is to *think about* and *understand* the situation. Read the problem several times until you are confident you understand what is being asked. Study carefully any diagrams, graphs, tables, or photographs that accompany the problem. Imagine a movie, running in your mind, of what happens in the problem.

- If a diagram is not provided, you should almost always make a quick drawing of the situation. Indicate any known values, perhaps in a table or directly on your sketch.

- Now focus on what algebraic or numerical information is given in the problem. In the problem statement, look for key phrases such as "starts from at rest" ($v_i = 0$), "stops" ($v_f = 0$), or "freely falls" ($a_y = -g = -9.80 \text{ m/s}^2$). Key words can help simplify the problem.

- Next focus on the expected result of solving the problem. Exactly what is the question asking? Will the final result be numerical or algebraic? If it is numerical, what units will it have? If it is algebraic, what symbols will appear in it?

- Incorporate information from your own experiences and common sense. What should a reasonable answer look like? What should its order of magnitude be? You wouldn't expect to calculate the speed of an automobile to be $5 \times 10^6$ m/s.

## Categorize

- Once you have a really good idea of what the problem is about, you need to *simplify* the problem. Remove the details that are not important to the solution. For example, you can often model a moving object as a particle. Key words should tell you whether you can ignore air resistance or friction between a sliding object and a surface.

- Once the problem is simplified, it is important to *categorize* the problem. How does it fit into a framework of ideas that you construct to understand the world? Is it a simple *plug-in problem,* such that numbers can be simply substituted into a definition? If so, the problem is likely to be finished when this substitution is done. If not, you face what we can call an *analysis problem*—the situation must be analyzed more deeply to reach a solution.

- If it is an analysis problem, it needs to be categorized further. Have you seen this type of problem before? Does it fall into the growing list of types of problems that you have solved previously? Being able to classify a problem can make it much easier to lay out a plan to solve it. For example, if your simplification shows that the problem can be treated as a particle moving under constant acceleration and you have already solved such a problem (such as the examples in Section 2.6), the solution to the new problem follows a similar pattern.

## Analyze

- Now, you need to analyze the problem and strive for a mathematical solution. Because you have already categorized the problem, it should not be too difficult to select relevant equations that apply to the type of situation in the problem. For example, if your categorization shows that the problem involves a particle moving under constant acceleration, Equations 2.9 to 2.13 are relevant.

- Use algebra (and calculus, if necessary) to solve symbolically for the unknown variable in terms of what is given. Substitute in the appropriate numbers, calculate the result, and round it to the proper number of significant figures.

## Finalize

- This final step is the most important part. Examine your numerical answer. Does it have the correct units? Does it meet your expectations from your conceptualization of the problem? What about the algebraic form of the result—before you substituted numerical values? Does it make sense? Try looking at the variables in it to see whether the answer would change in a physically meaningful way if they were drastically increased or decreased or even became zero. Looking at limiting cases to see whether they yield expected values is a very useful way to make sure that you are obtaining reasonable results.

- Think about how this problem compares with others you have done. How was it similar? In what critical ways did it differ? Why was this problem assigned? You should have learned something by doing it. Can you figure out what? Can you use your solution to expand, strengthen, or otherwise improve your framework of ideas? If it is a new category of problem, be sure you understand it so that you can use it as a model for solving future problems in the same category.

When solving complex problems, you may need to identify a series of subproblems and apply the problem-solving strategy to each. For very simple problems, you probably don't need this whole strategy. But when you are looking at a problem and you don't know what to do next, remember the steps in the strategy and use them as a guide.

Work on problems in this Study Guide yourself and compare your solutions with ours. Your solution does not have to look just like the one presented here. A problem can sometimes be solved in different ways, starting from different principles. If you wonder about the validity of an alternative approach, ask your instructor.

5. We suggest that you use this Study Guide to review the material covered in the text, and as a guide in preparing for exams. You can use the sections Chapter Review, Notes From Selected Chapter Sections, and Equations and Concepts to focus in on any points which require further study. The main purpose of this Study Guide is to improve upon the efficiency and effectiveness of your study hours and your overall understanding of physical concepts. However, it should not be regarded as a substitute for your textbook or for individual study and practice in problem solving.

# Chapter 16

# Temperature and the Kinetic Theory of Gases

## NOTES FROM SELECTED CHAPTER SECTIONS

### Section 16.1 Temperature and the Zeroth Law of Thermodynamics

Thermal physics is the study of the behavior of solids, liquids and gases, using the concepts of heat and temperature. Two approaches are commonly used:

- macroscopic approach, called thermodynamics, in which one explains the bulk thermal properties of matter.

- microscopic approach, called statistical mechanics, in which properties of matter are explained on an atomic scale.

*All thermal phenomena are manifestations of the laws of mechanics as we have learned them. For example, internal energy is actually a consequence of the random motions of a large number of particles making up the system.*

The concept of the temperature of a system can be understood in connection with a measurement, such as the reading of a thermometer. Temperature, a scalar quantity, is a property that can only be defined when the system is in thermal equilibrium with another system. Thermal equilibrium implies that two (or more) systems are at the same temperature.

The **zeroth law of thermodynamics** states that if two systems are in thermal equilibrium with a third system, they must be in thermal equilibrium with each other. The third system can be a calibrated thermometer whose reading determines whether or not the systems are in thermal equilibrium.

## Section 16.2 Thermometers and Temperature Scales

Thermometers are devices used to define and measure the temperature of a system. All thermometers make use of a change in some physical property with temperature.

Some of the physical properties used to establish a temperature scale are:

- change in volume of a liquid
- change in length of a solid
- change in pressure of a gas held at constant volume
- change in volume of a gas held at constant pressure
- change in electric resistance of a conductor
- change in color of a very hot object

The **gas thermometer** is a standard device for defining temperature. In the constant-volume gas thermometer, a low-density gas is placed in a flask, and its volume is kept constant while it is heated or cooled. The pressure is measured as the temperature of the gas is changed. Experimentally, one finds that the temperature is proportional to the absolute pressure.

The **thermodynamic temperature scale** is based on a scale for which the reference temperature is taken to be the **triple point of water**; that is, the temperature and pressure at which water, water vapor, and ice coexist in equilibrium. On this scale, the SI unit of temperature is the kelvin, defined as the fraction 1/273.16 of the temperature of the triple point of water.

## Section 16.5 The Kinetic Theory of Gases

A **microscopic model of an ideal gas** is based on the following assumptions:

- **The number of molecules is large, and the average separation between them is large** compared with their dimensions. Therefore, the molecules occupy a negligible volume compared with the volume of the container.

- **The molecules obey Newton's laws of motion, but the individual molecules move in a random fashion**. By random fashion, we mean that the molecules move in all directions with equal probability and with various speeds. This distribution of velocities does not change in time, despite the collisions between molecules.

- **The molecules undergo elastic collisions with each other**. Thus, the molecules are considered to be structureless, and for any pair of colliding molecules as a system both kinetic energy and momentum are conserved.

- **The forces between molecules are negligible except during a collision**. The forces between molecules are short-range, so that the only time the molecules interact with each other is during a collision.

- **The gas is in thermal equilibrium with the walls of the container**. The collisions of molecules with the walls are perfectly elastic.

- **The gas under consideration is a pure gas**. That is, all molecules are identical.

## EQUATIONS AND CONCEPTS

**Pressure** is force per unit area and has units of $\mathrm{N/m^2}$. The SI unit of pressure is the pascal (Pa). One atmosphere of pressure is atmospheric pressure at sea level.

$$1\ \mathrm{Pa} = 1\ \mathrm{N/m^2}$$

$$1\ \mathrm{atm} = 1.013\times10^5\ \mathrm{Pa}$$

The **Celsius temperature** $T_C$ is related to the absolute temperature $T$ (in kelvins) according to Equation 16.1, where $0°\mathrm{C}$ corresponds to 273.15 K.

$$T_C = T - 273.15 \quad (16.1)$$

The **Fahrenheit temperature** $T_F$ can be converted to degrees Celsius using Equation 16.2. Note that $0°\mathrm{C} = 32°\mathrm{F}$ and $100°\mathrm{C} = 212°\mathrm{F}$.

$$T_F = \frac{9}{5}T_C + 32°\mathrm{F} \quad (16.2)$$

$$\Delta T_F = \frac{9}{5}\Delta T_C \quad (16.3)$$

The **change in its length** (linear expansion) $\Delta L$, due to a change in temperature is proportional to the change in temperature and the original length. The proportionality constant $\alpha$ is called the **average coefficient of linear expansion**.

$$\Delta L = \alpha L_i \Delta T \quad (16.4)$$

$$\alpha = \frac{1}{L_i}\frac{\Delta L}{\Delta T}$$

The **change in its volume** is proportional to $\Delta T$ and the original volume. The constant of proportionality $\beta$ is the **average coefficient of volume expansion**.

$$\Delta V = V_f - V_i = \beta_i V \Delta T \quad (16.6)$$

$$\beta \approx 3\alpha$$

(for isotropic solid)

The **change in area** is proportional to the change in temperature and to the initial area.

$$\Delta A = \gamma A_i \Delta T \quad (16.7)$$

$$\gamma \approx 2\alpha$$

The **number of moles** in a sample of any substance equals the ratio of the mass of the sample to the molar mass characteristic of that particular substance.

$$n = \frac{m_{\text{substance}}}{M} \quad (16.8)$$

The **equation of state of an ideal gas**, relates the variables $P$, $V$, and $T$ at equilibrium and can be expressed in the form of Equation 16.9 or Equation 16.11. $R$ is the universal gas constant and $k_B$ is **Boltzmann's constant**.

$$PV = nRT \quad (16.9)$$

$n$ = number of moles

$$R = 8.314 \text{ J/mol}\cdot\text{K} = 0.0821 \text{ L}\cdot\text{atm/mol}\cdot\text{K}$$

$$k_B = \frac{R}{N_A} = 1.38\times10^{-23} \text{ J/K}$$

$$PV = N k_B T \quad (16.11)$$

$N$ = number of molecules

The **pressure of an ideal gas** is proportional to the number of molecules per unit volume and the average translational kinetic energy per molecule.

$$P = \frac{2}{3}\left(\frac{N}{V}\right)\left(\frac{1}{2}m_0\overline{v^2}\right) \quad (16.13)$$

$m_0$ = mass of a single molecule.

The **average molecular kinetic energy** of the molecules of an ideal gas and the total translational kinetic energy of the gas are proportional to the absolute temperature.

$$\tfrac{1}{2}m_0\overline{v^2} = \tfrac{3}{2}k_B T \quad (16.15)$$

$$E_{\text{total}} = N\left(\tfrac{1}{2}m_0\overline{v^2}\right) = \tfrac{3}{2}Nk_B T = \tfrac{3}{2}nRT \quad (16.17)$$

The expression for the **root-mean-square speed** of molecules shows that at a given temperature, lighter molecules move faster on the average than heavier molecules.

$$v_{\text{rms}} = \sqrt{\overline{v^2}} = \sqrt{\frac{3k_BT}{m_0}} = \sqrt{\frac{3RT}{M}} \qquad (16.19)$$

The **Maxwell-Boltzmann distribution** describes the distribution of speeds of gas molecules as a function of temperature.

$$N_v = 4\pi N \left(\frac{m_0}{2\pi k_B T}\right)^{3/2} v^2 e^{-m_0 v^2/2k_B T} \qquad (16.20)$$

The root-mean-square speed, the average speed, and the most probable speed are related as shown in Equations 16.21, 16.22, and 16.23: $v_{\text{rms}} > \bar{v} > v_{\text{mp}}$.

$$v_{\text{rms}} = 1.73\sqrt{\frac{k_BT}{m_0}} \qquad (16.21)$$

$$\bar{v} = \sqrt{\frac{8k_BT}{\pi m_0}} = 1.60\sqrt{\frac{k_BT}{m_0}} \qquad (16.22)$$

$$v_{\text{mp}} = \sqrt{\frac{2k_BT}{m_0}} = 1.41\sqrt{\frac{k_BT}{m_0}} \qquad (16.23)$$

## REVIEW CHECKLIST

✓ Understand the concepts of thermal equilibrium and thermal contact between two bodies, and state the zeroth law of thermodynamics.

✓ Discuss some physical properties of substances, which change with temperature, and the manner in which these properties are used to construct thermometers.

✓ Describe the operation of the constant-volume gas thermometer and how it is used to define the ideal gas temperature scale.

✓ Convert between the various temperature scales, especially the conversion from degrees Celsius into kelvins, degrees Fahrenheit into kelvins, and degrees Celsius into degrees Fahrenheit.

✓ Provide a qualitative description of the origin of thermal expansion of solids and liquids; define the linear expansion coefficient and volume expansion coefficient for an isotropic solid, and learn how to deal with these coefficients in practical situations involving expansion or contraction.

✓ State and understand the assumptions made in developing the molecular model of an ideal gas.

✓ Recognize that the temperature of an ideal gas is proportional to the average molecular kinetic energy. State the theorem of equipartition of energy, noting that each degree of freedom of a molecule contributes an equal amount of energy of magnitude $\frac{1}{2}k_B T$.

## ANSWERS TO SELECTED QUESTIONS

**Q16.1** A piece of copper is dropped into a beaker of water. If the water's temperature rises, what happens to the temperature of the copper? Under what conditions are the water and copper in thermal equilibrium?

**Answer** If the water's temperature increases, that means that energy is being transferred by heat to the water. This can happen either if the copper changes phase from liquid to solid, or if the temperature of the copper is above that of the water, and falling.

In this case, the copper is referred to as a "piece" of copper, so it is already in the solid phase; therefore, its temperature must be falling. When the temperature of the copper reaches that of the water, the copper and water will reach equilibrium, and the subsequent net energy transfer by heat between the two will be zero.

**Q16.11** When the metal ring and metal sphere in Figure Q16.11 are both at room temperature, the sphere can just be passed through the ring. After the sphere is heated, it cannot be passed through the ring. Explain. What if the ring is heated and the sphere is left at room temperature? Does the sphere pass through the ring?

**FIG. Q16.11**

**Answer** The hot sphere has expanded, so it no longer fits through the ring. When the ring is heated, the cool sphere fits through more easily.

Suppose a cool sphere is put through a cool ring and then heated so that it does not come back out. With the sphere still hot, you can separate the sphere and ring by heating the ring as shown in the figure. This more surprising result occurs because the thermal expansion of the ring is not like the inflation of a blood-pressure cuff. Rather, it is like a photographic enlargement; every linear dimension, including the hole diameter, increases by the same factor. The reason for this is that the atoms everywhere, including those around the inner circumference, push away from each other. The only way that the atoms can accommodate the greater distances is for the circumference—and corresponding diameter—to grow. This property was once used to fit metal rims to wooden wagon and horse-buggy wheels.

**Q16.13** When alcohol is rubbed on your body, it lowers your skin temperature. Explain this effect.

**Answer** As the alcohol evaporates, high-speed molecules leave the liquid. This reduces the average speed of the remaining molecules. Since the average speed is lowered, the temperature of the alcohol is reduced. This process helps to carry energy away from the skin of the patient, resulting in cooling of the skin. The alcohol plays the same role of evaporative cooling as does perspiration, but alcohol evaporates much more quickly than perspiration.

**Q16.15** If a helium-filled balloon initially at room temperature is placed in a freezer, will its volume increase, decrease, or remain the same?

**Answer** The helium is far from liquefaction. Therefore, we model it as an ideal gas, described by $PV = nRT$. In this case, the pressure stays nearly constant, being equal to 1 atm. Since the temperature may decrease by 10% (from 293 K to 263 K), the volume should also **decrease** by 10%. This process is called "isobaric cooling", or "isobaric contraction."

**Q16.17** What happens to a helium-filled balloon released into the air? Will it expand or contract? Will it stop rising at some height?

**Answer** Imagine the balloon rising into the air. The air cannot be uniform in pressure, because the lower layers support the weight of all the air above. Therefore, as the balloon rises, the pressure will decrease. At the same time, the temperature decreases slightly, but not enough to overcome the effects of the pressure.

By the ideal gas law, $PV = nRT$. Since the decrease in temperature has little effect, the decreasing pressure results in an increasing volume; thus the balloon expands quite dramatically. By the time the balloon reaches an altitude of 10 miles, its volume will be 90 times larger.

Long before that happens, one of two events will occur. The first possibility, of course, is that the balloon will break, and the pieces will fall to the earth. The other possibility is that the balloon will come to a spot where the average density of the balloon and its payload is equal to the average density of the air. At that point, buoyancy will cause the balloon to "float" at that altitude, until it loses helium and descends.

People who remember releasing balloons with pen-pal notes will perhaps also remember that replies most often came back when the balloons were low on helium, and were barely able to take off. Such balloons found a fairly low flotation altitude, and were the most likely to be found intact at the end of their journey.

## SOLUTIONS TO SELECTED PROBLEMS

**P16.1** A constant-volume gas thermometer is calibrated in dry ice (that is, evaporating carbon dioxide in the solid state, with a temperature of –80.0°C) and in boiling ethyl alcohol (78.0°C). The two pressures are 0.900 atm and 1.635 atm.

(a) What Celsius value of absolute zero does the calibration yield?

(b) What is the pressure at the freezing point of water?

(c) What is the pressure at the boiling point of water?

**Solution** Since we have a linear graph, the pressure is related to the temperatures as $P = A + BT$, where $A$ and $B$ are constants.

To find $A$ and $B$, we use the given data: $0.900 \text{ atm} = A + (-80.0°\text{C})B$

and $1.635 \text{ atm} = A + (78.0°\text{C})B$.

Solving these simultaneously, we find $A = 1.272 \text{ atm}$

and $B = 4.652 \times 10^{-3} \text{ atm/°C}$.

Therefore, $P = 1.272 \text{ atm} + \left(4.652 \times 10^{-3} \text{ atm/°C}\right)T$.

(a) At absolute zero, $P = 0 = 1.272 \text{ atm} + \left(4.652 \times 10^{-3} \text{ atm/°C}\right)T$

which gives $T = -273°\text{C}$. ◊

(b) At the freezing point of water, $P = 1.272 \text{ atm} + 0 = 1.27 \text{ atm}$. ◊

(c) At the boiling point of water,

$P = 1.272 \text{ atm} + \left(4.652 \times 10^{-3} \text{ atm/°C}\right)(100°\text{C}) = 1.74 \text{ atm}$. ◊

**P16.3** Liquid nitrogen has a boiling point of –195.81°C at atmospheric pressure. Express this temperature in

(a) degrees Fahrenheit, and

(b) in kelvins.

**Solution** (a) By Equation 16.2, $T_F = \frac{9}{5}T_C + 32°\text{F} = \frac{9}{5}(-195.81) + 32 = -320°\text{F}$ ◊

(b) Applying Equation 16.1, $273.15\text{ K} - 195.81\text{ K} = 77.3\text{ K}$ ◊

**Related Comment:** A convenient way to remember Equation 16.1 and 16.2 is to remember the freezing and boiling points of water, in each form:

$$T_{\text{freeze}} = 32.0°\text{F} = 0°\text{C} = 273.15\text{ K}$$
$$T_{\text{boil}} = 212°\text{F} = 100°\text{C}$$

To convert from Fahrenheit to Celsius, subtract 32 (the freezing point), and then adjust the scale by the liquid range of the water.

$$\text{Scale} = \frac{(100-0)°\text{C}}{(212-32)°\text{F}} = \frac{5°\text{C}}{9°\text{F}}$$

A kelvin is the same size change as a degree Celsius, but the kelvin scale takes its zero point at **absolute zero**, instead of the freezing point of water. Therefore, to convert from kelvin to Celsius, subtract 273.15 K.

**P16.9** The active element of a certain laser is made of a glass rod 30.0 cm long by 1.50 cm in diameter. If the temperature of the rod increases by 65.0°C, what is the increase in

(a) its length,

(b) its diameter,

(c) its volume? Assume that the average coefficient of linear expansion of the glass is $9.00 \times 10^{-6}\ °\text{C}^{-1}$.

*continued on next page*

**Solution** (a) $\Delta L = \alpha L_i \Delta T = \left(9.00\times10^{-6}\ °\text{C}^{-1}\right)(0.300\ \text{m})(65.0°\text{C}) = 1.76\times10^{-4}\ \text{m}$ ◊

(b) The diameter is a linear dimension, so the same equation applies:

$$\Delta D = \alpha D_i \Delta T = \left(9.00\times10^{-6}\ °\text{C}^{-1}\right)(0.015\,0\ \text{m})(65.0°\text{C}) = 8.78\times10^{-6}\ \text{m}$$ ◊

(c) The original volume is

$$V = \pi r^2 L = \frac{\pi}{4}(0.015\,0\ \text{m})^2(0.300\ \text{m}) = 5.30\times10^{-5}\ \text{m}^3$$

Using the volumetric coefficient of expansion, $\beta$,

$$\Delta V = \beta V_i \Delta T \approx 3\alpha V_i \Delta T$$
$$\Delta V \approx 3\left(9.00\times10^{-6}\ °\text{C}^{-1}\right)\left(5.30\times10^{-5}\ \text{m}^3\right)(65.0°\text{C}) = 93.0\times10^{-9}\ \text{m}^3$$ ◊

**Related Calculation:** The above calculation ignores $\Delta L^2$ and $\Delta L^3$ terms. Calculate the change in volume exactly, and compare your answer with the approximate solution above.

The volume will increase by a factor of $\dfrac{\Delta V}{V_i} = \left(1+\dfrac{\Delta D}{D_i}\right)^2\left(1+\dfrac{\Delta L}{L_i}\right) - 1$

$$\frac{\Delta V}{V_i} = \left(1+\frac{8.78\times10^{-6}\ \text{m}}{0.015\ \text{m}}\right)^2\left(1+\frac{1.76\times10^{-4}\ \text{m}}{0.300\ \text{m}}\right) - 1 = 1.76\times10^{-3}$$

$$\Delta V = \frac{\Delta V}{V_i}V_i$$

$$\Delta V = \left(1.76\times10^{-3}\right)\left(5.30\times10^{-5}\ \text{m}^3\right) = 93.1\times10^{-9}\ \text{m}^3 = 93.1\ \text{mm}^3$$ ◊

The answer is virtually identical; the approximation $\beta \approx 3\alpha$ is a good one.

**P16.13** A hollow aluminum cylinder 20.0 cm deep has an internal capacity of 2.000 L at 20.0°C. It is completely filled with turpentine and then slowly warmed to 80.0°C.

(a) How much turpentine overflows?

(b) If the cylinder is then cooled back to 20.0°C, how far below the cylinder's rim does the turpentine surface recede?

*continued on next page*

**Solution** When the temperature is increased from 20.0°C to 80.0°C, both the cylinder and the turpentine increase in volume by $\Delta V = \beta V_i \Delta T$ :

(a) The overflow,

$$V_{over} = \Delta V_{turp} - \Delta V_{Al}$$

$$V_{over} = (\beta V_i \Delta T)_{turp} - (\beta V_i \Delta T)_{Al} = V_i \Delta T(\beta_{turp} - 3\alpha_{Al})$$

$$V_{over} = (2.000 \text{ L})(60.0°\text{C})(9.00\times10^{-4} \text{ °C}^{-1} - 0.720\times10^{-4} \text{ °C}^{-1})$$

$$V_{over} = 0.099\,4 \text{ L}$$ ◊

(b) After warming the whole volume of the turpentine is

$$V' = 2\,000 \text{ cm}^3 + (9.00\times10^{-4} \text{ °C}^{-1})(2\,000 \text{ cm}^3)(60.0°\text{C}) = 2\,108 \text{ cm}^3$$

The fraction lost is $\frac{99.4 \text{ cm}^3}{2\,108 \text{ cm}^3} = 4.71\times10^{-2}$

This also is the fraction of the cylinder that will be empty after cooling:

$$\Delta h = (4.71\times10^{-2})(20.0 \text{ cm}) = 0.943 \text{ cm}$$ ◊

**P16.19** An auditorium has dimensions $10.0 \text{ m}\times20.0 \text{ m}\times30.0 \text{ m}$. How many molecules of air fill the auditorium at 20.0°C and a pressure of 101 kPa?

**Solution** **Conceptualize:** The given room conditions are close to Standard Temperature and Pressure (STP is 0°C and 101.3 kPa), so we can use the estimate that one mole of an ideal gas at STP occupies a volume of about 22 L. The volume of the auditorium is $6\times10^3 \text{ m}^3$ and $1 \text{ m}^3 = 1\,000 \text{ L}$, so we can estimate the number of molecules to be:

$$N \approx (6\,000 \text{ m}^3)\left(\frac{10^3 \text{ L}}{1 \text{ m}^3}\right)\left(\frac{1 \text{ mol}}{22 \text{ L}}\right)\left(\frac{6\times10^{23} \text{ molecules}}{1 \text{ mol}}\right) \approx 1.6\times10^{29} \text{ molecules of air}.$$

**Categorize:** The number of molecules can be found more precisely by applying $PV = nRT$ .

**Analyze:** The equation of state of an ideal gas is $PV = nRT$ . We need to solve for the number of moles to find *N*.

*continued on next page*

$$n = \frac{PV}{RT} = \frac{(1.01\times10^5\ \text{N/m}^2)[(10.0\ \text{m})(20.0\ \text{m})(30.0\ \text{m})]}{(8.314\ \text{J/mol}\cdot\text{K})(293\ \text{K})} = 2.49\times10^5\ \text{mol}$$

$$N = n(N_A) = (2.49\times10^5\ \text{mol})(6.022\times10^{23}\ \text{molecules/mol}) = 1.50\times10^{29}\ \text{molecules} \ \Diamond$$

**Finalize:** This result agrees quite well with our initial estimate. The numbers would match even better if the temperature of the auditorium were 0°C.

**P16.21** The mass of a hot-air balloon and its cargo (not including the air inside) is 200 kg. The air outside is at 10.0°C and 101 kPa. The volume of the balloon is $400\ \text{m}^3$. To what temperature must the air in the balloon be heated before the balloon will lift off? (Air density at 10.0°C is $1.25\ \text{kg/m}^3$.)

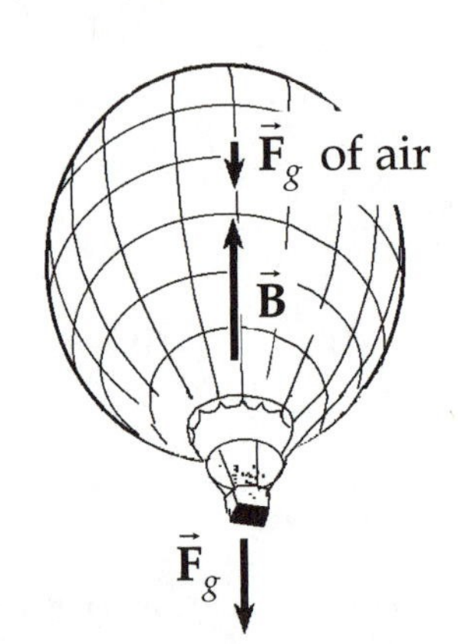


**FIG. P16.21**

**Solution** **Conceptualize:** The air inside the balloon must be significantly hotter than the outside air in order for the balloon to feel a net upward force, but the temperature must also be less than the melting point of the nylon used for the balloon's envelope. (Rip-stop nylon melts around 200°C.) Otherwise the results could be disastrous!

**Categorize:** The density of the air inside the balloon must be sufficiently low so that the buoyant force is equal to the weight of the balloon, its cargo, and the air inside. The temperature of the air required to achieve this density can be found from the equation of state of an ideal gas.

**Analyze:** The buoyant force equals the weight of the air at 10.0°C displaced by the balloon:

$B = m_{\text{air}}g = \rho_{\text{air}}Vg$: $\quad B = (1.25\ \text{kg/m}^3)(400\ \text{m}^3)(9.80\ \text{m/s}^2) = 4\,900\ \text{N}$.

The weight of the balloon and its cargo is

$$F_g = m_b g = (200\ \text{kg})(9.80\ \text{m/s}^2) = 1\,960\ \text{N}.$$

Since $B > F_g$, the balloon has a chance of lifting off as long as the weight of the air inside the balloon is less than the difference in these forces:

$$F_{\text{air}} < B - F_g = 4\,900\ \text{N} - 1\,960\ \text{N} = 2\,940\ \text{N}.$$

The mass of this air is $m_{\text{air}} = \dfrac{F_{\text{air}}}{g} = \dfrac{2\,940\ \text{N}}{9.80\ \text{m/s}^2} = 300\ \text{kg}$.

*continued on next page*

To find the required temperature of this air from $PV = nRT$, we must find the corresponding number of moles of air. Dry air is approximately 20% $O_2$, and 80% $N_2$. Using data from a periodic table, we can calculate the molar mass of the air to be approximately

$$M = (0.80)(28.0\ \text{g/mol}) + (0.20)(32.0\ \text{g/mol}) = 28.8\ \text{g/mol}$$

so the number of moles is $n = \dfrac{m}{M} = (300\ \text{kg})\left(\dfrac{1\,000\ \text{g/kg}}{28.8\ \text{g/mol}}\right) = 1.04\times10^4\ \text{mol}$.

The pressure of this air is the ambient pressure; from $PV = nRT$, we can find the minimum temperature required for lift off:

$$T = \frac{PV}{nR} = \frac{(1.01\times10^5\ \text{N/M}^2)(400\text{m}^3)}{(1.04\times10^4\text{mol})(8.314\ \text{J/mol}\cdot\text{K})} = 466\ \text{K} = 193^\circ\text{C}. \qquad \lozenge$$

**Finalize:** The average temperature of the air inside the balloon required for lift off appears to be close to the melting point of the nylon fabric, so this seems like a dangerous situation! A larger balloon would be better suited for the given weight of the balloon. (A quick check on the Internet reveals that this balloon is only about 1/10 the size of most sport balloons, which have a volume of about $3\,000\ \text{m}^3$).

If the buoyant force were less than the weight of the balloon and its cargo, the balloon would not lift off no matter how hot the air inside might be! If this were the case, then either the weight would have to be reduced or a bigger balloon would be required.

**P16.25** The pressure gauge on a tank registers the gauge pressure, which is the difference between the interior and exterior pressure. When the tank is full of oxygen $(O_2)$, it contains 12.0 kg of the gas at a gauge pressure of 40.0 atm. Determine the mass of oxygen that has been withdrawn from the tank when the pressure reading is 25.0 atm. Assume that the temperature of the tank remains constant.

**Solution** The ideal gas law states $P_1V_1 = n_1RT_1$ and $P_2V_2 = n_2RT_2$.

At constant volume and temperature, $\dfrac{P}{n} =$ constant or $\dfrac{P_1}{n_1} = \dfrac{P_2}{n_2}$

and $n_2 = \left(\dfrac{P_2}{P_1}\right)n_1$.

*continued on next page*

However, $n$ is proportional to $m$, so

$$m_2 = \left(\frac{P_2}{P_1}\right)m_1 = \left(\frac{26.0 \text{ atm}}{41.0 \text{ atm}}\right)(12.0 \text{ kg}) = 7.61 \text{ kg}.$$

The mass removed is $\Delta m = 12.0 \text{ kg} - 7.61 \text{ kg} = 4.39 \text{ kg}$. ◊

**P16.33** (a) How many atoms of helium gas fill a balloon having a diameter of 30.0 cm at 20.0°C and 1.00 atm?

(b) What is the average kinetic energy of the helium atoms?

(c) What is the root-mean-square speed of the helium atoms?

**Solution** (a) The volume is $V = \frac{4}{3}\pi r^3 = \frac{4}{3}\pi(0.150 \text{ m})^3 = 1.41 \times 10^{-2} \text{ m}^3$

Now $PV = nRT$:

$$n = \frac{PV}{RT} = \frac{\left(1.013 \times 10^5 \text{ N/m}^2\right)\left(1.41 \times 10^{-2} \text{ m}^3\right)}{(8.314 \text{ N} \cdot \text{m/mol} \cdot \text{K})(293 \text{ K})}$$

$$n = 0.588 \text{ mol}$$

$$N = nN_A$$

$$= (0.588 \text{ mol})\left(6.02 \times 10^{23} \text{ molecules/1 mol}\right)$$

$$N = 3.54 \times 10^{23} \text{ helium atoms}$$ ◊

(b) The kinetic energy is $\overline{K} = \frac{1}{2}m_0\overline{v^2} = \frac{3}{2}k_BT$

$$\overline{K} = \frac{3}{2}\left(1.38 \times 10^{-23} \text{ J/K}\right)(293 \text{ K}) = 6.07 \times 10^{-21} \text{ J}$$ ◊

(c) A He atom has mass $m_0 = \frac{M}{N_A} = \frac{4.002\,6 \text{ g/mol}}{6.02 \times 10^{23} \text{ molecules/mol}}$

$$m_0 = 6.65 \times 10^{-24} \text{ g} = 6.65 \times 10^{-27} \text{ kg}$$

So the kinetic energy is $\frac{1}{2}\left(6.65 \times 10^{-27} \text{ kg}\right)\overline{v^2} = 6.07 \times 10^{-21} \text{ J}$

and $v_{\text{rms}} = \sqrt{\overline{v^2}} = 1.35 \text{ km/s}$ ◊

**P16.35** A cylinder contains a mixture of helium and argon gas in equilibrium at 150°C.

(a) What is the average kinetic energy for each type of gas molecule?

(b) What is the root-mean-square speed of each type of molecule?

**Solution** (a) Both kinds of molecules have the same average kinetic energy.

$$\frac{1}{2}m_0\overline{v^2} = \frac{3}{2}k_BT = \frac{3}{2}\left(1.38\times10^{-23}\ \text{J/K}\right)(273+150)\ \text{K}$$

$$\overline{K} = 8.76\times10^{-21}\ \text{J}$$ ◊

(b) The root-mean-square velocity can be calculated from the kinetic energy:

$$v_{\text{rms}} = \sqrt{\overline{v^2}} = \sqrt{\frac{2\overline{K}}{m_0}}$$

These two gases are noble, and therefore monatomic. The masses of the molecules are:

$$m_{\text{He}} = \frac{(4.00\ \text{g/mol})\left(10^{-3}\ \text{kg/g}\right)}{6.02\times10^{23}\ \text{atoms/mol}} = 6.64\times10^{-27}\ \text{kg}$$

$$m_{\text{Ar}} = \frac{(39.9\ \text{g/mol})\left(10^{-3}\ \text{kg/g}\right)}{6.02\times10^{23}\ \text{atoms/mol}} = 6.63\times10^{-26}\ \text{kg}$$

Substituting these values,

$$v_{\text{rms, He}} = \sqrt{\frac{2\left(8.76\times10^{-21}\ \text{J}\right)}{6.64\times10^{-27}\ \text{kg}}} = 1.62\times10^{3}\ \text{m/s}$$

and

$$v_{\text{rms, Ar}} = \sqrt{\frac{2\left(8.76\times10^{-21}\ \text{J}\right)}{6.63\times10^{-26}\ \text{kg}}} = 514\ \text{m/s}$$ ◊

**P16.37** From the Maxwell-Boltzmann speed distribution, show that the most probable speed of a gas molecule is given by Equation 16.23. Note that the most probable speed corresponds to the point at which the slope of the speed distribution curve $\frac{dN_v}{dv}$ is zero.

*continued on next page*

**Solution** From the Maxwell speed distribution function,

$$N_v = 4\pi N\left(\frac{m_0}{2\pi k_B T}\right)^{3/2} v^2 e^{-m_0 v^2/2k_B T}$$

we locate the peak in the graph of $N_v$ versus $v$ by evaluating $\frac{dN_v}{dv}$ and setting it equal to zero, to solve for the most probable speed:

$$\frac{dN_v}{dv} = 4\pi N\left(\frac{m_0}{2\pi k_B T}\right)^{3/2}\left(\frac{-2m_0 v}{2k_B T}\right)v^2 e^{-m_0 v^2/2k_B T} + 4\pi N\left(\frac{m_0}{2\pi k_B T}\right)2ve^{-m_0 v^2/2k_B T} = 0\,.$$

This equation has solutions $v = 0$ and $v \to \infty$, but those correspond to minimum-probability speeds, so we divide by $v$ and by the exponential function.

$$v_{\text{mp}}\left(-\frac{m_0\left(2v_{\text{mp}}\right)}{2k_B T}\right) + 2 = 0 \qquad \text{or} \qquad v_{\text{mp}} = \sqrt{\frac{2k_B T}{m_0}}$$

This is Equation 16.23. ◊

**P16.45** A liquid has a density $\rho$.

(a) Show that the fractional change in density for a change in temperature $\Delta T$ is $\frac{\Delta\rho}{\rho} = -\beta\Delta T$. What does the negative sign signify?

(b) Fresh water has a maximum density of $1.000\,0 \text{ g/cm}^3$ at 4.0°C. At 10.0°C, its density is $0.999\,7 \text{ g/cm}^3$. What is $\beta$ for water over this temperature interval?

**Solution** We start with the two equations: $\rho = \frac{m}{V}$ and $\frac{\Delta V}{V} = \beta\Delta T$.

(a) Differentiating the first equation, $d\rho = -\frac{m}{V^2}dV$.

For very small changes in $V$ and $\rho$, this can be written

$$\Delta\rho = -\frac{m}{V}\frac{\Delta V}{V} = -\left(\frac{m}{V}\right)\frac{\Delta V}{V}.$$

*continued on next page*

Substituting both of our initial equations, we find that $\Delta\rho = -\rho\beta\Delta T$.

The negative sign means that if $\beta$ is positive, any increase in temperature causes the density to decrease and vice versa. ◊

(b) We apply the equation $\beta = -\dfrac{\Delta\rho}{\rho\Delta T}$ for the specific case of water:

$$\beta = -\frac{\left(1.000\,0\ \text{g/cm}^3 - 0.999\,7\ \text{g/cm}^3\right)}{\left(1.000\,0\ \text{g/cm}^3\right)(4.00°\text{C} - 10.0°\text{C})}.$$

Calculating, we find that $\beta = 5.00 \times 10^{-5}\ °\text{C}^{-1}$. ◊

**P16.47** A vertical cylinder of cross-sectional area $A$ is fitted with a tight-fitting, frictionless piston of mass $m$ (Fig. P16.47).

(a) If $n$ moles of an ideal gas are in the cylinder at a temperature of $T$, what is the height $h$ at which the piston is in equilibrium under its own weight?

(b) What is the value for $h$ if $n = 0.200$ mol, $T = 400$ K, $A = 0.008\,00\ \text{m}^2$, and $m = 20.0$ kg?

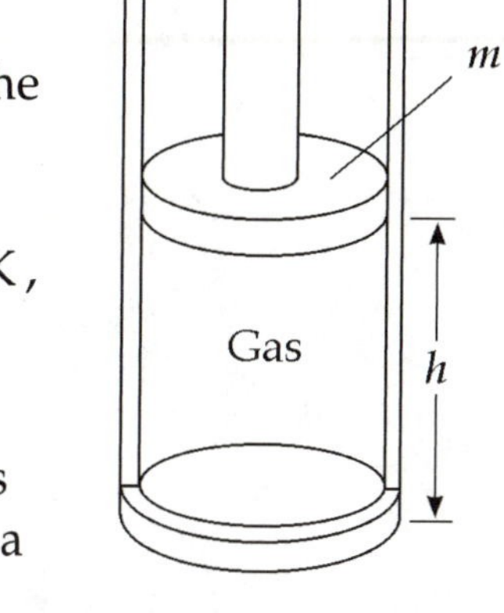


**FIG. P16.47**

**Solution** (a) We suppose that the air above the piston remains at atmospheric pressure, $P_0$. Model the piston as a particle in equilibrium.

$\sum F_y = ma_y$ yields $-P_0A - mg + PA = 0$

where $P$ is the pressure exerted by the gas contained.

Noting that $V = Ah$, and that $n$, $T$, $m$, $g$, $A$, and $P_0$ are given,

$PV = nRT$ becomes $P = \dfrac{nRT}{Ah}$

so $-P_0A - mg + \dfrac{nRT}{Ah}A = 0$

and $h = \dfrac{nRT}{P_0A + mg}$ ◊

*continued on next page*

(b) $$h = \frac{(0.200\ \text{mol})(8.314\ \text{J/mol}\cdot\text{K})(400\ \text{K})}{\left(1.013\times10^{5}\ \text{N/m}^2\right)\left(0.008\ 00\ \text{m}^2\right)+(20.0\ \text{kg})\left(9.80\ \text{m/s}^2\right)}$$

$$h = \frac{665\ \text{N}\cdot\text{m}}{810\ \text{N}+196\ \text{N}} = 0.661\ \text{m}$$ ◊

The equation derived in part (a) predicts the way the piston moves down as the weight of the piston increases. It also describes isobaric expansion, by showing that $h$ is directly proportional to the absolute temperature $T$. The equation is dimensionally correct, since its units work out as $\text{J/N} = \text{N}\cdot\text{m/N} = \text{m}$. The quantity $nRT$ on top in the expression is related (by a factor of $\frac{3}{2}$) to the translational kinetic energy content of the gas. The quantity $P_0A + mg$ on the bottom is the net downward force on the piston.

# Chapter 17

# Energy in Thermal Processes: The First Law of Thermodynamics

## NOTES FROM SELECTED CHAPTER SECTIONS

### Section 17.1 Heat and Internal Energy

When two systems at different temperatures are in contact with each other, energy will transfer between them until they reach the same temperature (that is, until they are in thermal equilibrium with each other). The term heat refers to energy transfer as a consequence of a temperature difference.

One unit of heat is the calorie (cal), defined as the amount of heat necessary to increase the temperature of 1 g of water from 14.5°C to 15.5°C. The mechanical equivalent of heat, first measured by Joule, is given by $1 \text{ cal} = 4.186 \text{ J}$.

### Section 17.2 Specific Heat

The **specific heat**, $c$, of any substance is defined as the amount of energy required to increase the temperature of 1 kg of that substance by one Celsius degree. Its units are $\text{J/kg}\cdot{}^\circ\text{C}$.

### Section 17.3 Latent Heat and Phase Changes

The **heat of fusion** is the amount of energy required to melt (or freeze) 1 kg of a specific substance, with no temperature change in the substance. The **heat of vaporization** characterizes the liquid-to-gas change in a similar manner. Both of these parameters have units of $\text{J/kg}$.

## Section 17.4 Work in Thermodynamic Processes

The work done on a gas as its volume changes from $V_i$ to $V_f$ is given by Eq. 17.7,

$$W = -\int_i^f P dV$$

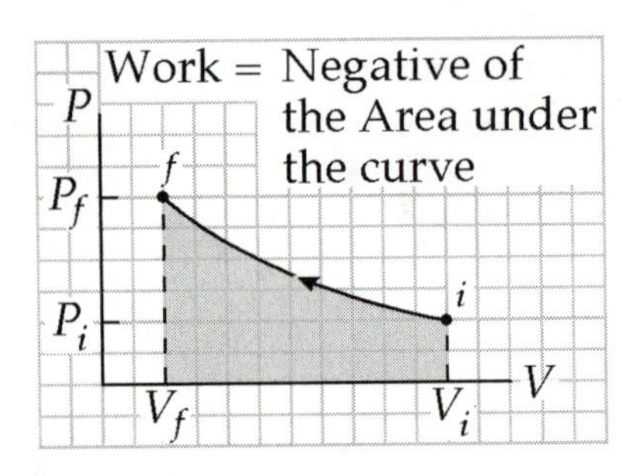


**FIG. 17.1**

A ***PV* diagram** is a graphical representation that shows the path followed by a gas as it progresses from an initial to a final state. The work done on a gas in a process that takes the gas between initial and final states is the **negative of the area** under the curve on a *PV* diagram. Figure 17.1 illustrates the case for compression of a gas.

- If a gas is compressed, $V_f < V_i$, the area is negative and the work on the gas is positive.
- If the gas expands, $V_f > V_i$, the area is positive, the work on the gas is negative, and the gas does positive work on the piston.
- If the gas expands at constant pressure, called an isobaric process, then the work done on the gas is $W = -P(V_f - V_i)$.

*The work done on a system depends on the process by which the system goes from the initial to the final state. In other words, the work done depends on the initial, final, and intermediate states of the system.*

## Section 17.5 The First Law of Thermodynamics

In the first law of thermodynamics, $\Delta E_{\text{int}} = Q + W$:

$\Delta E_{\text{int}}$ = change in internal energy of the system.
$Q$ = energy added to the system by heat.
$W$ = work done on the system.

**The initial and final states must be equilibrium states**; however, the intermediate states often are not equilibrium states since the thermodynamic coordinates may be impossible to determine during the thermodynamic process. For an infinitesimal change of the system, the first law of thermodynamics becomes $dE_{int} = dQ + dW$. We note that $dQ$ and $dW$ are not exact differentials, since both $Q$ and $W$ are not functions of the system coordinates. *Both Q and W depend on the path taken between the initial and final equilibrium states, during which time the system interacts with its environment.* On the other hand, $dE_{int}$ is an exact differential and the internal energy $E_{int}$ is a state function.

## Section 17.6 Some Applications of the First Law of Thermodynamics

An **isolated system** is one that does not interact with its surroundings. In such a system, $Q = W = 0$, so it follows from the first law that $\Delta E_{int} = 0$. That is, the internal energy of an isolated system cannot change.

A **cyclic process** is one that originates and ends up at the same state. In this situation, $\Delta E_{int} = 0$, so from the first law we see that $Q = -W$. That is, the work done per cycle equals the energy added by heat to the system per cycle. This is important to remember when dealing with heat engines in the next chapter.

An **adiabatic process** is a process in which no energy enters or leaves the system by heat; that is, $Q = 0$. The first law applied to this process gives $\Delta E_{int} = W$. A system may undergo an adiabatic process if it is thermally insulated from its surroundings, or if the process is so rapid that negligible energy has time to flow by heat.

An **isobaric process** is a process that occurs at constant pressure.

An **isovolumetric process** is one that occurs at constant volume. By definition, $W = 0$ for such a process (since $dV = 0$), so from the first law it follows that $\Delta E_{int} = Q$. That is, all of the energy added by heat to a system kept at constant volume goes into increasing the internal energy of the system.

An **isothermal process** is one that occurs at constant temperature. A plot of $P$ versus $V$ at constant temperature for an ideal gas yields a hyperbolic curve called an **isotherm**. The internal energy of an ideal gas is a function of temperature only. Hence, in an isothermal process of an ideal gas, $\Delta E_{int} = 0$.

## Section 17.10 Energy Transfer Mechanisms in Thermal Processes

There are three basic processes of thermal energy transfer: conduction, convection, and radiation.

**Conduction** is an energy transfer process that occurs in a substance when there is a temperature gradient across the substance. That is, conduction of energy occurs only when the temperature of the substance is not uniform. For example, if you position a metal rod with one end in a flame, energy will flow from the hot end to the colder end. The rate of flow of heat along the rod is proportional to the cross-sectional area of the rod, the temperature gradient, and $k$, the thermal conductivity of the material of which the rod is made.

**Convection** is a process of energy transfer by the motion of material, such as the mixing of hot and cold fluids. Convection can be due to a difference in density (natural convection) or due to movement of a substance by a fan or pump (forced convection). Convection heating is used in conventional hot-air and hot-water heating systems. Convection currents produce changes in weather conditions when warm and cold air masses mix in the atmosphere.

**Radiation** is energy transfer by the emission of electromagnetic radiation.

## EQUATIONS AND CONCEPTS

When two systems initially at different temperatures are placed in contact with each other, energy will be transferred from the system at higher temperature to the system at lower temperature until the two systems reach a common temperature (thermal equilibrium).

**Thermal equilibrium**

The **mechanical equivalent of heat** was first measured by Joule. This is the definition of the calorie as an energy unit.

$$1 \text{ cal} = 4.186 \text{ J} \tag{17.1}$$

**Specific heat** is a characteristic of a substance and is a measure of the quantity of energy required to change the temperature of 1 kg of the substance by 1°C.

$$c \equiv \frac{Q}{m\Delta T} \tag{17.2}$$

*In this and subsequent equations, Q represents the quantity of energy transferred by heat between a system and its environment.*

$$Q = mc\Delta T \tag{17.3}$$

The **energy $Q$ required to change the phase** of a mass $m$ depends on a property of the material called the **latent heat,** $L$. The phase change process occurs at constant temperature (as energy is transferred between the system and its environment). The value of $L$ depends on the nature of the phase change and the thermal properties of the substance. *Choose the positive sign when energy is transferred into the system to change solid to liquid or liquid to gas. Choose the negative sign when energy leaves the system during condensation or freezing.*

$$Q = \pm mL \qquad (17.5)$$

$L_f$ = heat of fusion (when phase change involves melting or freezing)

$L_v$ = heat of vaporization (when phase change involves boiling or condensing)

The **work done on a gas** that undergoes an expansion or compression from volume $V_i$ to volume $V_f$ depends on the path between the initial and final states. The pressure is generally not constant, so you must exercise care in evaluating $W$ using this equation. *In general, the work done is the negative of the area under the PV curve bounded by $V_i$ and $V_f$, and the pressure function P.*

$$W = -\int_{V_i}^{V_f} P\,dV \qquad (17.7)$$

The **first law of thermodynamics** is a version of the law of conservation of energy that describes changes in internal energy. It states that the change in internal energy of a system, $\Delta E_{\text{int}}$, equals the quantity $Q + W$, where $Q$ is the energy transferred by heat to the system and $W$ is the work done on it. *The quantity $Q + W$ is independent of the path taken between the initial and final states.*

$$\Delta E_{\text{int}} = Q + W \qquad (17.8)$$

$Q$ is positive when energy is transferred to a system.

$W$ is positive when work is done on a system

The **following definitions describing thermodynamic processes** will be important in applications of the laws of thermodynamics.

**Thermodynamic processes**

In an **isolated system** $Q = W = 0$, so that $\Delta E_{\text{int}} = 0$. *The internal energy of an isolated system remains constant.*

**Isolated system**

A system that does not interact with its surroundings.

In a **cyclic process,** $\Delta E_{int} = 0$ and $Q = -W$. *The net work done per cycle by the gas equals the area enclosed by the path representing the process on a PV diagram.*

**Cyclic process**
A process for which the initial and final states are the same.

At **constant volume,** $\Delta V = 0$, $W = 0$, and $\Delta E_{int} = Q$. *The net energy added by heat at constant volume goes into increasing the internal energy.*

**Isovolumetric process**
A process which occurs at constant volume.

$$\Delta E_{int} = Q \qquad (17.11)$$

$$Q = nC_V \Delta T \qquad (17.13)$$

$$C_V = \frac{3}{2}R \text{ (monatomic gas)} \qquad (17.17)$$

In an **isothermal process,** $\Delta E_{int} = 0$ and $W = -Q$. *The internal energy is constant. Any energy transferred to the gas by work is transferred by heat from the gas to its surroundings.*

**Isothermal process**
A process that occurs at constant temperature.
$W = -Q$

$$W = -Q = -nRT \ln\left(\frac{V_f}{V_i}\right) \qquad (17.12)$$

*In an **isobaric process,** the work done and the energy transferred by heat are both nonzero.*

**Isobaric process**
A process that occurs at constant pressure.

$$Q = nC_p \Delta T \qquad (17.14)$$

$$C_p = \frac{5}{2}R \text{ (monatomic gas)}$$

In an **adiabatic process,** $Q = 0$ and, therefore, the change in internal energy equals the work done on the system. *A system can undergo an adiabatic process if it is thermally insulated from its surroundings.*

**Adiabatic process**

A process in which no energy is transferred by heat.

$$\Delta E_{\text{int}} = W \quad (17.10)$$

$$PV^{\gamma} = \text{constant} \quad (17.24)$$

$$P_i V_i^{\gamma} = P_f V_f^{\gamma} \quad (17.25)$$

$$T_i V_i^{\gamma-1} = T_f V_f^{\gamma-1} \quad (17.26)$$

$$W = \frac{1}{\gamma - 1}\left(P_f V_f - P_i V_i\right) \quad (17.27)$$

The **internal energy** $E_{\text{int}}$ of $N$ molecules (or $n$ moles) of a monatomic ideal gas is proportional to the absolute temperature.

$$E_{\text{int}} = \frac{3}{2} N k_B T = \frac{3}{2} nRT \quad (17.15)$$

The **ratio of the specific heat** $\gamma$ is a dimensionless quantity that is equal to 1.67 for a monatomic ideal gas.

$$C_P - C_V = R \text{ (any ideal gas)} \quad (17.21)$$

The molar specific heat at constant pressure is greater than the molar specific heat at constant volume by an amount $R$.

$$\gamma = \frac{C_P}{C_V} \text{ (always)} \quad (17.22)$$

$$\gamma = \frac{C_P}{C_V} = 1.67 \text{ (monatomic gas)}$$

This is the basic **law of thermal conduction**. The constant $k$ is called the thermal conductivity and is characteristic of a particular material. Conduction of energy occurs only when there is a temperature gradient (or temperature difference between two points in the conducting material).

$$\mathcal{P} = kA\left(\frac{T_h - T_c}{L}\right) \quad (17.35)$$

**Stefan's law** expresses the rate of emission of heat energy by **radiation** (the power radiated). *The radiated power is proportional to the fourth power of the absolute temperature.*

$$\mathcal{P} = \sigma A e T^4 \tag{17.36}$$

$\sigma = 5.696 \times 10^{-8}\ \text{W/m}^2 \cdot \text{K}^4$

$T$ = temperature in kelvins

$e$ = emissivity (values from 0 to 1)

**Net radiated** ($T > T_0$) **or absorbed** ($T < T_0$) **power** when an object at temperature $T$ in surroundings at temperature $T_0$. *At thermal equilibrium ($T = T_0$), an object radiates and absorbs energy at the same rate and the temperature of the object remains constant.*

$$\mathcal{P}_{\text{net}} = \sigma A e\left(T^4 - T_0^{\,4}\right) \tag{17.37}$$

## SUGGESTIONS, SKILLS, AND STRATEGIES

Many applications of the first law of thermodynamics deal with the work done on a system that undergoes a change in state. As a gas is taken from a state with initial pressure $P_i$ and volume $V_i$, to a state with final pressure $P_f$ and volume $V_f$, the work can be calculated if the process can be drawn on a $PV$ diagram as in Figure 17.2. The work input during the expansion is given by the integral expression

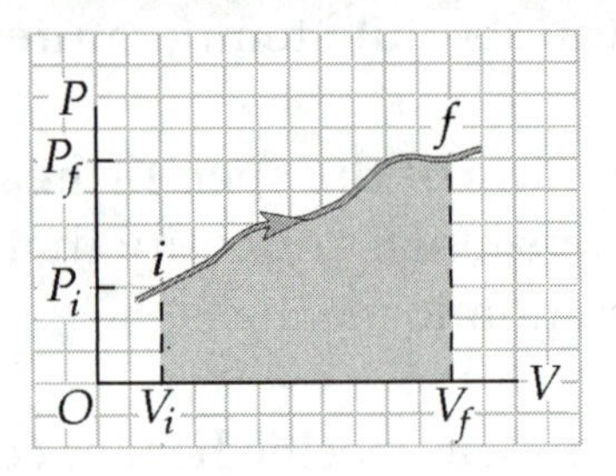


**FIG. 17.2**

$$W = -\int_{V_i}^{V_f} P\,dV$$

which numerically represents the **negative of the area** under the $PV$ curve (the shaded region) shown in Figure 17.2. It is important to recognize that the **work depends on the path taken as the gas goes from *i* to *f*.** That is, $W$ depends on the specific manner in which the pressure $P$ changes during the process.

## Problem-Solving Strategy: Calorimetry Problems

If you are having difficulty with calorimetry problems, consider the following factors:

- Be sure your units are consistent throughout. For instance, if you are using specific heats in cal/g·°C, be sure that masses are in grams and temperatures are Celsius throughout.

- Energy losses and gains are found by using $Q = mc\Delta T$ only for those intervals in which no phase changes are occurring. The equations $Q = \pm mL_f$ and $Q = \pm mL_v$ are to be used only when phase changes **are** taking place.

- Often sign errors occur in calorimetry equations. Remember the negative sign in the equation $Q_{\text{cold}} = -Q_{\text{hot}}$.

## Calculating Specific Heats

It is important to remember that for gases, two different values of molar heat capacity can be calculated: $C_P$, molar heat capacity at constant pressure and $C_V$, molar heat capacity at constant volume. For any ideal gas $C_P - C_V = R$, the universal gas constant and the ratio of $\frac{C_P}{C_V} = \gamma$ has a value that depends on the nature of the gas (see Table 17.3 of the text). The two equations above can be combined to find the following expressions for $C_V$ and $C_P$:

$$C_V = \frac{R}{\gamma - 1} \quad \text{and} \quad C_P = \frac{\gamma R}{\gamma - 1}$$

These may be used for ideal gases whether or not they are monatomic.

## REVIEW CHECKLIST

✓ Understand the concepts of heat, internal energy, and thermodynamic processes.

✓ Define and discuss the calorie, specific heat, and latent heat.

✓ Understand how work is defined when a system undergoes a change in state, and the fact that work (like heat) depends on the path taken by the system. You should also know how to sketch processes on a *PV* diagram, and calculate work using these diagrams.

✓ State the first law of thermodynamics ($\Delta E_{int} = Q + W$), and explain the meaning of the three energy terms related by this statement.

✓ Discuss the implications of the first law of thermodynamics as applied to (a) an isolated system, (b) a cyclic process, (c) an adiabatic process, and (d) an isothermal process.

✓ Recognize that the internal energy of an ideal gas is proportional to the absolute temperature, and be able to derive the specific heat of an ideal gas at constant volume from the first law of thermodynamics.

✓ Define an adiabatic process, and be able to derive the expression $PV^{\gamma} = \text{constant}$, which applies to an adiabatic process for an ideal gas.

## ANSWERS TO SELECTED QUESTIONS

**Q17.5** What is wrong with this statement, "Given any two objects, the one with the higher temperature contains more heat?"

**Answer** The statement shows a misunderstanding of the concept of heat. Heat is a process by which energy is transferred, not a form of energy that is held or contained. If you wish to speak of energy that is "contained", you speak of **internal energy**, not **heat**.

Further, even if the statement used the term "internal energy", it would still be incorrect, since the effects of specific heat and mass are both ignored. A 1-kg mass of water at 20°C has more internal energy than a 1-kg mass of air at 30°C. Similarly, the earth has far more internal energy than a drop of molten titanium metal.

Correct statements would be: (1) "given any two bodies in thermal contact, the one with the higher temperature will transfer energy to the other by heat", and (2) "given any two bodies of equal mass, the one with the higher product of absolute temperature and specific heat contains more internal energy".

**Q17.7** The air temperature above coastal areas is profoundly influenced by the large specific heat of water. One reason is that the energy released when $1\ \text{m}^3$ of water cools by 1°C will raise the temperature of a much larger volume of air by 1°C. Find this volume of air. The specific heat of air is approximately $1\ \text{kJ/kg}\cdot°\text{C}$. Take the density of air to be $1.3\ \text{kg/m}^3$.

*continued on next page*

**Answer** The mass of one cubic meter of water is specified by its density,

$$m = \rho V = \left(1.00\times10^3 \text{ kg/m}^3\right)\left(1 \text{ m}^3\right) = 1\times10^3 \text{ kg}.$$

When one cubic meter of water cools by 1°C it releases energy

$$Q_c = mc\Delta T = \left(1\times10^3 \text{ kg}\right)\left(4\,186 \text{ J/kg}\cdot{}^\circ\text{C}\right)\left(-1^\circ\text{C}\right) = -4\times10^6 \text{ J}$$

where the negative sign represents the heat output. When $+4\times10^6$ J is transferred to the air, raising its temperature 1°C, the volume of the air is given by $Q_c = mc\Delta T = \rho V c\Delta T$:

$$V = \frac{Q_c}{\rho c\Delta T} = \frac{4\times10^6 \text{ J}}{\left(1.3 \text{ kg/m}^3\right)\left(1\times10^3 \text{ J/kg}\cdot{}^\circ\text{C}\right)\left(1^\circ\text{C}\right)} = 3\times10^3 \text{ m}^3.$$

The volume of the air is thousands of times larger than the volume of the water.

**Q17.9** Using the first law of thermodynamics, explain why the **total** energy of an isolated system is always constant.

**Answer** The first law of thermodynamics says that the net change in internal energy of a system is equal to the energy added by heat, plus the work done on the system.

$$\Delta E_{\text{int}} = Q + W$$

However, an isolated system is defined as an object or set of objects for which there is no exchange of energy with its surroundings. In the case of the first law of thermodynamics, this means that $Q = W = 0$, so the change in internal energy of the system at all times must be zero.

As it stays constant in amount, an isolated system's energy may change from one form to another or move from one object to another within the system. For example, a "bomb calorimeter" is a closed system that consists of a sturdy insulated steel container, a water bath, an item of food, and oxygen. The food is burned in the presence of an excess of oxygen, and chemical energy is converted to internal energy. In the process of oxidation, energy is transferred by heat to the water bath, raising its temperature. The change in temperature of the water bath is used to determine the caloric content of the food. The process works specifically **because** the total energy remains unchanged in a closed system.

## SOLUTIONS TO SELECTED PROBLEMS

**P17.5** A 1.50-kg iron horseshoe initially at 600°C is dropped into a bucket containing 20.0 kg of water at 25.0°C. What is the final temperature? (Ignore the heat capacity of the container and assume that a negligible amount of water boils away.)

**Solution** **Conceptualize:** Even though the horseshoe is much hotter than the water, the mass of the water is significantly greater, so we might expect the water temperature to rise less than 10°C.

**Categorize:** The heat lost by the iron will be gained by the water, and from this energy transfer, the change in water temperature can be found.

**Analyze:** $Q_{\text{iron}} = -Q_{\text{water}}$ or $(mc\Delta T)_{\text{iron}} = -(mc\Delta T)_{\text{water}}$

$$m_{\text{Fe}}c_{\text{Fe}}(T - 600°\text{C}) = -m_w c_w (T - 25.0°\text{C})$$

$$T = \frac{m_w c_w (25.0°\text{C}) + m_{\text{Fe}}c_{\text{Fe}}(600°\text{C})}{m_{\text{Fe}}c_{\text{Fe}} + m_w c_w}$$

$$T = \frac{(20.0\text{ kg})(4\,186\text{ J/kg}\cdot°\text{C})(25.0°\text{C}) + (1.50\text{ kg})(448\text{ J/kg}\cdot°\text{C})(600°\text{C})}{(1.50\text{ kg})(448\text{ J/kg}\cdot°\text{C}) + (20.0\text{ kg})(4\,186\text{ J/kg}\cdot°\text{C})} = 29.6°\text{C} \quad \Diamond$$

**Finalize:** The temperature only rose about 5°C, so our answer seems reasonable. The specific heat of the water is about 10 times greater than the iron, so this effect also reduces the change in water temperature. In this problem, we assumed that a negligible amount of water boiled away, but in reality, the final temperature of the water would be somewhat less than we calculated, since some of the heat energy would be used to vaporize a bit of water.

**P17.13** A 3.00-kg lead bullet at 30.0°C is fired at a speed of 240 m/s into a large block of ice at 0°C, in which it becomes embedded. What quantity of ice melts?

**Solution** **Conceptualize:** The amount of ice that melts is probably small, maybe only a few grams based on the size, speed, and initial temperature of the bullet.

**Categorize:** We will assume that all of the initial kinetic energy and excess internal energy of the bullet goes into internal energy to melt the ice, the mass of which can be found from the latent heat of fusion. Because the ice does not all melt, in the final state everything is at 0°C.

*continued on next page*

**Analyze:** At thermal equilibrium, the energy lost by the bullet equals the energy gained by the ice: $|\Delta K_b| + |Q_b| = Q_{\text{ice}}$ gives

$$\frac{1}{2}m_b v_b^2 + m_b c_{\text{Pb}} = m_{\text{ice}} L_f \qquad \text{or} \qquad m_{\text{ice}} = m_b \left( \frac{\left(\frac{1}{2}\right) v_b^2 + c_{\text{Pb}} |\Delta T|}{L_f} \right)$$

$$m_{\text{ice}} = \left(3.00 \times 10^{-3}\ \text{kg}\right) \left( \frac{\left(\frac{1}{2}\right)(240\ \text{m/s})^2 + (128\ \text{J/kg}\cdot{}^\circ\text{C})(30.0^\circ\text{C})}{3.33 \times 10^5\ \text{J/kg}} \right)$$

$$m_{\text{ice}} = \frac{86.4\ \text{J} + 11.5\ \text{J}}{3.33 \times 10^5\ \text{J/kg}} = 2.94 \times 10^{-4}\ \text{kg} = 0.294\ \text{g} \qquad \Diamond$$

**Finalize:** The amount of ice that melted is less than a gram, which agrees with our prediction. It appears that most of the energy used to melt the ice comes from the kinetic energy of the bullet (88%), while the excess internal energy of the bullet only contributes 12% to melt the ice. Small chips of ice probably fly off when the bullet makes impact. Some of the energy is transferred to their kinetic energy, so in reality, the amount of ice that would melt should be less than what we calculated. If the block of ice were colder than 0°C (as is usually the case), then the melted ice would refreeze.

**P17.15** In an insulated vessel, 250 g of ice at 0°C is added to 600 g of water at 18.0°C.

(a) What is the final temperature of the system?

(b) How much ice remains when the system reaches equilibrium?

**Solution** (a) When 250 g of ice is melted,

$$Q_f = mL_f = (0.250\ \text{kg})\left(3.33 \times 10^5\ \text{J/kg}\right) = 83.3\ \text{kJ}.$$

The energy released when 600 g of water cools from 18.0°C to 0°C is

$$|Q| = |mc\Delta T| = (0.600\ \text{kg})(4\,186\ \text{J/kg}\cdot{}^\circ\text{C})(18.0^\circ\text{C}) = 45.2\ \text{kJ}.$$

Since the energy required to melt 250 g of ice at 0°C **exceeds** the energy released by cooling 600 g of water from 18.0°C to 0°C, the final temperature of the system (water + ice) must be 0°C. ◊

*continued on next page*

(b) The energy released by the water (45.2 kJ) will melt a mass of ice $m$, where $Q = mL_f$.

Solving for the mass, $m = \frac{Q}{L_f} = \frac{45.2 \times 10^3 \text{ J}}{3.33 \times 10^5 \text{ J/kg}} = 0.136 \text{ kg}$.

Therefore, the ice remaining is $m' = 0.250 \text{ kg} - 0.136 \text{ kg} = 0.114 \text{ kg}$. ◊

**P17.19** A sample of ideal gas is expanded to twice its original volume of $1.00 \text{ m}^3$ in a quasi-static process for which $P = \alpha V^2$, with $\alpha = 5.00 \text{ atm/m}^6$, as shown in Figure P17.19. How much work is done on the expanding gas?

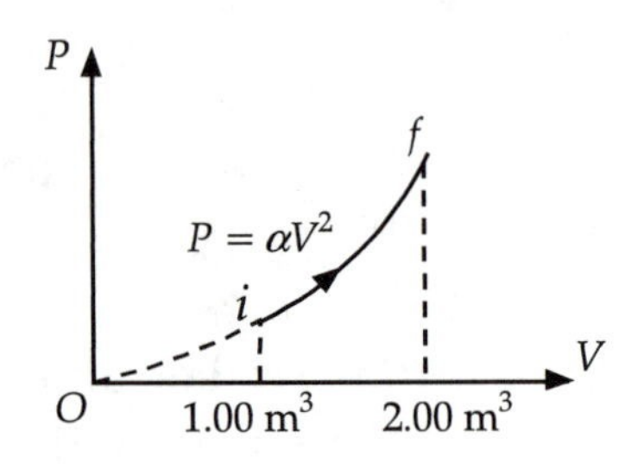


**FIG. P17.19**

**Solution** $W = -\int_{V_i}^{V_f} P dV$ and $V_f = 2V_i = 2(1.00 \text{ m}^3) = 2.00 \text{ m}^3$.

The work done on the gas is the negative of the area under the curve $P = \alpha V^2$, from $V_i$ to $V_f$.

$$W = -\int_{V_i}^{V_f} \alpha V^2 dV = -\frac{1}{3}\alpha\left(V_f^3 - V_i^3\right)$$

$$W = -\frac{1}{3}(5.00 \text{ atm/m}^6)(1.013 \times 10^5 \text{ Pa/atm})\left[(2.00 \text{ m}^3)^3 - (1.00 \text{ m}^3)^3\right]$$

$$W = -1.18 \times 10^6 \text{ J}$$ ◊

**P17.23** A thermodynamic system undergoes a process in which its internal energy decreases by 500 J. At the same time, 220 J of work is done on the system. Find the energy transferred to or from it by heat.

**Solution** $\Delta E_{\text{int}} = Q + W$, where $W$ is positive, because work is done **on** the system:

$$Q = \Delta E_{\text{int}} - W = -500 \text{ J} - 220 \text{ J} = -720 \text{ J}$$

+720 J of energy is transferred **from** the system by heat. ◊

**P17.27** An ideal gas initially at 300 K undergoes an isobaric expansion at 2.50 kPa. If the volume increases from 1.00 $m^3$ to 3.00 $m^3$ and 12.5 kJ is transferred to the gas by heat, what are

(a) the change in its internal energy and

(b) its final temperature?

**Solution** We use the energy version of the nonisolated system model.

(a) $\Delta E_{\text{int}} = Q + W$ where $W = -P\Delta V$

so that $\Delta E_{\text{int}} = Q - P\Delta V$

$$\Delta E_{\text{int}} = 1.25\times10^4 \text{ J} - \left(2.50\times10^3 \text{ N/m}^2\right)\left(3.00 \text{ m}^3 - 1.00 \text{ m}^3\right) = 7\,500 \text{ J}$$ ◊

(b) Since $\dfrac{V_1}{T_1} = \dfrac{V_2}{T_2}$, $T_2 = \left(\dfrac{V_2}{V_1}\right)T_1 = \left(\dfrac{3.00 \text{ m}^3}{1.00 \text{ m}^3}\right)(300 \text{ K}) = 900 \text{ K}$ ◊

**P17.31** A 2.00-mol sample of helium gas initially at 300 K and 0.400 atm is compressed isothermally to 1.20 atm. Noting that the helium behaves as an ideal gas, find

(a) the final volume of the gas,

(b) the work done on the gas, and

(c) the energy transferred by heat.

**Solution** (a) Rearranging $PV = nRT$,

we get $V_i = \dfrac{nRT}{P_i}$

$$V_i = \frac{(2.00 \text{ mol})(8.314 \text{ J/mol}\cdot\text{K})(300 \text{ K})}{(0.400 \text{ atm})\left(1.013\times10^5 \text{ Pa/atm}\right)}\left(\frac{1 \text{ Pa}}{\text{N/m}^2}\right) = 0.123 \text{ m}^3.$$

For isothermal compression, *PV* is constant, so $P_iV_i = P_fV_f$:

$$V_f = V_i\left(\frac{P_i}{P_f}\right) = \left(0.123 \text{ m}^3\right)\left(\frac{0.400 \text{ atm}}{1.20 \text{ atm}}\right) = 0.041\,0 \text{ m}^3$$ ◊

*continued on next page*

(b) $W = -\int P dV : W = -\int \frac{nRT}{V} dV = -nRT \ln\left(\frac{V_f}{V_i}\right) = -(4\,988 \text{ J}) \ln\left(\frac{1}{3}\right) = +5.48 \text{ kJ}$ ◊

(c) $\Delta E_{\text{int}} = 0 = Q + W : Q = -5.48 \text{ kJ}$ ◊

**P17.33** A 1.00-mol sample of hydrogen gas is heated at constant pressure from 300 K to 420 K. Calculate

(a) the energy transferred to the gas by heat,

(b) the increase in its internal energy, and

(c) the work done on the gas.

**Solution** Since this is a constant-pressure process, $Q = nC_P\Delta T$.

(a) The temperature rises by $\Delta T = 420 \text{ K} - 300 \text{ K} = 120 \text{ K}$

$$Q = (1.00 \text{ mol})(28.8 \text{ J/mol}\cdot\text{K})(120 \text{ K}) = 3.46 \text{ kJ}$$ ◊

(b) For any gas $\Delta E_{\text{int}} = nC_V\Delta T$

so $\Delta E_{\text{int}} = (1.00 \text{ mol})(20.4 \text{ J/mol}\cdot\text{K})(120 \text{ K}) = 2.45 \text{ kJ}$

(c) $\Delta E_{\text{int}} = Q + W$ so $W = \Delta E_{\text{int}} - Q = 2.45 \text{ kJ} - 3.46 \text{ kJ} = -1.01 \text{ kJ}$ ◊

**P17.41** A 2.00-mol sample of a diatomic ideal gas expands slowly and adiabatically from a pressure of 5.00 atm and a volume of 12.0 L to a final volume of 30.0 L.

(a) What is the final pressure of the gas?

(b) What are the initial and final temperatures?

(c) Find Q, W, and $\Delta E_{\text{int}}$.

*continued on next page*

**Solution** (a) $P_iV_i^\gamma = P_fV_f^\gamma: \; P_f = P_i\left(\frac{V_i}{V_f}\right)^\gamma = (5.00 \text{ atm})\left(\frac{12.0 \text{ L}}{30.0 \text{ L}}\right)^{1.40} = 1.39 \text{ atm}$ ◊

(b) $T_i = \frac{P_iV_i}{nR} = \frac{(5.00 \text{ atm})(1.013\times10^5 \text{ Pa/atm})(12.0\times10^{-3} \text{ m}^3)}{(2.00 \text{ mol})(8.314 \text{ N}\cdot\text{m/mol}\cdot\text{K})} = 366 \text{ K}$ ◊

$T_f = \frac{P_fV_f}{nR} = 253 \text{ K}$ ◊

(c) This is an adiabatic process, so by the definition $Q = 0$. ◊

For any process, $\Delta E_{\text{int}} = nC_V\Delta T$.

And for this gas, $C_V = \frac{R}{\gamma - 1} = \frac{5}{2}R$.

Thus, $\Delta E_{\text{int}} = \frac{5}{2}(2.00 \text{ mol})(8.314 \text{ J/mol}\cdot\text{K})(253 \text{ K} - 366 \text{ K}) = -4\,660 \text{ J}$. ◊

Now, $W = \Delta E_{\text{int}} - Q = -4\,660 \text{ J} - 0 = -4\,660 \text{ J}$. ◊

Note that in this case the work done on the gas is negative, so positive work is done by the gas.

**P17.47** The *heat capacity* of a sample of a substance is the product of the mass of the sample and the specific heat of the substance. Consider 2.00 mol of an ideal diatomic gas.

(a) Find the total heat capacity of the gas at constant volume and at constant pressure, assuming that the molecules rotate but do not vibrate.

(b) Repeat the problem, assuming that the molecules both rotate and vibrate.

*continued on next page*

**Solution** We use the kinetic theory structural model of an ideal gas.

(a) Count the degrees of freedom. A diatomic molecule oriented along the $y$ axis can possess energy by moving in the $x$, $y$, and $z$ directions and by rotating around the $x$ and $z$ axes. Rotation around the $y$ axis does not represent an energy contribution because the moment of inertia of the molecule about this axis is essentially zero. The molecule will have an average energy $\frac{1}{2}k_BT$ for each of these five degrees of freedom.

The gas will have internal energy

$$E_{int} = N\left(\frac{5}{2}\right)k_BT = nN_A\left(\frac{5}{2}\right)\left(\frac{R}{N_A}\right)T = \frac{5}{2}nRT$$

so the constant-volume heat capacity of the whole sample is

$$\frac{\Delta E_{int}}{\Delta T} = \frac{5}{2}nR = \frac{5}{2}(2.00 \text{ mol})(8.314 \text{ J/mol}\cdot\text{K}) = 41.6 \text{ J/K}. \quad \Diamond$$

For one mole, $C_P = C_V + R$. For the sample, $nC_P = nC_V + nR$.

With $P$ constant, $nC_P = 41.6 \text{ J/K} + (2.00 \text{ mol})(8.314 \text{ J/mol}\cdot\text{K}) = 58.2 \text{ J/K}$. ◊

(b) Vibration adds a degree of freedom for kinetic energy and a degree of freedom for elastic energy. Now the molecule's average energy is $\frac{7}{2}k_BT$,

the sample's internal energy is $N\left(\frac{7}{2}\right)k_BT = \frac{7}{2}nRT$

and the sample's constant-volume heat capacity is

$$\frac{7}{2}nR = \frac{7}{2}(2.00 \text{ mol})(8.314 \text{ J/mol}\cdot\text{K}) = 58.2 \text{ J/K}. \quad \Diamond$$

At constant pressure, its heat capacity is

$$nC_V + nR = 58.2 \text{ J/K} + (2.00 \text{ mol})(8.314 \text{ J/mol}\cdot\text{K}) = 74.8 \text{ J/K}. \quad \Diamond$$

**P17.49** A bar of gold is in thermal contact with a bar of silver of the same length and area (Fig. P17.49). One end of the compound bar is maintained at 80.0°C and the opposite end is at 30.0°C. When the energy transfer reaches steady state, what is the temperature at the junction?

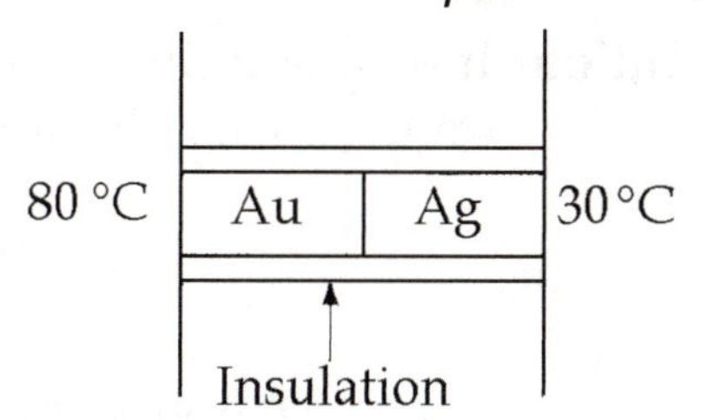


**FIG. P17.49**

**Solution** Call the gold bar Object 1 and the silver bar Object 2. Each is a nonisolated system in steady state. When energy transfer by heat reaches a steady state, the flow rate through each will be the same:

$$\mathscr{P}_1 = \mathscr{P}_2 \quad \text{or} \quad \frac{k_1 A_1 \Delta T}{L_1} = \frac{k_2 A_2 \Delta T}{L_2}.$$

In this case, $L_1 = L_2$ and $A_1 = A_2$

so $k_1 \Delta T_1 = k_2 \Delta T_2$.

Let $T_3$ be the temperature at the junction; then $k_1(80.0°\text{C} - T_3) = k_2(T_3 - 30.0°\text{C})$.

Rearranging, we find

$$T_3 = \frac{(80.0°\text{C})k_1 + (30.0°\text{C})k_2}{k_1 + k_2}$$

$$T_3 = \frac{(80.0°\text{C})(314\ \text{W/m}\cdot°\text{C}) + (30.0°\text{C})(427\ \text{W/m}\cdot°\text{C})}{(314\ \text{W/m}\cdot°\text{C}) + (427\ \text{W/m}\cdot°\text{C})} = 51.2°\text{C}. \quad \Diamond$$

**P17.65** A solar cooker consists of a curved reflecting surface that concentrates sunlight onto the object to be warmed (Fig. P17.65). The solar power per unit area reaching the Earth's surface at the location is $600\ \text{W/m}^2$. The cooker faces the Sun and has a diameter of 0.600 m. Assume that 40.0% of the incident energy is transferred to 0.500 L of water in an open container, initially at 20.0°C. How long does it take to completely boil away the water? (Ignore the heat capacity of the container.)

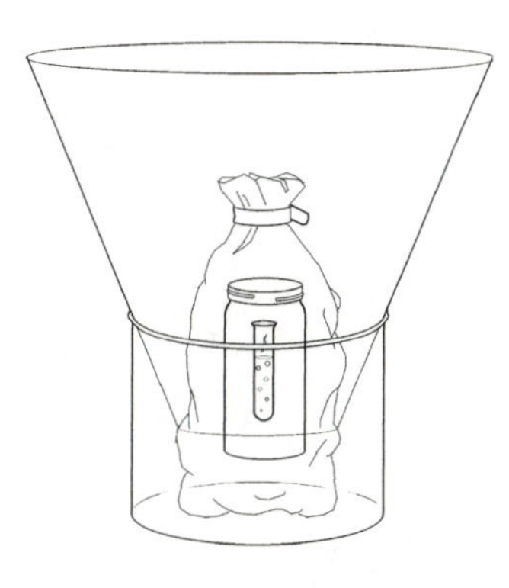

**FIG. P17.65**

*continued on next page*

**Solution** If we point the axis of the reflecting surface toward the Sun, the power incident on the solar collector is

$$\mathcal{P}_i = IA = \left(600\ \mathrm{W/m^2}\right)\left[\pi(0.300\ \mathrm{m})^2\right] = 170\ \mathrm{W}.$$

For a 40.0%-efficient reflector, the collected power is

$$\mathcal{P}_i = (0.400)(170\ \mathrm{W}) = 67.9\ \mathrm{W} = 67.9\ \mathrm{J/s}.$$

The total energy required to increase the temperature of the water to the boiling point and to evaporate it is

$$Q = mc\Delta T + mL_v$$

$$Q = (0.500\ \mathrm{kg})(4\,186\ \mathrm{J/kg{\cdot}^\circ C})(80.0^\circ\mathrm{C}) + (0.500\ \mathrm{kg})\left(2.26\times10^6\ \mathrm{J/kg}\right)$$

$$Q = 1.30\times10^6\ \mathrm{J}.$$

The time required is

$$\Delta t = \frac{Q}{\mathcal{P}_c} = \frac{1.30\times10^6\ \mathrm{J}}{67.9\ \mathrm{J/s}} = 1.91\times10^4\ \mathrm{s} = 5.31\ \mathrm{h}.$$ ◊

Chapter 18

# Heat Engines, Entropy, and the Second Law of Thermodynamics

## NOTES FROM SELECTED CHAPTER SECTIONS

### Section 18.1 Heat Engines and the Second Law of Thermodynamics

A heat engine is a device that takes in energy as heat and puts out energy in other useful forms, such as mechanical or electrical energy. A practical heat engine carries some working substance through a cyclic process during which (1) energy is absorbed by heat from a source at a high temperature, (2) work is done by the engine, and (3) energy is expelled by heat from the engine to a reservoir at a lower temperature.

The engine absorbs a quantity of energy, $|Q_h|$, from a hot reservoir, does work $|W| = W_{\text{eng}}$, and then gives up energy $|Q_c|$ to a cold reservoir. Because the working substance goes through a cycle, its initial and final internal energies are equal, so $\Delta E_{\text{int}} = 0$. From the first law we see that the net work done by a heat engine equals the net energy absorbed by the engine: $W_{\text{eng}} = |Q_h| - |Q_c|$.

*If the working substance is a gas, the net work done for a cyclic process is the area enclosed by the curve representing the process on a PV diagram.*

The **thermal efficiency**, *e*, of a heat engine is the ratio of the net work done by the engine to the energy absorbed by heat at the higher temperature during one cycle.

The **second law of thermodynamics** can be stated in several ways:

- **Clausius Statement:** Energy will not flow spontaneously by heat from a cold object to a hot object. *No thermodynamic process can occur whose only result is to transfer energy from a colder to a hotter body by heat.* Such a process (refrigeration) is only possible if work is done on the system.

- **Kelvin-Planck Statement:** It is impossible for a cyclic thermodynamic process to occur whose only result is the complete conversion of energy extracted by heat from a hot reservoir into work. *That is, it is impossible to construct a heat engine that, operating in a cycle, produces no other effect than the absorption of energy by heat from a reservoir and the performance of an equal amount of work.* A heat engine must also release energy by heat into a cold reservoir. The efficiency of a heat engine must be less than 100%. The engine's energy output by heat is often called heat exhaust, wasted heat, rejected heat, expelled heat, or thermal pollution.

- **Entropy statement:** The second law of thermodynamics can be stated in terms of entropy as follows: *The total entropy of an isolated system always increases in time if the system undergoes an irreversible process.* If an isolated system undergoes a reversible process, the total entropy remains constant.

## Section 18.2 Reversible and Irreversible Processes

A process is **reversible** if the system passes from the initial to the final state through a succession of equilibrium states. Then, the process can be made to run in the opposite direction at any point by an infinitesimal change in conditions. Otherwise the process is irreversible.

## Section 18.3 The Carnot Engine

An ideal reversible cyclic process, called the **Carnot cycle**, is described in the *PV* diagram of Figure 18.1. The Carnot cycle consists of two adiabatic and two isothermal processes, all being reversible.

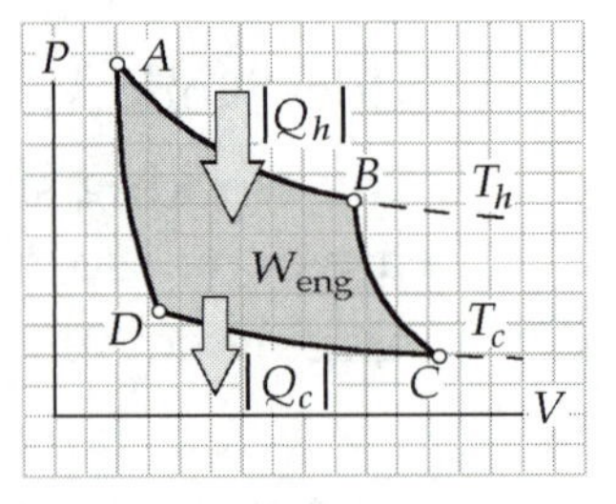


**FIG. 18.1**

- The process $A \rightarrow B$ is an isotherm (constant *T*), during which time the gas expands at constant temperature $T_h$, and absorbs energy $|Q_h|$ by heat from the hot reservoir.

- The process $B \rightarrow C$ is an adiabatic expansion ($Q = 0$), during which time the gas expands and cools to a temperature $T_c$.

- The process $C \rightarrow D$ is a second isotherm, during which time the gas is compressed at constant temperature $T_c$, and gives up energy $|Q_c|$ by heat to the cold reservoir.

- The final process $D \rightarrow A$ is an adiabatic compression in which the gas temperature increases to a final temperature of $T_h$.

No working engine is 100% efficient, even when losses such as friction are neglected. One can determine the theoretical limit on the efficiency of a real engine by comparison with the ideal Carnot engine. A reversible engine is one that will operate with the same efficiency in the forward and reverse directions. The Carnot engine is one example of a reversible engine.

**Carnot's theorems**, which can be proved from the first and second laws of thermodynamics, can be stated as follows:

- **Theorem I.** No real (irreversible) engine can have an efficiency greater than that of a reversible engine operating between the same two temperatures.

- **Theorem II.** All reversible engines operating between $T_h$ and $T_c$ have the same efficiency, given by Equation 18.4.

A schematic diagram of a heat engine is shown in Figure 18.2a, where $|Q_h|$ is the energy extracted from the hot reservoir by heat at temperature $T_h$, $|Q_c|$ is the energy rejected to the cold reservoir by heat at temperature $T_c$, and $W_{\text{eng}}$ is the work done by the engine.

## Section 18.4 Heat Pumps and Refrigerators

A refrigerator is a heat engine operating in reverse, as shown in Figure 18.2b. During one cycle of operation, the refrigerator absorbs energy $|Q_c|$ by heat from the cold reservoir, expels energy $|Q_h|$ by heat to the hot reservoir, and the work done on the system is $W = |Q_h| - |Q_c|$.

Hot Reservoir at $T_h$

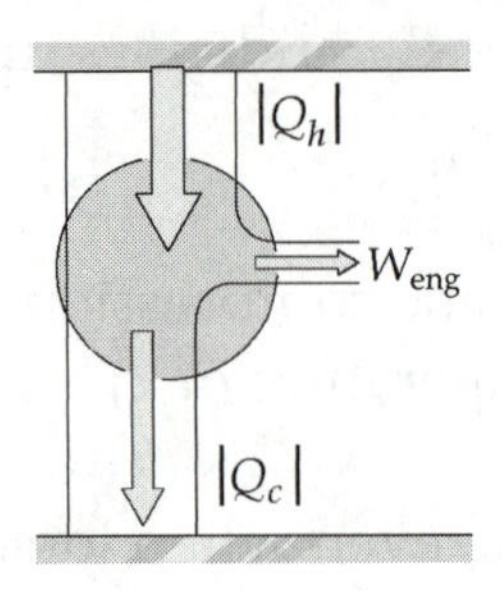


Cold Reservoir at $T_c$

(a) Heat engine

Hot Reservoir at $T_h$

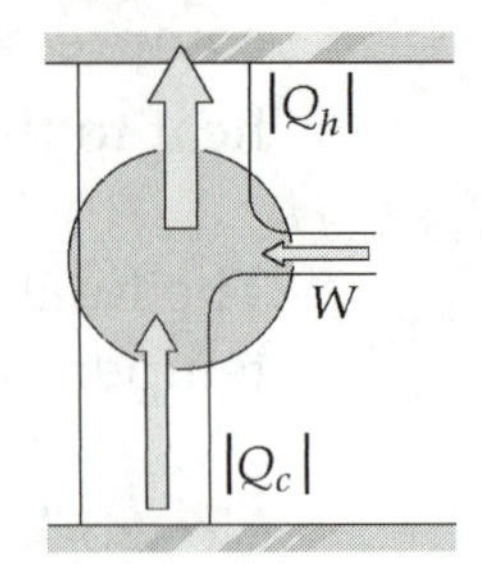


Cold Reservoir at $T_c$

(b) Refrigerator (heat pump)

**FIG. 18.2**

## Section 18.6 Entropy

## Section 18.7 Entropy and the Second Law of Thermodynamics

## Section 18.8 Entropy Changes in Irreversible Processes

Entropy is a quantity used to measure the degree of disorder in a system. For example, the molecules of a gas in a container at a high temperature are in a more disordered state (higher entropy) than the same molecules at a lower temperature.

When energy is added by heat to a system in an incremental reversible process, $dQ_r$ is positive and the entropy increases. When energy is removed, $dQ_r$ is negative and the entropy decreases. Note that only changes in entropy are defined by Equation 18.8; therefore, the concept of entropy is most useful when a system undergoes a change in its state.

When using Equation 18.12 to calculate entropy changes, note that $\Delta S$ may be obtained even if the process is irreversible, since $\Delta S$ depends only on the initial and final equilibrium states, not on the path. In order to calculate $\Delta S$ for an irreversible process, you must devise a reversible process (or sequence of reversible processes) between the initial and final states, and compute $\dfrac{dQ_r}{T}$ for the reversible process. The entropy change for the irreversible process is the same as that of the reversible process between the same initial and final equilibrium states.

# EQUATIONS AND CONCEPTS

The net **work done by a heat engine** during one cycle equals the net energy absorbed by the engine. During each cycle:

$$W_{\text{eng}} = |Q_h| - |Q_c| \tag{18.1}$$

- energy is transferred from a reservoir at a high temperature,
- work is done by the engine, and
- energy is expelled by the engine to a reservoir at a low temperature.

$Q_h$ = quantity of energy absorbed from the high temperature reservoir.

$Q_c$ = quantity of energy expelled to the low temperature reservoir.

The **thermal efficiency**, $e$, of a heat engine is defined as the ratio of the net work done to the energy absorbed as heat during one cycle of the process.

$$e = \frac{W_{\text{eng}}}{|Q_h|} = 1 - \frac{|Q_c|}{|Q_h|} \tag{18.2}$$

The **Carnot efficiency** is the limiting efficiency for an engine operating reversibly in a Carnot cycle between two given temperatures.

$$e_C = 1 - \frac{T_c}{T_h} \tag{18.4}$$

The **coefficient of performance**, COP, of a heat pump, operating in the heating mode, is the ratio of the energy transferred as heat at the high temperature to the work done on the pump.

$$\text{COP (heating mode)} = \frac{|Q_h|}{W} \tag{18.5}$$

The **coefficient of performance** of a refrigerator (or a heat pump operating in the cooling mode) is defined by the ratio of the energy absorbed by heat $Q_c$, to the work done. A good refrigerator has a high coefficient of performance.

$$\text{COP (cooling mode)} = \frac{|Q_c|}{W} \tag{18.6}$$

The **change in entropy** as a system changes from one equilibrium state to another, under a reversible, quasi-static process, is defined by Equation 18.8.

$$dS = \frac{dQ_r}{T} \tag{18.8}$$

$dQ_r$ = quantity of energy added (or removed)
$T$ = absolute temperature

**Entropy**, $S$, is a thermodynamic variable that characterizes the degree of disorder in a system. *All physical processes tend toward a state of increasing entropy; the entropy of the Universe increases in all natural processes.*

$$S \equiv k_B \ln W \tag{18.9}$$

$W$ = the number of possible microstates of the system

The **change in entropy** of a system that undergoes a reversible process between the states $i$ and $f$ depends only on the properties of the initial and final equilibrium states.

$$\Delta S = \int_i^f \frac{dQ_r}{T} \text{ (reversible path)} \tag{18.10}$$

For any **arbitrary reversible cycle** the change in entropy of a system is identically zero.

$$\oint \frac{dQ_r}{T} = 0 \tag{18.11}$$

During a **free expansion**, the entropy of a gas increases.

$$\Delta S = nR \ln\left(\frac{V_f}{V_i}\right) \tag{18.12}$$

## REVIEW CHECKLIST

✓ Understand the basic principle of the operation of a heat engine, and be able to define and discuss the thermal efficiency of a heat engine.

✓ State the second law of thermodynamics, and discuss the difference between reversible and irreversible processes. Discuss the importance of the first and second laws of thermodynamics as they apply to various forms of energy conversion and thermal pollution.

✓ Describe the processes that take place in an ideal heat engine taken through a Carnot cycle.

✓ Calculate the efficiency of a Carnot engine, and note that the efficiency of real heat engines is always less than the Carnot efficiency.

✓ Calculate entropy changes for reversible processes (such as one involving an ideal gas).

✓ Calculate entropy changes for irreversible processes, recognizing that the entropy change for an irreversible process is equivalent to that of a reversible process between the same two equilibrium states.

## ANSWERS TO SELECTED QUESTIONS

**Q18.1** What are some factors that affect the efficiency of automobile engines?

**Answer** A gasoline engine does not exactly fit the definition of a heat engine. It takes in energy by mass transfer. Nevertheless, it converts this chemical energy into internal energy, so it can be modeled as taking in energy by heat. Therefore Carnot's limit on the efficiency of a heat engine applies to a gasoline engine. The fundamental limit on its efficiency is $1-\frac{T_c}{T_h}$, set by the maximum temperature the engine block can stand and the temperature of the surroundings into which exhaust heat must be dumped.

The Carnot efficiency can only be attained by a reversible engine. To run with any speed, a gasoline engine must carry out irreversible processes and have an efficiency below the Carnot limit. In order for the compression and power strokes to be adiabatic, we would like to minimize irreversibly losing heat to the engine block. We would like to have the processes happen very quickly, so that there is negligible time for energy to be conducted away. But in a standard piston engine the compression and power strokes must take one-half of the total cycle time, and the angular speed of the crankshaft is limited by the time required to open and close the valves to get the fuel into each cylinder and the exhaust out.

Other limits on efficiency are imposed by irreversible processes like friction, both inside and outside the engine block. If the ignition timing is off, then the effective compressed volume will not be as small as it could be. If the burning of the gasoline is not complete, then some of the chemical energy will never enter the process.

The approximation that the intake and exhaust gases have the same value of $\gamma$, hides other possibilities. Combustion breaks up the gasoline and oxygen molecules into a larger number of simpler water and $CO_2$ molecules, raising $\gamma$. By increasing $\gamma$ and $N$, combustion slightly improves the efficiency. For this reason a gasoline engine can be more efficient than a methane engine, and also more efficient after the gasoline is refined to remove double carbon bonds (which would raise $\gamma$ for the fuel).

**Q18.3** A steam-driven turbine is one major component of an electric power plant. Why is it advantageous to have the temperature of the steam as high as possible?

**Answer** The most optimistic limit of efficiency in a steam engine is the ideal (Carnot) efficiency. This can be calculated from the high and low temperatures of the steam, as it expands:

$$e_C = \frac{T_h - T_c}{T_h} = 1 - \frac{T_c}{T_h}$$

The engine will be most efficient when the low temperature is extremely low, and the high temperature is extremely high. However, since the electric power plant is typically placed on the surface of the earth, there is a limit to how much the steam can expand, and how the lower temperature can be. The only way to further increase the efficiency, then, is to raise the temperature of the hot steam, as high as possible.

**Q18.9** The device shown in Figure Q18.9, called a thermoelectric converter, uses a series of semiconductor cells to convert internal energy to electric potential energy. In the figure shown, both legs of the device are at the same temperature and no electric potential energy is produced. When one leg is at a higher temperature than the other, however, as shown in the figure on the right, electric potential energy is produced as the device extracts energy from the hot reservoir and drives a small electric motor.

**FIG. Q18.9**

(a) Why does the temperature differential produce electric potential energy in this demonstration?

(b) In what sense does this intriguing experiment demonstrate the second law of thermodynamics?

**Answer** (a) The semiconductor converter operates essentially like a thermocouple, which is a pair of wires of different metals, with a junction at each end. When the junctions are at different temperatures, a small voltage appears around the loop, so that the device can be used to measure temperature or (here) to drive a small motor.

*continued on next page*

(b) The second law states than an engine operating in a cycle cannot absorb energy by heat from one reservoir and expel it entirely by work. This exactly describes the first situation, where both legs are in contact with a single reservoir, and the thermocouple fails to produce electrical work. To expel energy by work, the device must transfer energy by heat from a hot reservoir to a cold reservoir, as in the second situation.

**Q18.11** Discuss the change in entropy of a gas that expands

(a) at constant temperature and

(b) adiabatically.

**Answer** (a) The expanding gas is doing work. If it is ideal, its constant temperature implies constant internal energy, and it must be taking in energy by heat equal in amount to the work it is doing. As energy enters the gas by heat its entropy increases. The derivation of Equation 18.12 shows that the change in entropy is $\Delta S = nR\ln\left(\frac{V_f}{V_i}\right)$.

(b) In a reversible adiabatic expansion there is no entropy change. We can say this is because the heat input is zero, or we can say it is because the temperature drops to compensate for the volume increase. In an irreversible adiabatic expansion the entropy increases. Equation 18.12 shows the entropy change in a free expansion.

## SOLUTIONS TO SELECTED PROBLEMS

**P18.3** A particular heat engine has a useful power output of 5.00 kW and an efficiency of 25.0%. The engine expels 8 000 J of exhaust energy in each cycle. Find

(a) the energy taken in during each cycle and

(b) the time interval for each cycle.

**Solution** We are given that $|Q_c| = 8\,000\text{ J}$

(a) We have $$e = \frac{W_{\text{eng}}}{|Q_h|} = \frac{|Q_h| - |Q_c|}{|Q_h|} = 1 - \frac{|Q_c|}{|Q_h|} = 0.250.$$

Isolating $|Q_h|$, we have $$|Q_h| = \frac{|Q_c|}{1-e} = \frac{8\,000\text{ J}}{1-0.250} = 10.7\text{ kJ}.$$ ◊

*continued on next page*

(b) The work per cycle is $W_{\text{eng}} = |Q_h| - |Q_c| = 2\,667 \text{ J}$.

From $\mathscr{P} = \dfrac{W_{\text{eng}}}{\Delta t}$, $\quad \Delta t = \dfrac{W_{\text{eng}}}{\mathscr{P}} = \dfrac{2\,667 \text{ J}}{5\,000 \text{ J/s}} = 0.533 \text{ s}$. ◊

**P18.5** One of the most efficient heat engines ever built is a steam turbine in the Ohio valley, operating between 430°C and 1 870°C on energy from West Virginia coal to produce electricity for the Midwest.

(a) What is its maximum theoretical efficiency?

(b) The actual efficiency of the engine is 42.0%. How much useful power does the engine deliver if it takes in $1.40 \times 10^5$ J of energy each second from its hot reservoir?

**Solution** The engine is a steam turbine in an electric generating station:

$$T_c = 430°\text{C} = 703 \text{ K} \qquad \text{and} \qquad T_h = 1\,870°\text{C} = 2\,143 \text{ K}$$

(a) $e_C = \dfrac{\Delta T}{T_h} = \dfrac{1\,440 \text{ K}}{2\,143 \text{ K}} = 0.672 \text{ or } 67.2\%$ ◊

(b) $|Q_h| = 1.40 \times 10^5 \text{ J}$ $\qquad W_{\text{eng}} = 0.420|Q_h| = 5.88 \times 10^4 \text{ J}$

$$\mathscr{P} = \frac{W_{\text{eng}}}{\Delta t} = \frac{5.88 \times 10^4 \text{ J}}{1 \text{ s}} = 58.8 \text{ kW}$$ ◊

**P18.7** An ideal gas is taken through a Carnot cycle. The isothermal expansion occurs at 250°C, and the isothermal compression takes place at 50.0°C. The gas takes in 1 200 J of energy from the hot reservoir during the isothermal expansion. Find

(a) the energy expelled to the cold reservoir in each cycle and

(b) the net work done by the gas in each cycle.

**Solution** (a) For a Carnot cycle, $\quad e_C = 1 - \dfrac{T_c}{T_h}$.

For any engine, $\quad e = \dfrac{W_{\text{eng}}}{|Q_h|} = 1 - \dfrac{|Q_c|}{|Q_h|}$.

Therefore, for a Carnot engine, $\quad 1 - \dfrac{T_c}{T_h} = 1 - \dfrac{|Q_c|}{|Q_h|}$.

*continued on next page*

Then, since $|Q_c| = |Q_h|\left(\frac{T_c}{T_h}\right)$, $|Q_c| = (1\,200\text{ J})\left(\frac{323\text{ K}}{523\text{ K}}\right) = 741\text{ J}.$ ◊

(b) The work is calculated as $W_{\text{eng}} = |Q_h| - |Q_c| = 1\,200\text{ J} - 741\text{ J} = 459\text{ J}.$ ◊

**P18.17** An ideal refrigerator or ideal heat pump is equivalent to a Carnot engine running in reverse. That is, energy $Q_c$ is taken in from a cold reservoir and energy $|Q_h|$ is rejected to a hot reservoir.

(a) Show that the work that must be supplied to run the refrigerator or heat pump is $W = \left(\frac{T_h - T_c}{T_c}\right)Q_c$.

(b) Show that the coefficient of performance of the ideal refrigerator is $\text{COP} = \frac{T_c}{T_h - T_c}$.

**Solution** (a) For a complete cycle $\Delta E_{\text{int}} = 0$, and $W = |Q_h| - |Q_c| = |Q_c|\left(\frac{|Q_h|}{|Q_c|} - 1\right)$.

For a Carnot cycle (and only for a Carnot cycle), $\frac{|Q_h|}{|Q_c|} = \frac{T_h}{T_c}$.

Then, $W = |Q_c|\frac{T_h - T_c}{T_c}.$ ◊

(b) For a refrigerator $\text{COP} = \frac{|Q_c|}{W}$, so $\text{COP} = \frac{T_c}{T_h - T_c}.$ ◊

**P18.21** How much work does an ideal Carnot refrigerator require to remove 1.00 J of energy from helium at 4.00 K and reject this energy to a room-temperature (293-K) environment?

**Solution** $(\text{COP})_{\text{Carnot, refrig}} = \frac{T_c}{\Delta T} = \frac{4.00\text{ K}}{293\text{ K} - 4.00\text{ K}} = 0.013\,8 = \frac{|Q_c|}{W}$

$W = \frac{|Q_c|}{\text{COP}} = \frac{1.00\text{ J}}{0.013\,8} = 72.2\text{ J}$ ◊

**P18.23** Calculate the change in entropy of 250 g of water warmed slowly from 20.0°C to 80.0°C. (*Suggestion:* Note that $dQ = mc\,dT$.)

**Solution** We use the energy version of the nonisolated system model. To do the heating reversibly, put the water pot successively into contact with reservoirs at temperatures $20.0°\text{C} + \delta$, $20.0°\text{C} + 2\delta$, … 80.0°C, where $\delta$ is some small increment.

Then, $$\Delta S = \int_i^f \frac{dQ}{T} = \int_{T_i}^{T_f} mc\frac{dT}{T}.$$

Here $T$ means the absolute temperature. We would ordinarily think of $dT$ as the change in the Celsius temperature, but one Celsius degree of temperature change is the same size as one kelvin of change, so $dT$ is also the change in absolute $T$.

Then $$\Delta S = mc\ln T\Big|_{T_i}^{T_f} = mc\ln\left(\frac{T_f}{T_i}\right)$$

$$\Delta S = (0.250\text{ kg})(4\,186\text{ J/kg}\cdot\text{K})\ln\left(\frac{353\text{ K}}{293\text{ K}}\right) = 195\text{ J/K}.$$ ◊

**P18.29** A bag contains 50 red marbles and 50 green marbles.

(a) You draw a marble at random from the bag, notice its color, return it to the bag, and repeat the process for a total of three draws. You record the result as the number of red marbles and the number of green marbles in the set of three. Construct a table listing each possible macrostate and the number of microstates within it. For example, RRG, RGR, and GRR are the three microstates constituting the macrostate 1G, 2R.

(b) Construct a table for the case in which you draw five marbles instead of three.

**Solution** (a)

| Result | Possible combinations | Total |
|---|---|---|
| All Red | RRR | 1 |
| 2R, 1G | RRG, RGR, GRR | 3 |
| 1R, 2G | RGG, GRG, GGR | 3 |
| All Green | GGG | 1 ◊ |

*continued on next page*

(b)

| Result | Possible combinations | Total |
|---|---|---|
| All Red | RRRRR | 1 |
| 4R, 1G | RRRRG, RRRGR, RRGRR, RGRRR, GRRRR | 5 |
| 3R, 2G | RRRGG, RRGRG, RGRRG, GRRRG, RRGGR, RGRGR, GRRGR, RGGRR, GRGRR, GGRRR | 10 |
| 2R, 3G | GGGRR, GGRGR, GRGGR, RGGGR, GGRRG, GRGRG, RGGRG, GRRGG, RGRGG, RRGGG | 10 |
| 1R, 4G | GGGGR, GGGRG, GGRGG, GRGGG, RGGGG | 5 |
| All Green | GGGGG | 1 ◊ |

Conclusion: The macrostates with roughly equal numbers of red and green marbles are the high-entropy states. The irreversible process of shaking the whole bag and emptying half of it onto the floor will likely result in such a state of disorder.

**P18.31** A 1 500-kg car is moving at 20.0 m/s. The driver brakes to a stop. The brakes cool off to the temperature of the surrounding air, which is nearly constant at 20.0°C. What is the total entropy change?

**Solution** The original kinetic energy of the car,

$$K = \frac{1}{2}mv^2 = \frac{1}{2}(1\,500 \text{ kg})(20.0 \text{ m/s})^2 = 300 \text{ kJ}$$

becomes irreversibly 300 kJ of extra internal energy in the brakes, the car, and its surroundings. Since their total heat capacity is so large, their equilibrium temperature will be approximately 20.0°C. To carry them reversibly to this same final state, imagine putting 300 kJ into the car and its environment from a heater at 20.001°C.

Then $$\Delta S = \int_i^f \frac{dQ}{T} = \frac{1}{T}\int dQ_r = \frac{Q}{T} = \frac{300 \text{ kJ}}{293 \text{ K}}$$

$$\Delta S = 1.02 \text{ kJ/K}$$ ◊

**P18.39** A house loses energy through the exterior walls and roof at a rate of $5\,000 \text{ J/s} = 5.00 \text{ kW}$ when the interior temperature is 22.0°C and the outside temperature is $-5.00°\text{C}$. Calculate the electric power required to maintain the interior temperature at 22.0°C for the following two cases.

(a) The electric power is used in electric resistance heaters (which convert all the energy transferred in by electrical transmission into internal energy).

*continued on next page*

(b) Assume instead that the electric power is used to drive an electric motor that operates the compressor of a heat pump, which has a coefficient of performance equal to 60.0% of the Carnot-cycle value.

**Solution** **Conceptualize:** The electric heater should be 100% efficient, so $\mathcal{P} = 5$ kW in part (a). It sounds as if the heat pump is only 60% efficient, so we might expect $\mathcal{P} = 9$ kW in part (b).

**Categorize:** Power is the amount of energy transferred per unit of time, so we can find the power in each case by examining the energy input as heat required for the house as a nonisolated system in steady state.

**Analyze:**

(a) As in Section 6.6, let $T_{ET}$ represent the energy transferred by electric transmission. We know that $\mathcal{P}_{\text{electric}} = \dfrac{T_{ET}}{\Delta t}$, so if all of the energy transferred into the heater by electrical transmission is stored in the heater as internal energy, then

$$\mathcal{P}_{\text{electric}} = \frac{T_{ET}}{\Delta t} = 5.00 \text{ kW} \qquad \Diamond$$

Now we let $T$ represent temperature.

(b) For a heat pump,

$$(\text{COP})_{\text{Carnot}} = \frac{T_h}{\Delta T} = \frac{295 \text{ K}}{27.0 \text{ K}} = 10.93$$

$$\text{Actual COP} = (0.600)(10.93) = \frac{|Q_h|}{W} = \frac{|Q_h|/\Delta t}{W/\Delta t}$$

Therefore, to bring 5 000 W of heat into the house only requires input power

$$P_{\text{heat pump}} = \frac{W}{\Delta t} = \frac{|Q_h|/\Delta t}{\text{COP}} = \frac{5\,000 \text{ W}}{6.56} = 763 \text{ W} \qquad \Diamond$$

**Finalize:** The result for the electric heater's power is consistent with our prediction, but the heat pump actually requires **less** power than we expected. Since both types of heaters use electricity to operate, we can now see why it is more cost effective to use a heat pump even though it is less than 100% efficient!

**P18.43** In 1816, Robert Stirling, a Scottish clergyman, patented the *Stirling engine* which has found a wide variety of applications ever since. Fuel is burned externally to warm one of the engine's two cylinders. A fixed quantity of inert gas moves cyclically between the cylinders, expanding in the hot one and contracting in the cold one. Figure P18.43 represents a model for its thermodynamic cycle. Consider $n$ mol of an ideal monatomic gas being taken once through the cycle, consisting of two isothermal processes at temperatures $3T_i$ and $T_i$ and two constant-volume processes. Determine, in terms of $n$, $R$, and $T_i$,

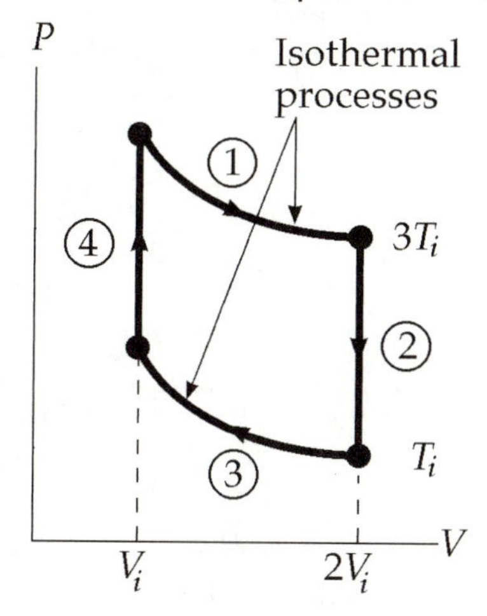


**FIG. P18.43**

(a) the net energy transferred by heat to the gas and

(b) the efficiency of the engine.

A Stirling engine is easier to manufacture than an internal combustion engine or a turbine. It can run on burning garbage. It can run on the energy of sunlight and produce no material exhaust.

**Solution** The Stirling engine need have no material exhaust, but it has energy exhaust.

(a) For an isothermal process, $Q = nRT\ln\left(\frac{V_f}{V_i}\right)$.

Therefore, $Q_1 = nR(3T_i)\ln 2$ and $Q_3 = nRT_i\ln\left(\frac{1}{2}\right)$.

The internal energy of a monatomic ideal gas is $E_{\text{int}} = \frac{3}{2}nRT$.

In the constant-volume processes, $Q_2 = \Delta E_{\text{int},2} = \frac{3}{2}nR(T_i - 3T_i)$

and $Q_4 = \Delta E_{\text{int},4} = \frac{3}{2}nR(3T_i - T_i)$.

The net energy transferred by heat is then $Q = Q_1 + Q_2 + Q_3 + Q_4$,

or $Q = 2nRT_i\ln 2$. ◊

*continued on next page*

(b) $|Q_h|$ is the sum of the positive contributions to $Q$.

$$|Q_h| = Q_1 + Q_4 = 3nRT_i(1+\ln 2)$$

Since the change in temperature for the complete cycle is zero,

$$\Delta E_{\text{int}} = 0 \text{ and } W_{\text{eng}} = Q.$$

Therefore, the efficiency is $e = \dfrac{W_{\text{eng}}}{|Q_h|} = \dfrac{Q}{|Q_h|} = \dfrac{2\ln 2}{3(1+\ln 2)} = 0.273 = 27.3\%$. ◊

**P18.51** A 1.00-mol sample of a monatomic ideal gas is taken through the cycle shown in Figure P18.51. At point $A$, the pressure, volume, and temperature are $P_i$, $V_i$, and $T_i$, respectively. In terms of $R$ and $T_i$, find

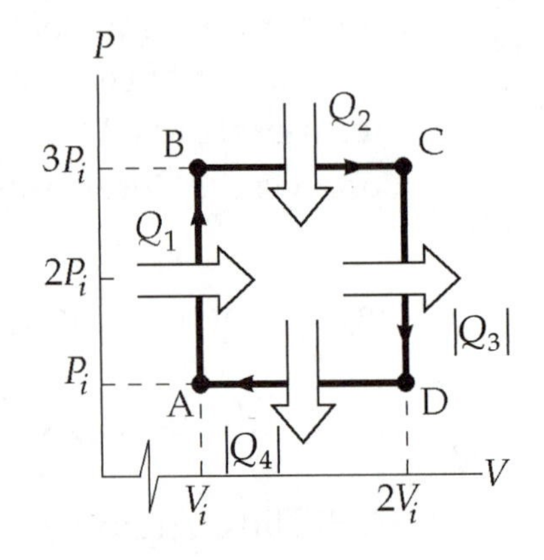


**FIG. P18.51**

(a) the total energy entering the system by heat per cycle,

(b) the total energy leaving the system by heat per cycle,

(c) the efficiency of an engine operating in this cycle, and

(d) the efficiency of an engine operating in a Carnot cycle between the same temperature extremes.

**Solution** At point $A$, $P_iV_i = nRT_i$ with $n = 1.00$ mol.

At point $B$, $3P_iV_i = nRT_B$ so $T_B = 3T_i$.

At point $C$, $(3P_i)(2V_i) = nRT_C$ and $T_C = 6T_i$.

At point $D$, $P_i(2V_i) = nRT_D$ and $T_D = 2T_i$.

We find the energy transfer by heat for each step in the cycle using

$$C_V = \frac{3}{2}R \quad \text{and} \quad C_P = \frac{5}{2}R$$

$$Q_1 = Q_{AB} = C_V(3T_i - T_i) = 3RT_i \qquad Q_2 = Q_{BC} = C_P(6T_i - 3T_i) = 7.5RT_i$$

$$Q_3 = Q_{CD} = C_V(2T_i - 6T_i) = -6RT_i \qquad Q_4 = Q_{DA} = C_P(T_i - 2T_i) = -2.5RT_i.$$

*continued on next page*

(a) Therefore, $Q_{\text{in}} = |Q_h| = Q_{AB} + Q_{BC} = 10.5RT_i$. ◊

(b) $Q_{\text{out}} = |Q_c| = |Q_{CD} + Q_{DA}| = 8.5RT_i$ ◊

(c) $e = \dfrac{|Q_h| - |Q_c|}{|Q_h|} = 0.190 = 19.0\%$ ◊

(d) Carnot efficiency, $e_C = 1 - \dfrac{T_c}{T_h} = 1 - \dfrac{T_i}{6T_i} = 0.833 = 83.3\%$. ◊

**P18.53** A system consisting of $n$ mol of an ideal gas undergoes two reversible processes. It starts with pressure $P_i$ and volume $V_i$, expands isothermally, and then contracts adiabatically to reach a final state with pressure $P_i$ and volume $3V_i$.

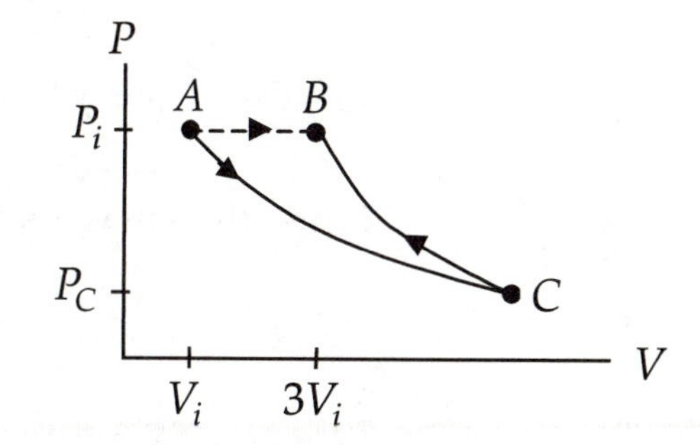


**FIG. P18.53**

(a) Find its change in entropy in the isothermal process. The entropy does not change in the adiabatic process.

(b) Explain why the answer to part (a) must be the same as the answer to Problem 18.52.

**Solution** The diagram shows the isobaric process considered in Problem 18.52 as $AB$. The processes considered in this problem are $AC$ and $CB$.

(a) For isotherm ($AC$), $P_A V_A = P_C V_C$.

For adiabat ($CB$) $P_C V_C^{\gamma} = P_B V_B^{\gamma}$.

Combining these gives

$$V_C = \left(\frac{P_B V_B^{\gamma}}{P_A V_A}\right)^{1/(\gamma-1)} = \left[\left(\frac{P_i}{P_i}\right)\frac{(3V_i)^{\gamma}}{V_i}\right]^{1/(\gamma-1)} = \left(3^{\gamma/(\gamma-1)}\right)V_i.$$

Therefore, $$\Delta S_{AC} = nR\ln\left(\frac{V_C}{V_A}\right) = nR\ln\left[3^{\gamma/(\gamma-1)}\right] = \frac{nR\gamma\ln 3}{\gamma - 1}.$$ ◊

*continued on next page*

(b) Since the change in entropy is path independent, $\Delta S_{AB} = \Delta S_{AC} + \Delta S_{CB}$.

But because ($CB$) is adiabatic and reversible, $\Delta S_{CB} = 0$.

Then $\Delta S_{AB} = \Delta S_{AC}$.

The answer to Problem 18.52 was stated as $\Delta S_{AB} = nC_P \ln 3$.

Because $\gamma = \frac{C_P}{C_V}$, $C_V = \frac{C_P}{\gamma}$, and $C_P - C_V = R$

gives $C_P - \frac{C_P}{\gamma} = R$

so $\gamma C_P - C_P = \gamma R$ and $C_P = \frac{\gamma R}{\gamma - 1}$.

Thus, the answers to problem Problems 18.52 and 18.53 are, in fact, equal.

# Chapter 19

# Electric Forces and Electric Fields

## NOTES FROM SELECTED CHAPTER SECTIONS

### Section 19.2 Properties of Electric Charges

**Electric charge** has the following important properties:

- There are two kinds of charges in nature, positive and negative, with the property that unlike charges attract one another and like charges repel one another.
- Charge is conserved.
- Charge is quantized.

### Section 19.3 Insulators and Conductors

**Conductors** are materials in which electric charges move freely under the influence of an electric field; **insulators** are materials that do not readily transport charge.

### Section 19.4 Coulomb's Law

Experiments show that an electric force between a pair of point charges is:

- of equal magnitude and opposite direction on the two charges
- attractive if the charges are of opposite sign and repulsive if the charges have the same sign

- inversely proportional to the square of the separation, $r$, between the two charges and is along the line joining them

- proportional to the product of the magnitudes of the charges.

## Section 19.5 Electric Fields

An electric field exists at some point if a test charge at rest placed at that point experiences an electrical force.

The electric field vector $\vec{\mathbf{E}}$ at some point in space is defined as the electric force $\vec{\mathbf{F}}_e$ acting on a positive test charge placed at that point divided by the magnitude of the test charge $q_0$.

At any point the total electric field created by a group of discrete point charges equals the vector sum of the electric fields due to each of the individual charges.

## Section 19.6 Electric Field Lines

A convenient aid for visualizing electric field patterns is to draw lines pointing in the direction of the electric field vector at every point. These lines are called electric field lines.

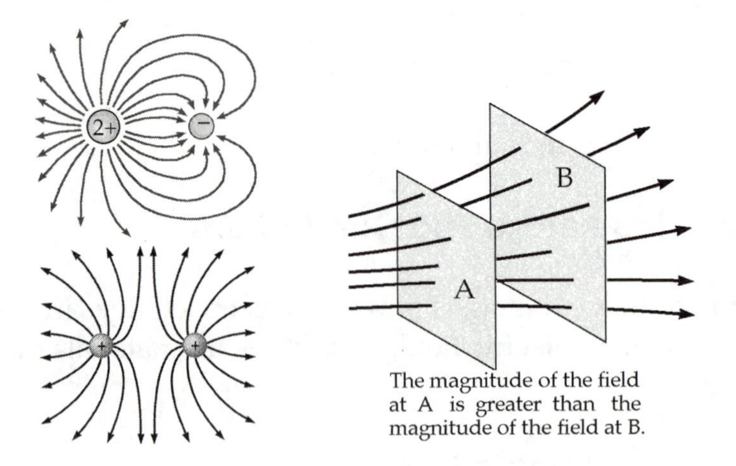


The magnitude of the field at A is greater than the magnitude of the field at B.

Lines are drawn so that:

- Lines begin on a positive charge (or at infinity) and terminate on a negative charge (or at infinity).

- The number of lines leaving a positive charge (or approaching a negative charge) is proportional to the magnitude of the charge.

- No two field lines can cross.

Electric field lines are related to the electric field in the following manner:

- The electric field vector is tangent to an electric field line at each point.
- The number of lines per unit area through a surface perpendicular to the field lines is proportional to the magnitude of the electric field in that region.
- Thus, $\vec{\mathbf{E}}$ is large when the field lines are close together and small when they are far apart.

## Section 19.9 Gauss's Law

Gauss's law states that the net electric flux through a closed gaussian surface is equal to the net charge inside the surface divided by $\epsilon_0$.

When using Gauss's law to calculate an electric field, choose the gaussian surface so that it has the same symmetry as the charge distribution.

## Section 19.11 Conductors in Electrostatic Equilibrium

A conductor in electrostatic equilibrium has the following properties:

- The electric field is zero everywhere inside the conductor.
- Any excess charge on an isolated conductor resides entirely on its surface.
- The electric field just outside a charged conductor is perpendicular to the conductor's surface and has a magnitude $\frac{\sigma}{\epsilon_0}$, where $\sigma$ is the charge per unit area at that point.
- On an irregularly shaped conductor, charge tends to accumulate at locations where the radius of curvature of the surface is the smallest, that is, at sharp points.

## EQUATIONS AND CONCEPTS

The **magnitude of the electrostatic force** between two stationary point charges, $q_1$ and $q_2$, separated by a distance $r$ is given by **Coulomb's law.**

$$F_e = k_e \frac{|q_1||q_2|}{r^2} \qquad (19.1)$$

$$\text{where } k_e = \frac{1}{4\pi\epsilon_0} = 8.99\times10^9 \ \text{N}\cdot\text{m}^2/\text{C}^2$$

In calculations, an approximate value for $k_e$ may be used.

$\epsilon_0 = 8.85\times10^{-12}\ \text{C}^2/\text{N}\cdot\text{m}^2$
**permittivity of space**

The **direction of the electrostatic force** on each charge is determined from the experimental observation that like sign charges experience forces of mutual repulsion and unlike sign charges attract each other. *By virtue of Newton's third law, the magnitude of the force on each of the two charges is the same regardless of the relative magnitude of the values of $q_1$ and $q_2$.*

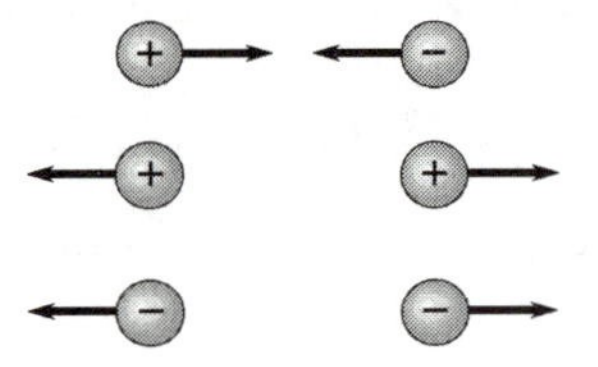

The **vector form of Coulomb's law** includes a unit vector. $\vec{\mathbf{F}}_{12}$ is the force on $q_2$ due to $q_1$. The unit vector $\hat{\mathbf{r}}_{12}$ is directed from $q_1$ to $q_2$. *Coulomb's law applies exactly only to point charges. Regardless of the relative magnitudes of the two charges, $\vec{\mathbf{F}}_{21} = -\vec{\mathbf{F}}_{12}$; this follows from Newton's third law.*

$$\vec{\mathbf{F}}_{12} = k_e \frac{q_1 q_2}{r^2}\,\hat{\mathbf{r}}_{12} \qquad (19.2)$$

The **resultant force** on any one charge within a group of charges is the vector sum of the forces exerted on that charge by the remaining individual charges present. *The principle of superposition applies.*

The **electric field** at any point in space is defined as the ratio of electric force per charge exerted on a small positive test charge placed at the point where the field is to be determined.

$$\vec{\mathbf{E}} \equiv \frac{\vec{\mathbf{F}}_e}{q_0} \qquad (19.3)$$

The **electric field a distance *r* from a point charge** is given by Equation 19.5. The unit vector $\hat{\mathbf{r}}$ is directed away from $q$ and toward the point where the field is to be calculated. *The direction of the electric field is radially outward from a positive point charge and radially inward toward a negative point charge.*

$$\vec{\mathbf{E}} = k_e \frac{q}{r^2}\hat{\mathbf{r}} \quad (19.5)$$

The **superposition principle** holds when the electric field at a point is due to a number of point charges.

$$\vec{\mathbf{E}} = k_e \sum_i \frac{q_i}{r_i^2}\hat{\mathbf{r}}_i \quad (19.6)$$

(vector sum)

The **electric field** of a **continuous charge distribution** is found by integrating over the entire region that contains the charge. This is a vector operation and can usually be carried out easily when the charge is distributed along a line, over a surface or throughout a volume.

$$\vec{\mathbf{E}} = \lim_{\Delta q_i \to 0} k_e \sum_i \frac{\Delta q_i}{r_i^2}\hat{\mathbf{r}}_i = k_e \int \frac{dq}{r^2}\hat{\mathbf{r}} \quad (19.7)$$

The concept of **charge density** is utilized to perform the integration described above. It is convenient to represent a charge increment $dq$ as the product of an element of length, area, or volume and the charge density over that region.

For an element of length $dx$

$$dq = \lambda dx$$

For an element of area $dA$

$$dq = \sigma dA$$

For an element of volume $dV$

$$dq = \rho dV$$

For **uniform charge distributions** the volume charge density ($\rho$), the surface charge density ($\sigma$), and the linear charge density ($\lambda$) can be calculated from the total charge and the total volume, area, or length.

$$\rho \equiv \frac{Q}{V} \quad (19.8)$$

$$\sigma \equiv \frac{Q}{A} \quad (19.9)$$

**For non-uniform distributions**, the densities $\lambda$, $\sigma$, and $\rho$ must be stated as functions of position.

$$\lambda \equiv \frac{Q}{\ell} \quad (19.10)$$

**Electric flux** is a measure of the number of electric field lines that penetrate a surface. The flux equals the product of the magnitude of the field and the projection of the area onto a plane perpendicular to the direction of the field. Equation 19.18 gives the value of the flux through a plane area when a uniform field makes an angle $\theta$ with the normal to the surface. *Electric flux is a scalar quantity and has SI units of* $\text{N}\cdot\text{m}^2/\text{C}$.

$$\Phi_E = EA\cos\theta \qquad (19.18)$$

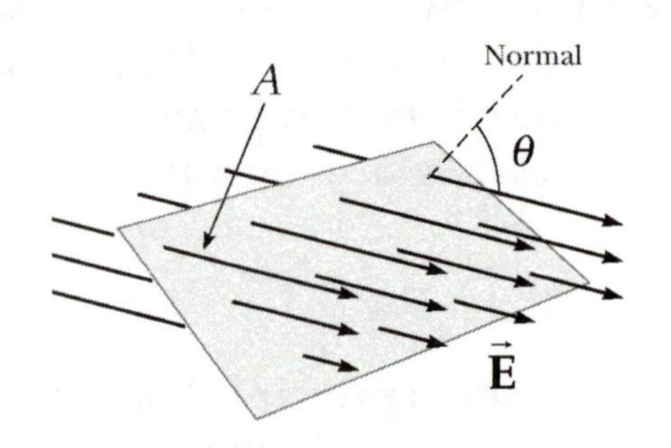


**For a surface of arbitrary shape or a nonuniform field,** the flux is calculated by integrating the normal component of the field over the surface in question. *The integrand is the "dot" product of two vectors and the integral must be evaluated over the entire surface in question.*

$$\Phi_E = \oint \vec{\mathbf{E}} \cdot d\vec{\mathbf{A}} = \oint E_n dA \qquad (19.20)$$

**Gauss's law** states that the net flux through any closed surface surrounding a net charge $q$ equals the net charge enclosed by the surface divided by the constant $\epsilon_0$. The net flux is independent of the shape and size of the surface. *The surface is called a "gaussian" surface.*

$$\Phi_E = \oint \vec{\mathbf{E}} \cdot d\vec{\mathbf{A}} = \frac{q_{\text{in}}}{\epsilon_0} \qquad (19.22)$$

The **electric field due to symmetric charge distributions** can be determined by evaluating Equation 19.22 for the respective charge distributions. Some examples are:

a **distance $r$ from a point charge $q$**

$$E = k_e \frac{q}{r^2}$$

a point **exterior to a uniformly charged insulating sphere** of radius $a$ and charge $Q$

$$E = k_e \frac{Q}{r^2} \text{ for } r > a$$

a point **interior to a uniformly charged insulating sphere** with total charge $Q$

$$E = k_e \left(\frac{Q}{a^3}\right) r \text{ for } r < a$$

a point **outside a thin uniformly charged spherical shell** of radius $a$ and charge $Q$

$$E = k_e \frac{Q}{r^2} \text{ for } r > a$$

a **distance $r$ from an infinitely long uniform line of charge** with linear charge density $\lambda$ $$E = 2k_e \frac{\lambda}{r}$$

any distance from an **infinite plane of charge** with surface charge density $\sigma$ $$E = \frac{\sigma}{2\epsilon_0}$$

a point **just outside the surface of a charged conductor** in equilibrium with surface charge density $\sigma$ $$E_n = \frac{\sigma}{\epsilon_0} \qquad (19.25)$$

## SUGGESTIONS, SKILLS, AND STRATEGIES

### Problem-Solving Hints for Electric Forces and Fields

- **Units:** When performing calculations that involve the use of the Coulomb constant $k_e$ that appears in Coulomb's law, charges must be in coulombs and distances in meters. If they are given in other units, you must convert them to SI.

- **Applying Coulomb's law to point charges:** It is important to use the superposition principle properly when dealing with a collection of interacting point charges. When several charges are present, the resultant force on any one of them is found by finding the individual force that every other charge exerts on it and then finding the vector sum of all these forces. The magnitude of the force that any charged object exerts on another is given by Coulomb's law, and the direction of the force is found by noting that the forces are repulsive between like charges and attractive between unlike charges.

- **Calculating the electric field due to a group of point charges:** Remember that the superposition principle can also be applied to electric fields, which are also vector quantities. To find the total electric field at a given point, first calculate the electric field at the point due to each individual charge. The resultant field at the point is the vector sum of the fields due to the individual charges.

- **Calculating the electric field due to a continuous charge distribution:** To evaluate the electric field of a continuous charge distribution, it is convenient to employ the concept of charge density. Charge density can be written in different ways: charge per unit volume, $\rho$; charge per unit area, $\sigma$; or charge per unit length, $\lambda$. The total charge distribution is then subdivided into a small element of volume $dV$, area $dA$, or length $dx$. Each element contains an increment of charge $dq$ (equal to $\rho dV$, $\sigma dA$, or $\lambda dx$). If the charge is nonuniformly distributed over the region, then the charge densities must be written as functions of position. For example, if the charge density along a line or long bar is proportional to the distance from one end of the bar, then the linear charge density could be written as $\lambda = bx$ and the charge increment $dq$ becomes $dq = bxdx$.

- **Symmetry:** Whenever dealing with either a distribution of point charges or a continuous charge distribution, take advantage of any symmetry in the system to simplify your calculations.

## Problem-Solving Hints for Gauss's Law

Gauss's law is a very powerful theorem, which relates any charge distribution to the resulting electric field at any point in the vicinity of the charge. In this chapter you should learn how to apply Gauss's law to those cases in which the charge distribution has a sufficiently high degree of symmetry. As you review the examples of Gauss's law presented in the text, observe how each of the following steps has been included in the application of the equation $\Phi_E = \oint \vec{\mathbf{E}} \cdot d\vec{\mathbf{A}} = \frac{q_{\text{in}}}{\epsilon_0}$ to that particular situation.

- The **shape of the gaussian surface** should be chosen to have the same symmetry as the charge distribution.

- The **dimensions of the gaussian surface** must be such that the surface includes the point where the electric field is to be calculated.

- From the **symmetry of the charge distribution**, identify the direction of the electric field vector, $\vec{\mathbf{E}}$, relative to the direction of an element of surface area vector, $d\vec{\mathbf{A}}$, (which points outward from the surface) over each region of the gaussian surface.

- Write $\vec{\mathbf{E}} \cdot d\vec{\mathbf{A}}$ as $EdA\cos\theta$, and divide the surface into separate regions such that over each region the electric field has a constant value and can therefore be removed from the integral. Each of the separate regions should satisfy one or more of the following conditions:

    1. The value of the electric field can be argued by symmetry to be constant over the surface.

    2. The dot product in Equation 19.22 can be expressed as a simple algebraic product *EdA* because $\vec{\mathbf{E}}$ and $d\vec{\mathbf{A}}$ are parallel.

    3. The dot product in Equation 19.22 is zero because $\vec{\mathbf{E}}$ and $d\vec{\mathbf{A}}$ are perpendicular.

    4. The field can be argued to be zero everywhere on the surface.

- The total charge enclosed by the gaussian surface is that portion of the charge inside the gaussian surface.

- After the left and right sides of Gauss's law have each been evaluated, you can calculate the electric field on the gaussian surface, assuming the charge distribution is given in the problem. Conversely, if the electric field is known, you can calculate the charge distribution that produces the field.

## REVIEW CHECKLIST

✓ Describe the fundamental properties of electric charge and the nature of electrostatic forces between charged bodies.

✓ Use Coulomb's law to determine the net electrostatic force on a point electric charge due to a known distribution of a finite number of point charges.

✓ Calculate the electric field $\vec{\mathbf{E}}$ (magnitude and direction) at a specified location in the vicinity of a group of point charges.

✓ Calculate the electric field due to a continuous charge distribution. The charge may be distributed uniformly or nonuniformly along a line, over a surface, or throughout a volume.

✓ Calculate the electric flux through a surface; in particular, find the net electric flux through a closed surface.

✓ Understand that a gaussian surface is a closed mathematical surface; the surface may be on or within a conductor, an insulator, or in space. Also remember that the net electric flux through a closed gaussian surface is equal to the net charge enclosed by the surface divided by the constant $\epsilon_0$.

✓ Use Gauss's law to evaluate the electric field at points in the vicinity of charge distributions, which exhibit spherical, cylindrical, or planar symmetry.

## ANSWERS TO SELECTED QUESTIONS

**Q19.5** A balloon is negatively charged by rubbing and then clings to a wall. Does that mean that the wall is positively charged? Why does the balloon eventually fall?

**Answer** No. The balloon induces polarization of the molecules in the wall, so that a layer of positive charge exists near the balloon. This is just like the situation in Figure 19.6a, except that the signs of the charges are reversed. The attraction between these charges and the negative charges on the balloon is stronger than the repulsion between the negative charges on the balloon and the negative charges in the polarized molecules (because they are farther from the balloon), so that there is a net attractive force toward the wall. Ionization processes in the air surrounding the balloon provide ions to which excess electrons in the balloon can transfer, reducing the charge on the balloon and eventually causing the attractive force to be insufficient to support the weight of the balloon.

**Q19.9** Would life be different if the electron were positively charged and the proton were negatively charged? Does the choice of signs have any bearing on physical and chemical interactions? Explain.

**Answer** No, life would not be different. The character and effect of electric forces are defined by (1) the fact that there are only two types of electric charge—positive and negative, and (2) the fact that opposite charges attract, whereas like charges repel. The choice of signs is completely arbitrary.

As a related exercise, you might consider what would happen in a world where there were three types of electric charge, or in a world where opposite charges repelled, and like charges attracted.

**Q19.15** If the total charge inside a closed surface is known but the distribution of the charge is unspecified, can you use Gauss's law to find the electric field? Explain.

**Answer** No. If we wish to use Gauss's law to find the electric field, we must be able to bring the electric field, $\vec{\mathbf{E}}$, out of the integral. This can be done, in some cases—when the field is constant, for example. However, since we do not know the charge distribution, we cannot claim that the field is constant, and thus cannot find the electric field.

To illustrate this point, consider a sphere that contained a net charge of 100 $\mu$C. The charges could be located near the center, or they could all be grouped at the northernmost point within the sphere. In either case, the net electric flux would be the same, but the electric field would vary greatly.

**Q19.17** A person is placed in a large, hollow, metallic sphere that is insulated from ground. If a large charge is placed on the sphere, will the person be harmed upon touching the inside of the sphere? Explain what will happen if the person also has an initial charge whose sign is opposite that of the charge on the sphere.

**Answer** The metallic sphere is a good conductor, so any excess charge on the sphere will reside on the outside of the sphere. From Gauss's law, we know that the field inside the sphere will then be zero. As a result, when the person touches the inside of the sphere, no charge will be exchanged between the person and the sphere, and the person will not be harmed.

What happens, then, if the person has an initial charge? Regardless of the sign of the person's initial charge, the charges in the conducting surface will redistribute themselves to maintain a net zero charge within the **conducting metal**. Thus, if the person has a 5.00 $\mu$C charge on his skin, exactly –5.00 $\mu$C will gather on the inner surface of the sphere; so the electric field within the metallic material will be zero. When the person touches the metallic sphere then, he will receive a shock due to the charge on his own skin.

**Q19.19** A common demonstration involves charging a rubber balloon, which is an insulator, by rubbing it on your hair and touching the balloon to a ceiling or wall, which is also an insulator. The electrical attraction between the charged balloon and the neutral wall results in the balloon sticking to the wall. Imagine now that we have two infinitely large flat sheets of insulating material. One is charged and the other is neutral. If these sheets are brought into contact, will an attractive force exist between them, as there was for the balloon and the wall?

*continued on next page*

**Answer** There will not be an attractive force. There are two factors to consider in the attractive force between a balloon and a wall, or between any pair of charged and neutral objects. The first factor is that the molecules in the wall will orient themselves with their negative ends toward the balloon, and their positive ends pointing away from the balloon. The second factor to consider is that the balloon is of finite curved dimensions, and thus the molecules in the wall are in a nonuniform electric field. Therefore the nearby "negative ends" of the molecules in the wall will experience an attractive electrostatic force that will be greater in magnitude than the repulsive force exerted on the more distant "positive ends" of the molecules. The net result is an overall force of attraction.

Now consider the infinite sheets brought into contact. The polarization of the molecules in the neutral sheet will indeed occur, as in the wall. But the electric field from the charged sheet is **uniform**, and therefore is independent of the distance from the sheet. Thus, both the negative and positive charges in the neutral sheet will experience the same electric field and the same magnitude of electric force. The attractive force on the negative charges will cancel with the repulsive force on the positive charges, and there will be no net force.

## SOLUTIONS TO SELECTED PROBLEMS

**P19.3** Nobel laureate Richard Feynman once said that if two persons stood at arm's length from each other and each person had 1% more electrons than protons, the force of repulsion between them would be enough to lift a "weight" equal to that of the entire Earth. Carry out an order-of-magnitude calculation to substantiate this assertion.

**Solution** Suppose each person has mass 70 kg. In terms of elementary charges, each person consists of precisely equal numbers of protons and electrons and a nearly equal number of neutrons. The electrons comprise very little of the mass, so we find the number of protons-and-neutrons in each person:

$$(70 \text{ kg})\left(\frac{1 \text{ u}}{1.66\times10^{-27} \text{ kg}}\right)=4\times10^{28} \text{ u}$$

Of these, nearly one half, $2\times10^{28}$, are protons, and 1% of this is $2\times10^{26}$, constituting a charge of $\left(2\times10^{26}\right)\left(1.60\times10^{-19} \text{ C}\right)=3\times10^{7} \text{ C}$.

Thus, Feynman's force is $F=\dfrac{k_e q_1 q_2}{r^2}=\dfrac{\left(8.99\times10^{9} \text{ N}\cdot\text{m}^2/\text{C}^2\right)\left(3\times10^{7} \text{ C}\right)^2}{(0.5 \text{ m})^2}\sim10^{26} \text{ N}$

*continued on next page*

where we have used a half-meter arm's length. According to the particle in a gravitational field model, if the Earth were in an externally-produced uniform gravitational field of magnitude $9.80\ \text{m/s}^2$,

it would weigh $F_g = mg = (6\times10^{24}\ \text{kg})(10\ \text{m/s}^2) \sim 10^{26}\ \text{N}$

Thus, the forces are of the same order of magnitude. ◊

**P19.5** Three point charges are located at the corners of an equilateral triangle as shown in Figure P19.5. Calculate the resultant electric force on the 7.00-$\mu$C charge.

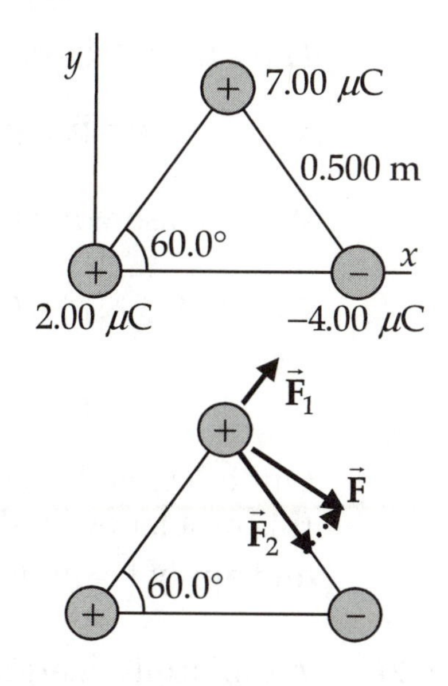

**FIG. P19.5**

**Solution** The 7.00-$\mu$C charge experiences a repulsive force $\vec{\mathbf{F}}_1$ due to the 2.00-$\mu$C charge, and an attractive force $\vec{\mathbf{F}}_2$ due to the −4.00-$\mu$C charge, where $F_2 = 2F_1$. If we sketch vectors representing $\vec{\mathbf{F}}_1$ and $\vec{\mathbf{F}}_2$ and the resultant, $\vec{\mathbf{F}}$ (see figure at lower right) we find that the resultant appears to be about the same magnitude as $F_2$ and is directed to the right about 30.0° below the horizontal.

We can find the net electric force by adding the two separate forces acting on the 7.00-$\mu$C charge. These individual forces can be found by applying Coulomb's law to each pair of charges.

The force on the 7.00-$\mu$C charge by the 2.00-$\mu$C charge is

$$\vec{\mathbf{F}}_1 = k_e \frac{q_1 q_2}{r^2}\hat{\mathbf{r}}$$

$$= \frac{(8.99\times10^9\ \text{N}\cdot\text{m}^2/\text{C}^2)(7.00\times10^{-6}\ \text{C})(2.00\times10^{-6}\ \text{C})}{(0.500\ \text{m})^2}(\cos 60°\,\hat{\mathbf{i}} + \sin 60°\,\hat{\mathbf{j}})$$

$$\vec{\mathbf{F}}_1 = (0.252\hat{\mathbf{i}} + 0.436\hat{\mathbf{j}})\ \text{N}$$

Similarly, the force on the 7.00-$\mu$C charge by the −4.00-$\mu$C charge is

*continued on next page*

$$\vec{\mathbf{F}}_2 = k_e \frac{q_1 q_3}{r^2}\hat{\mathbf{r}}$$

$$= -\frac{\left(8.99\times10^9\ \text{N}\cdot\text{m}^2/\text{C}^2\right)\left(7.00\times10^{-6}\ \text{C}\right)\left(-4.00\times10^{-6}\ \text{C}\right)}{(0.500\ \text{m})^2}\left(\cos 60°\hat{\mathbf{i}} - \sin 60°\hat{\mathbf{j}}\right)$$

$$\vec{\mathbf{F}}_2 = \left(0.503\hat{\mathbf{i}} - 0.872\hat{\mathbf{j}}\right)\text{ N}$$

Thus, the total force on the 7.00-$\mu$C charge, expressed as a set of components, is

$$\vec{\mathbf{F}} = \vec{\mathbf{F}}_1 + \vec{\mathbf{F}}_2 = \left(0.755\hat{\mathbf{i}} - 0.436\hat{\mathbf{j}}\right)\text{ N}$$ ◊

We can also write the total force as:

$$\vec{\mathbf{F}} = \sqrt{(0.755\ \text{N})^2 + (0.436\ \text{N})^2}\ \text{ at } \tan^{-1}\left(\frac{0.436\ \text{N}}{0.755\ \text{N}}\right) \text{ below the } +x \text{ axis}$$

$$\vec{\mathbf{F}} = 0.872\ \text{N at } 30.0° \text{ below the } +x \text{ axis}$$ ◊

Our calculated answer agrees with our initial estimate. An equivalent approach to this problem would be to find the net electric field due to the two lower charges and apply $\vec{\mathbf{F}} = q\vec{\mathbf{E}}$ to find the force on the upper charge in this electric field.

**P19.21** A uniformly charged insulating rod of length 14.0 cm is bent into the shape of a semicircle as shown in Figure P19.21. The rod has a total charge of $-7.50\ \mu$C. Find the magnitude and direction of the electric field at O, the center of the semicircle.

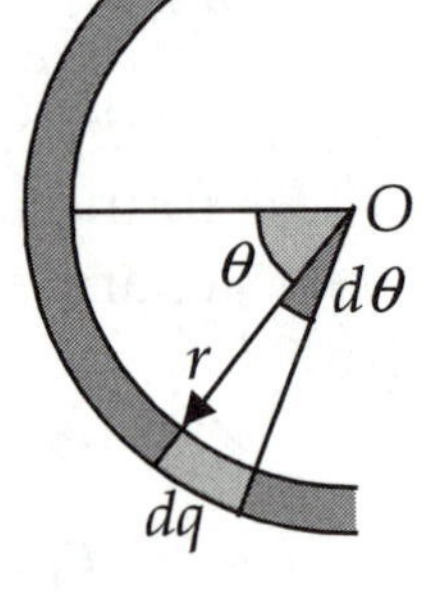


**FIG. P19.21**

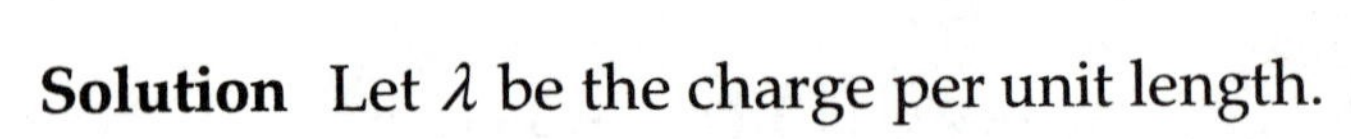

**Solution** Let $\lambda$ be the charge per unit length.

Then $\quad dq = \lambda ds = \lambda r d\theta$ and $dE = \dfrac{k_e dq}{r^2}$

In component form, $\quad E_y = 0$ (from symmetry) $dE_x = dE\cos\theta$

Integrating, $$E_x = \int dE_x = \int \frac{k_e \lambda r \cos\theta}{r^2} d\theta = \frac{k_e\lambda}{r}\int_{-\pi/2}^{\pi/2}\cos\theta d\theta = \frac{2k_e\lambda}{r}$$

But $\quad Q_{\text{total}} = \lambda\ell$, where $\ell = 0.140$ m, and $r = \dfrac{\ell}{\pi}$

Thus, $$E_x = \frac{2\pi k_e Q}{\ell^2} = \frac{2\pi\left(8.99\times10^9\ \text{N}\cdot\text{m}^2/\text{C}^2\right)\left(-7.50\times10^{-6}\ \text{C}\right)}{(0.140\ \text{m})^2}$$

$$\vec{\mathbf{E}} = \left(-2.16\times10^7\ \text{N/C}\right)\hat{\mathbf{i}}$$ ◊

**P19.25** A negatively charged rod of finite length carries charge with a uniform charge per unit length. Sketch the electric field lines in a plane containing the rod.

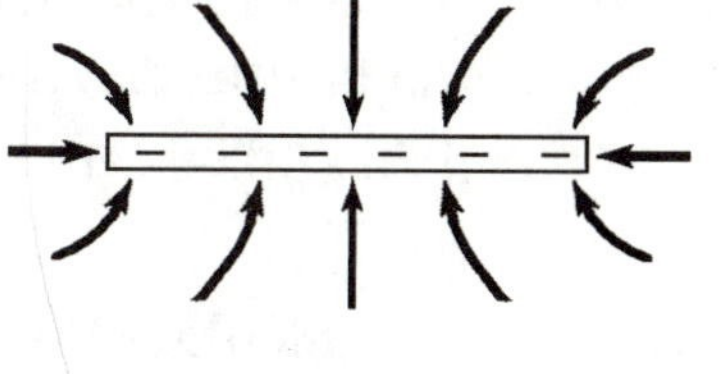
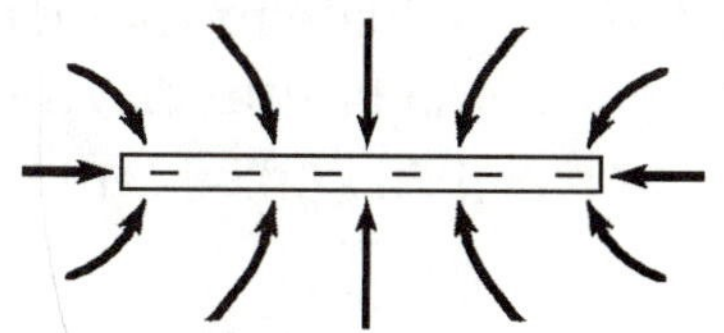

**FIG. P19.25**

**Solution** Since the rod has negative charge, field lines point inwards. Any field line points nearly toward the center of the rod at large distances, where the rod would look like just a point charge.

The lines curve to reach the rod perpendicular to its surface, where they end at equally-spaced points. ◊

**P19.27** A proton accelerates from rest in a uniform electric field of 640 N/C. At some later instant, its speed is $1.20\times10^{6}$ m/s (nonrelativistic, since $v$ is much less than the speed of light).

(a) Find the acceleration of the proton.

(b) After what time interval does the proton reach this speed?

(c) How far does the proton move in this time interval?

(d) What is its kinetic energy at the end of this time interval?

**Solution** (a) We use the particle in an electric field model and the particle under net force model.

$$a=\frac{F}{m}=\frac{qE}{m}=\frac{\left(1.602\times10^{-19}\text{ C}\right)\left(640\text{ N/C}\right)}{1.67\times10^{-27}\text{ kg}}=6.14\times10^{10}\text{ m/s}^2$$ ◊

(b) We use the particle under constant acceleration model.

$$\Delta t=\frac{\Delta v}{a}=\frac{1.20\times10^{6}\text{ m/s}}{6.14\times10^{10}\text{ m/s}^2}=19.5\ \mu\text{s}$$ ◊

(c) $$\Delta x=v_it+\frac{1}{2}at^2=0+\frac{1}{2}\left(6.14\times10^{10}\text{ m/s}^2\right)\left(19.5\times10^{-6}\text{ s}\right)^2=11.7\text{ m}$$ ◊

(d) $$K=\frac{1}{2}mv^2=\frac{1}{2}\left(1.67\times10^{-27}\text{ kg}\right)\left(1.20\times10^{6}\text{ m/s}\right)^2=1.20\times10^{-15}\text{ J}$$ ◊

**P19.31** A 40.0-cm-diameter loop is rotated in a uniform electric field until the position of maximum electric flux is found. The flux in this position is measured to be $5.20 \times 10^5\ \text{N} \cdot \text{m}^2/\text{C}$. What is the magnitude of the electric field?

**Solution** We calculate the flux as $\Phi = \vec{\mathbf{E}} \cdot \vec{\mathbf{A}} = EA\cos\theta$

The maximum value of the flux occurs when $\theta = 0$

Therefore, we can calculate the field strength at this point as

$$E = \frac{\Phi_{\max}}{A} = \frac{\Phi_{\max}}{\pi r^2}$$

$$E = \frac{5.20 \times 10^5}{\pi (0.200\ \text{m})^2} = 4.14 \times 10^6\ \text{N/C}$$ ◊

**P19.33** A point charge $Q$ is located just above the center of the flat face of a hemisphere of radius $R$ as shown in Figure P19.33. What is the electric flux

(a) through the curved surface and

(b) through the flat face?

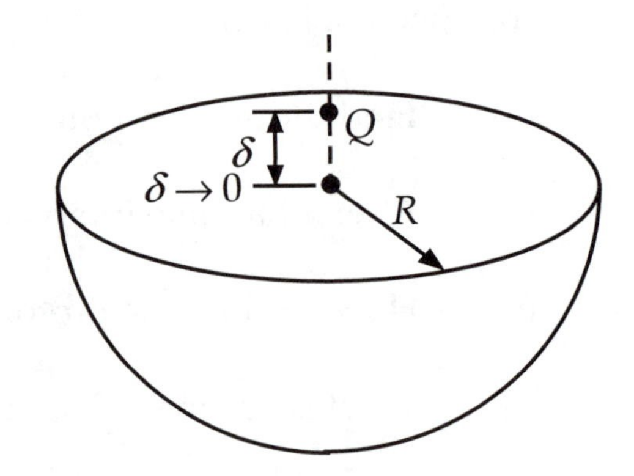


**FIG. P19.33**

**Solution** **Conceptualize:** From Gauss's law, the flux through a sphere with a point charge in it should be $\frac{Q}{\epsilon_0}$, so we should expect the electric flux through a hemisphere to be half this value:

$$\Phi_{\text{curved}} = \frac{Q}{2\epsilon_0}$$

Because the flat section appears like an infinite plane to a point just above its surface, so that half of all the field lines from the point charge are intercepted by the flat surface, the flux through this section should also equal $\frac{Q}{2\epsilon_0}$.

**Categorize:** We can apply the definition of electric flux directly for part (a) and then use Gauss's law to find the flux for part (b).

*continued on next page*

**Analyze:** (a) With $\delta$ very small, all points on the hemisphere are nearly at distance $R$ from the charge, so the field everywhere on the curved surface is $\frac{k_e Q}{R^2}$ radially outward (normal to the surface). Therefore, the flux is this field strength times the area of half a sphere:

$$\Phi_{\text{curved}} = \int \vec{\mathbf{E}} \cdot d\vec{\mathbf{A}} = E_{\text{local}} A_{\text{hemisphere}}$$

$$\Phi_{\text{curved}} = \left(k_e \frac{Q}{R^2}\right)\left(\frac{1}{2}\right)\left(4\pi R^2\right) = \frac{1}{4\pi \epsilon_0} Q(2\pi) = \frac{Q}{2\epsilon_0} \qquad \Diamond$$

(b) The closed surface encloses zero charge so Gauss's law gives

$$\Phi_{\text{curved}} + \Phi_{\text{flat}} = 0 \qquad \text{or} \qquad \Phi_{\text{flat}} = -\Phi_{\text{curved}} = \frac{-Q}{2\epsilon_0} \qquad \Diamond$$

**Finalize:** The direct calculations of the electric flux agree with our predictions, except for the negative sign in part (b), which comes from the fact that the area unit vector is defined as pointing outward from an enclosed surface, and in this case, the electric field has a component in the opposite direction (down).

**P19.39** Consider a long cylindrical charge distribution of radius $R$ with a uniform charge density $\rho$. Find the electric field at distance $r$ from the axis where $r < R$.

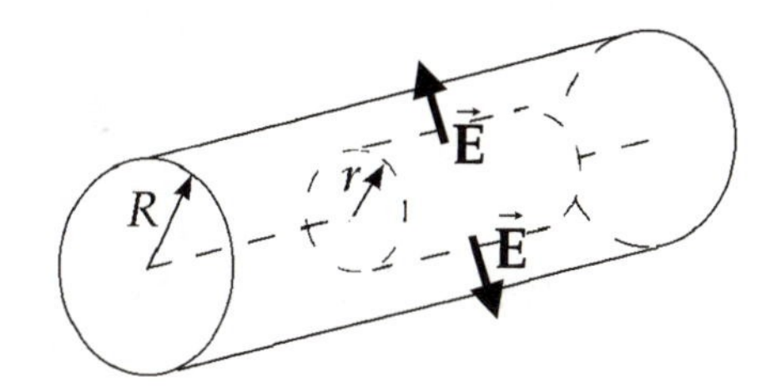


FIG. P19.39

**Solution** If $\rho$ is positive, the field must everywhere be radially outward. Choose as the gaussian surface a cylinder of length $L$ and radius $r$, contained inside the charged rod. Its volume is $\pi r^2 L$ and it encloses charge $\rho \pi r^2 L$. The circular end caps have no electric flux through them; there $\vec{\mathbf{E}} \cdot d\vec{\mathbf{A}} = EdA\cos 90.0° = 0$. The curved surface has $\vec{\mathbf{E}} \cdot d\vec{\mathbf{A}} = EdA\cos 0°$, and $E$ must be the same strength everywhere over the curved surface.

Then $\oint \vec{\mathbf{E}} \cdot d\vec{\mathbf{A}} = \frac{q}{\epsilon_0}$ becomes $E \int_{\substack{\text{Curved}\\\text{Surface}}} dA = \frac{\rho \pi r^2 L}{\epsilon_0}$.

Noting that $2\pi r L$ is the lateral surface area of the cylinder, $E(2\pi r)L = \frac{\rho \pi r^2 L}{\epsilon_0}$.

Thus, $\vec{\mathbf{E}} = \frac{\rho r}{2\epsilon_0}$ radially away from the axis. $\Diamond$

*continued on next page*

The field is proportional to the charge density. At the axis of the filament, $r = 0$, and the field is zero. It must be zero, because it has no preferred direction to point in. At the surface, $E = \dfrac{\rho R}{2\in_0}$ agrees with the field outside, $E = \dfrac{\lambda}{2\pi R\in_0}$ because

$\lambda L = \rho\pi R^2 L$ so $\lambda = \rho\pi R^2$ and $\dfrac{\lambda}{2\pi R\in_0} = \dfrac{\rho\pi R^2}{2\pi R\in_0} = \dfrac{\rho R}{2\in_0}$.

**P19.49** A thin, square conducting plate 50.0 cm on a side lies in the *xy* plane. A total charge of $4.00 \times 10^{-8}$ C is placed on the plate. Find

(a) the charge density on the plate,

(b) the electric field just above the plate, and

(c) the electric field just below the plate. You may assume that the charge density is uniform.

**Solution** In this problem ignore "edge" effects and assume that the total charge distributes uniformly over each side of the plate (one half the total charge on each side).

(a) $\sigma = \dfrac{q}{A} = \dfrac{4.00 \times 10^{-8}\ \text{C}}{2(0.500\ \text{m})^2} = 8.00 \times 10^{-8}\ \text{C/m}^2$ ◊

(b) Just above the plate, $E = \dfrac{\sigma}{\in_0} = \dfrac{8.00 \times 10^{-8}\ \text{C/m}^2}{8.85 \times 10^{-12}\ \text{C}^2/\text{N}\cdot\text{m}^2} = 9.04 \times 10^3\ \text{N/C}$

upward. ◊

(c) Just below the plate, $E = \dfrac{\sigma}{\in_0} = 9.04 \times 10^3\ \text{N/C}$ downward. ◊

**P19.59** Two small spheres of mass $m$ are suspended from strings of length $\ell$ that are connected at a common point. One sphere has charge $Q$, and the other has charge $2Q$. The strings make angles $\theta_1$ and $\theta_2$ with the vertical.

(a) How are $\theta_1$ and $\theta_2$ related?

(b) Assume that $\theta_1$ and $\theta_2$ are small. Show that the distance $r$ between the spheres is given by

$$r \approx \left(\frac{4k_e Q^2 \ell}{mg}\right)^{1/3}.$$

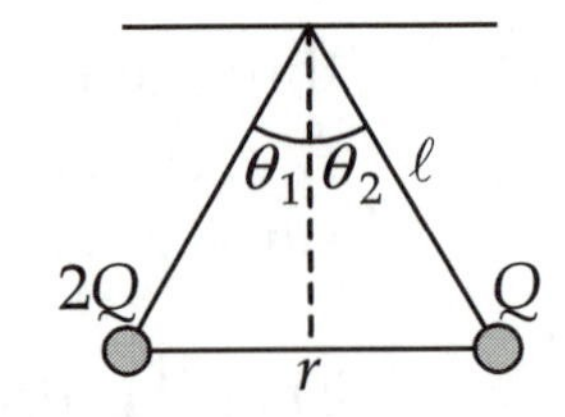


**FIG. P19.59**

*continued on next page*

**Solution** We use the particle in equilibrium model.

(a) The spheres have different charges, but each exerts an equal force on the other, given by $F_e = \dfrac{k_e(Q)(2Q)}{r^2}$, where $r$ is the distance between them. Since their masses are equal, $\theta_2 = \theta_1$. ◊

(b) For equilibrium, $\sum F_y = 0$: $T\cos\theta - mg = 0$, thus $T = \dfrac{mg}{\cos\theta}$

$\sum F_x = 0$: $F_e - T\sin\theta = 0$.

Substituting for $T$,

$$F_e = \frac{mg\sin\theta}{\cos\theta} = mg\tan\theta.$$

For small angles, $\tan\theta \approx \sin\theta = \dfrac{r}{2\ell}$.

Therefore,

$$F_e \approx mg\frac{r}{2\ell}.$$

The force $F_e$ is

$$\frac{k_e Q(2Q)}{r^2} \approx mg\frac{r}{2\ell}$$

so that

$$4k_e Q^2 \ell \approx mgr^3 \text{ and } r \approx \left(\frac{4k_e Q^2 \ell}{mg}\right)^{1/3}.$$ ◊

The value of $r$ increases with increasing string length $\ell$. It increases as the charge $Q$ increases and decreases as the mass increases. To check for dimensional correctness, we evaluate the units of the right-hand side:

$$\left(\frac{\text{N}\cdot\text{m}^2\ \text{C}^2\ \text{m s}^2}{\text{C}^2\ \text{kg m}}\right)^{1/3} = \left(\frac{\text{kg m}\cdot\text{m}^2\ \text{s}^2}{\text{s}^2\ \text{kg}}\right)^{1/3} = \text{m, as required.}$$

**P19.63** Repeat the calculations for Problem 19.62 when both sheets have *positive* uniform surface charge densities of value $\sigma$.

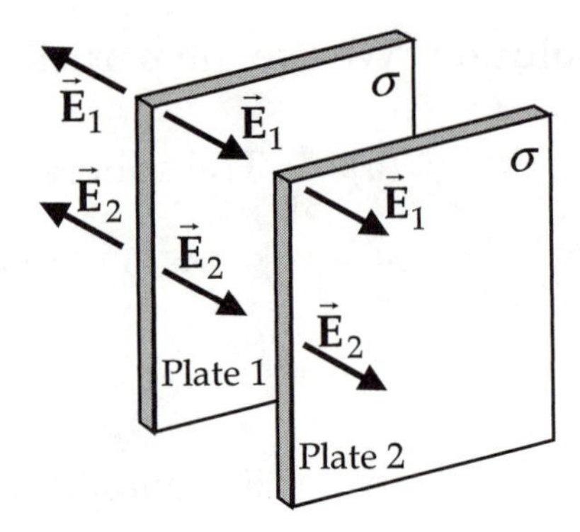


**FIG. P19.63**

**Solution** The new, modified problem statement reads:

"Two infinite, nonconducting sheets of charge are parallel to each other, as shown in Figure P19.63. The sheet on the left has a uniform surface charge density $\sigma$, and the one on the right has a uniform charge density $\sigma$."

(a) Calculate the electric field at points to the left of the two sheets.

(b) Calculate the electric field at points in between the two sheets.

(c) Calculate the electric field at points to the right of the two sheets.

**Conceptualize:** When both sheets have the same charge density, a positive test charge at a point midway between them will experience the same force in opposite directions from each sheet. Therefore, the electric field here will be zero. (We should ask: can we also conclude that the test charge will experience equal and oppositely directed forces *everywhere* in the region between the plates?)

Outside the sheets the electric field will point away and should be twice the strength due to one sheet of charge, so $E = \frac{\sigma}{\epsilon_0}$ in these regions.

**Categorize:** The principle of superposition can be applied to add the electric field vectors due to each sheet of charge.

**Analyze:** For each sheet, the electric field at any point is $\left|\vec{\mathbf{E}}\right| = \frac{\sigma}{2\epsilon_0}$ directed away from the sheet.

(a) At a point to the left of the two parallel sheets

$$\vec{\mathbf{E}} = E_1(-\hat{\mathbf{i}}) + E_2(-\hat{\mathbf{i}}) = 2E(-\hat{\mathbf{i}}) = -\frac{\sigma}{\epsilon_0}\hat{\mathbf{i}}. \qquad \Diamond$$

(b) At a point between the two sheets $\quad \vec{\mathbf{E}} = E_1\hat{\mathbf{i}} + E_2(-\hat{\mathbf{i}}) = 0. \qquad \Diamond$

*continued on next page*

(c) At a point to the right of the two parallel sheets

$$\vec{\mathbf{E}} = E_1\hat{\mathbf{i}} + E_2\hat{\mathbf{i}} = 2E\hat{\mathbf{i}} = \frac{\sigma}{\epsilon_0}\hat{\mathbf{i}}. \qquad \Diamond$$

**Finalize:** We essentially solved this problem in the Conceptualize step, so it is no surprise that these results are what we expected. A better check is to confirm that the results are complementary to the case where the plates are oppositely charged (Problem 19.62).

**P19.65** A solid, insulating sphere of radius $a$ has a uniform charge density $\rho$ and a total charge $Q$. Concentric with this sphere is an uncharged, conducting hollow sphere whose inner and outer radii are $b$ and $c$ as shown in Figure P19.65.

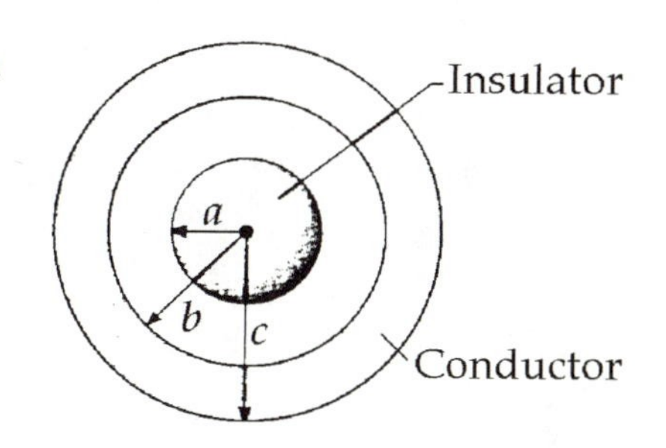


**FIG. P19.65**

(a) Find the magnitude of the electric field in the regions $r < a$, $a < r < b$, $b < r < c$, and $r > c$.

(b) Determine the induced charge per unit area on the inner and outer surfaces of the hollow sphere.

**Solution** (a) Choose as the gaussian surface a concentric sphere of radius $r$. The electric field will be perpendicular to its surface, and will be uniform in strength over its surface.

The surface of radius $r < a$ encloses charge $\rho\left(\frac{4}{3}\pi r^3\right)$

so $\Phi = \frac{q}{\epsilon_0}$ becomes $E\left(4\pi r^2\right) = \frac{\rho\left[(4/3)\pi r^3\right]}{\epsilon_0}$ and $E = \frac{\rho r}{3\epsilon_0}$. $\Diamond$

For $a < r < b$, we have $E\left(4\pi r^2\right) = \frac{4}{3}\rho\frac{\pi a^3}{\epsilon_0} = \frac{Q}{\epsilon_0}$ and $E = \frac{\rho a^3}{3\epsilon_0 r^2} = \frac{Q}{4\pi\epsilon_0 r^2}$. $\Diamond$

For $b < r < c$, we must have $E = 0$ $\Diamond$

because any nonzero field would be moving charges in the metal. Free charges did move in the metal to deposit charge $-Q_b$ on its inner surface, at radius $b$, leaving charge $+Q_c$ on its outer surface, at radius $c$.

Since the shell as a whole is neutral, $Q_c - Q_b = 0$.

For $r > c$, $\Phi = \frac{q}{\epsilon_0}$ becomes $E\left(4\pi r^2\right) = \frac{Q + Q_c - Q_b}{\epsilon_0}$ and $E = \frac{Q}{4\pi\epsilon_0 r^2}$. $\Diamond$

*continued on next page*

(b) For a gaussian surface of radius $b < r < c$, we have $0 = \frac{Q - Q_b}{\epsilon_0}$.

So $Q_b = Q$ and the charge density on the inner surface is $\frac{-Q_b}{A} = \frac{-Q}{4\pi b^2}$. ◊

Then $Q_c = Q_b = Q$ and the charge density on the outer surface is $+\frac{Q}{4\pi c^2}$. ◊

# Chapter 20

# Electric Potential and Capacitance

## NOTES FROM SELECTED CHAPTER SECTIONS

### Section 20.1 Potential Difference and Electric Potential

The electrostatic force is conservative; therefore, it is possible to define an electric potential energy function associated with this force. The change in potential between two points is proportional to the change in potential energy of a charge as it moves between the two points.

Electrical potential difference (a scalar quantity) is the work done to move a charge from point $A$ to point $B$ divided by the magnitude of the charge. Thus, the SI units of potential are joules per coulomb, and are called volts (V).

***A positive charge gains electrical potential energy when it is moved opposite the direction of the electric field.*** When a positive charge is released in an electric field, it moves in the direction of the field, from a point of high potential to a point of lower potential.

### Section 20.2 Potential Differences in a Uniform Electric Field

Electric field lines always point in the direction of decreasing electric potential. When a positive charge moves along the direction of the electric field, the electric potential energy of the charge-field system decreases. An equipotential surface is any surface consisting of a continuous distribution of points having the same electric potential. Equipotential surfaces are perpendicular to electric field lines and no work is required to move any charge between two points on an equipotential surface.

## Section 20.3 Electric Potential and Potential Energy Due to Point Charges

In electric circuits, a point of zero potential is often defined by grounding (connecting to Earth) some point in the circuit. In the case of a point charge, the point of zero potential is taken to be at an infinite distance from the charge. The electric potential at a given point in space due to a point charge $q$ depends only on the value of the charge and the distance, $r$, from the charge to the specified point in space. An electric potential can exist at a point in space whether or not a test charge exists at that point.

**Electric potential is a scalar quantity.** ***When using the superposition principle to determine the value of the electric potential at a point due to several point charges, the algebraic sum of the individual electric potentials must be used. Be careful not to confuse electric potential difference with electric potential energy.***

- Electric potential is a scalar property of the region (electric field) surrounding an electric charge. It does not depend on the presence of a test charge in the field.

- Electric potential energy is a characteristic of a charge-field system. It is due to the interaction of the field and a charge located within the field.

- In the figure to the right an electric field, due to the presence of +Q, is present in the region surrounding the charge. This field due to a point charge has the properties described in Chapter 19.

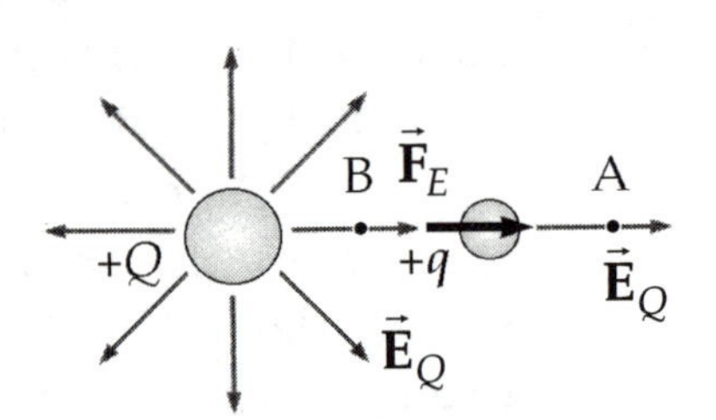


Refer to the figure as you consider the following statements about electric potential difference and electric potential energy.

- Electric field lines are directed along outward radials from the charge Q, and points "A" and "B" are shown along one of the field lines.

- If a positive test charge $q$ is placed at some point between A and B, the electric field will exert a force $F_E = qE$ on $q$ directed along the field line and away from Q.

- If the test charge $q$ is moved from point A to point B (where the potential is greater), the electric field will do work, $W = -q\Delta V = -q(V_B - V_A)$, and the electric potential energy of the two-charge system ($Q$ and $q$) changes by an amount equal to $q\Delta V$. ***The electric potential energy of a positive charge increases when the charge moves to a point closer to another isolated positive point charge (or along a direction opposite the direction of the electric field).***

- ***Since the electrostatic force is a conservative force, the work done by the electric field as the test charge q moves from point A to point B will be the same regardless of the actual path taken.***

You should consider the consequences of changing either or both $Q$ and $q$ from a positive to a negative charge.

## Section 20.4 Obtaining Electric Field From Electric Potential

If the electric potential (which is a scalar) is known as a function of coordinates ($x$, $y$, $z$), the components of the electric field (a vector quantity) can be obtained by taking the negative derivative of the potential with respect to the coordinates.

## Section 20.5 Potentials and Charged Conductors

## Section 20.6 Equipotential Surfaces

No work is required to move a charge between two points that are at the same potential. That is, $W = 0$ when $V_B = V_A$.

The electric potential is a constant everywhere on the surface of a charged conductor in electrostatic equilibrium.

The electric potential is constant everywhere inside a conductor and equal to its value at the surface. This is true because the electric field is zero inside a conductor and no work is required to move a charge between two points inside the conductor.

The electron volt (eV) is defined as the energy that an electron (or proton) gains when accelerated through a potential difference of 1 V.

A surface on which every point has the same value of electric potential is an equipotential surface. The electric field at every point on an equipotential surface is perpendicular to the surface; otherwise a parallel component of E would imply work would be required to move a charge between two points on the surface.

## Section 20.7 Capacitance

A capacitor is a device consisting of a pair of conductors (plates) separated by insulating material. A charged capacitor acts as a storehouse of charge and energy that can be reclaimed when needed for a specific application. The capacitance of a capacitor depends on the physical characteristics of the device (size, shape, plate separation, and the nature of the dielectric medium filling the region between the plates).

The capacitance, C, of a capacitor is defined as the ratio of the magnitude of the charge on either conductor to the magnitude of the potential difference between the conductors. Capacitance has SI units of coulombs per volt, called farads (F). The farad is a very large unit of capacitance. In practice, most capacitors have capacitances ranging from picofarads to microfarads.

## Section 20.8 Combinations of Capacitors

Two or more capacitors can be connected in a circuit in several possible combinations. For example, three capacitors (each assumed to have the same value of capacitance, C) can be combined as illustrated in the figure at right to achieve four different values of equivalent capacitance.

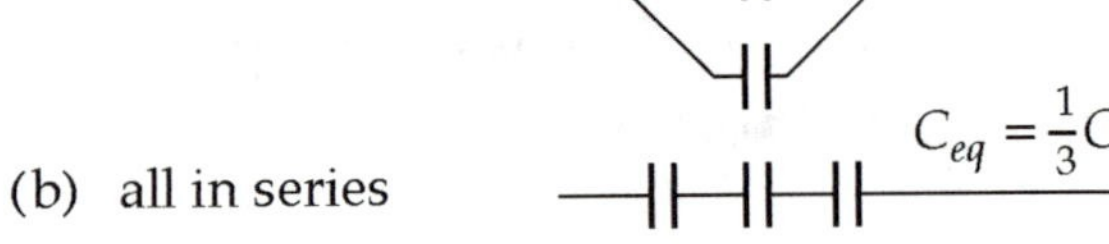


Can you determine how many different values of equivalent capacitance could be achieved by combining the three capacitors if they each had a different capacitance, $C_1$, $C_2$, and $C_3$?

*Note in the figure above that when a group of capacitors are connected in series they are arranged "end-to-end" and only adjacent capacitors have a circuit point in common. When connected in parallel, each capacitor in the group has two circuit points in common with each of the others.*

**For capacitors connected in series:**

- The reciprocal of the equivalent capacitance equals the sum of the reciprocals of the individual capacitances. See Equation 20.29.

- The potential difference across the series group equals the sum of the potential differences across the individual capacitors.

- Each capacitor of a group in series has the same charge (even if their capacitances are not the same) and this is also the value of the total charge on the series group.

**For capacitors connected in parallel:**

- The equivalent capacitance equals the sum of the individual capacitances. See Equation 20.25.

- The potential difference across each capacitor has the same value.

- The total charge on the parallel group equals the sum of the charges on the individual capacitors.

## Section 20.9 Energy Stored in a Charged Capacitor

The electrostatic potential energy stored in a charged capacitor equals the work done in the charging process—moving charges from one conductor at a lower potential to another conductor at a higher potential.

Each capacitor has a limiting voltage that depends on the capacitor's physical characteristics. When the potential difference between the plates of a capacitor exceeds that limiting voltage, a discharge will occur through the insulating material between the two plates. This is known as electrical breakdown. Therefore, the maximum energy that can be stored in a capacitor is limited by the breakdown voltage.

## Section 20.10 Capacitors with Dielectrics

A dielectric is an insulating material, such as rubber, glass, or waxed paper. When a dielectric is inserted between the plates of a capacitor, the capacitance increases. If the dielectric completely fills the space between the plates, the capacitance is multiplied by a dimensionless factor $\kappa$, called the dielectric constant. The dielectric constant is a property of the dielectric material and has a value of 1.000 for vacuum.

In general, the use of a dielectric has the following effects:

- It increases the capacitance.

- It increases the maximum operating voltage.

- It provides mechanical support for the two conductors.

## EQUATIONS AND CONCEPTS

The **potential difference** between two points, $\Delta V = V_B - V_A$, is defined as the change in the potential energy of a charge-field system when a test charge, $q_0$, is moved from point $A$ to point $B$ divided by the test charge $q_0$. $\Delta V$ can be evaluated by integrating $\vec{\mathbf{E}} \cdot d\vec{\mathbf{s}}$ along any path from $A$ to $B$. *Electric potential is a characteristic of an electric field, independent of any charges that may be placed in the field. The SI unit of electric potential is the volt (V).*

$$\Delta V = \frac{\Delta U}{q_0} = -\int_A^B \vec{\mathbf{E}} \cdot d\vec{\mathbf{s}} \quad (20.3)$$

$$1 \text{ V} = 1 \text{ J/C}$$

$$1 \text{ N/C} = 1 \text{ V/m}$$

The **electron volt** (eV) is a unit of energy that is defined as the energy that a charge-field system gains or loses when a charge of magnitude $e$ is moved through a potential difference of 1 V.

$$1 \text{ eV} = 1.60 \times 10^{-19} \text{ J} \quad (20.5)$$

In a **uniform electric field**, the potential difference between two points depends on the displacement $d$ along the direction parallel to $\vec{\mathbf{E}}$ and on the magnitude of $E$. *Electric field lines always point in the direction of decreasing electric potential. In the figure, the potential at point B is lower than that at point A.*

$$\Delta V = -E\int_A^B ds = -Ed \quad (20.6)$$

(when $d \| \vec{\mathbf{E}}$)

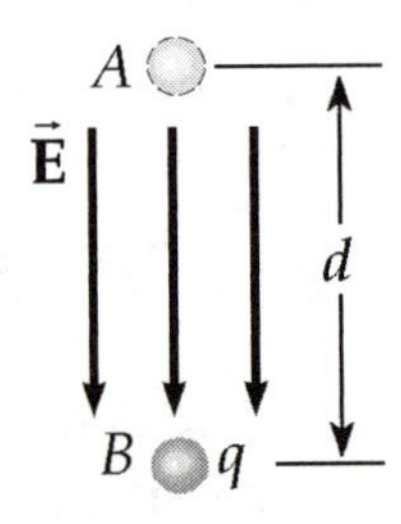


The **change in potential energy**, $\Delta U$, of a charge-field system as a charge moves from point $A$ to point $B$ in an electric field depends on the sign and magnitude of the charge as well as on the change in potential, $\Delta V$. *The electric potential energy of a charge-field system decreases when a positive charge moves in the direction of the field. See figure above.*

$$\Delta U = q_0 \Delta V = -q_0 Ed \quad (20.7)$$

The **electric potential due to a single point charge**, at a distance *r* from the charge, depends inversely on the distance from the charge. The potential is given by Equation 20.11 when the zero reference level for potential is taken to be at infinity.

$$V = k_e \frac{q}{r} \tag{20.11}$$

The **electric potential in the vicinity of several point charges** is the scalar sum of the potentials due to the individual charges.

$$V = k_e \sum_i \frac{q_i}{r_i} \tag{20.12}$$

The **electric potential energy associated with a pair of charges** separated by a distance *r* represents the work required to assemble the charges from an infinite separation. Hence, the negative of the potential energy equals the minimum work required to separate them by an infinite distance. *The electric potential energy associated with a system of two charged particles is positive if the two charges have the same sign, and negative if they are of opposite sign.*

$$U = k_e \frac{q_1 q_2}{r_{12}} \tag{20.13}$$

The **total potential energy of a group of charges** is found by calculating $U$ for each pair of charges and summing the terms algebraically.

$$U = \frac{k_e}{2} \sum_{i=1}^{N} \sum_{j=1}^{N} \frac{q_i q_j}{r_{ij}}$$

when $i \neq j$

The **components of the electric field** are the negative partial derivatives of the electric potential. The **components of the electric field** in rectangular coordinates are given here in terms of partial derivatives of the potential.

$$E_x = -\frac{\partial V}{\partial x} \tag{20.16}$$

$$E_y = -\frac{\partial V}{\partial y}$$

$$E_z = -\frac{\partial V}{\partial z}$$

In the case of a **spherical symmetric charge distribution**, the charge density depends only on the radial distance, *r*, and the electric field is radial.

$$E_r = -\frac{dV}{dr} \tag{20.15}$$

**For a continuous charge distribution, the electric potential** (relative to zero at infinity) can be calculated by integrating the contribution due to a charge element $dq$ over the line, surface, or volume that contains all the charge. It is convenient to represent $dq$ in terms of the appropriate charge density (line, area, or volume).

$$V = k_e \int_{\text{All Charge}} \frac{dq}{r} \qquad (20.18)$$

$dq = \lambda dx$ for a charged line
$dq = \sigma dA$ for a charged surface
$dq = \rho dV$ for a charged volume

The **capacitance** of a capacitor is defined as the ratio of the charge on either conductor (or plate) to the magnitude of the potential difference between the conductors. *Capacitance is always a positive quantity and has SI units of farads (F).*

$$C \equiv \frac{Q}{\Delta V} \qquad (20.19)$$

The **capacitance of an air-filled parallel-plate capacitor** is proportional to the area of the plates and inversely proportional to the separation of the plates.

$$C = \frac{\epsilon_0 A}{d} \qquad (20.21)$$

When a **dielectric material** (insulator) completely fills the region between the plates of a capacitor, the capacitance increases by a factor $\kappa$. *Kappa, called the dielectric constant, is dimensionless and is characteristic of a particular material.* Equation 20.34 is for the case of a parallel-plate capacitor.

$$C = \kappa \frac{\epsilon_0 A}{d} \qquad (20.34)$$

**For capacitors connected in parallel:**

- Each capacitor has two circuit points in common with each of the other capacitors in the group.

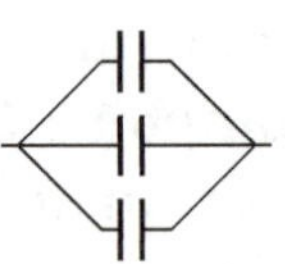

If the value of each capacitor is $C$, $C_{eq} = 3C$.

- The **equivalent capacitance** equals the algebraic sum of the individual capacitances.

$$C_{eq} = C_1 + C_2 + C_3 + \ldots \qquad (20.25)$$

- The **total charge** is the sum of charges on the individual capacitors.

$$Q_{total} = Q_1 + Q_2 + Q_3 + \ldots$$

- The **potential difference** across each capacitor is the same.

$$\Delta V_1 = \Delta V_2 = \Delta V_3 = \ldots$$

**For capacitors connected in series:**

- Adjacent capacitors have one circuit point in common.

If each capacitor has value $C$, $C_{eq} = \frac{C}{3}$

- The inverse of the **equivalent capacitance** equals the algebraic sum of the inverses of the capacitances of the individual capacitors.

$$\frac{1}{C_{eq}} = \frac{1}{C_1} + \frac{1}{C_2} + \frac{1}{C_3} + \ldots \qquad (20.29)$$

- The **potential difference** across a series group of capacitors equals the sum of the potential differences across the individual capacitors.

$$\Delta V_{total} = \Delta V_1 + \Delta V_2 + \Delta V_3 \ldots$$

- The **total charge** on the group is equal to the charge on each capacitor.

$$Q_{total} = Q_1 = Q_2 = Q_3 \ldots$$

A **series-parallel combination** of capacitors can be arranged using three or more capacitors. Using Equations 26.8 and 26.10 in sequence, each combination can be reduced to a single equivalent capacitor.

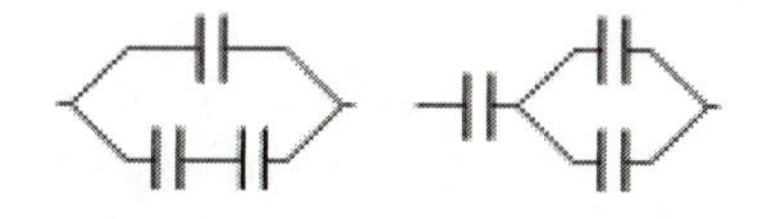

If each capacitor has value $C$, then in the first circuit $C_{eq} = \frac{3}{2C}$ and in the second circuit $C_{eq} = \frac{2}{3C}$.

The **electrostatic energy** stored in the electrostatic field of a charged capacitor equals the work done (by a battery or other source) in charging the capacitor from $q = 0$ to $q = Q$.

$$U = \frac{Q^2}{2C} = \frac{1}{2}Q\Delta V = \frac{1}{2}C(\Delta V)^2 \qquad (20.30)$$

The **energy density**, energy per unit volume, at any point in an electrostatic field is proportional to the square of the electric field intensity at that point.

$$u_E = \frac{1}{2}\epsilon_0 E^2 \qquad (20.32)$$

## SUGGESTIONS, SKILLS, AND STRATEGIES

The vector expression giving the components of the electric field over a region can be obtained from the scalar function that describes the electric potential, $V$, over the region by using Equation 20.15 and similar expressions for the $y$ and $z$ components:

$$E_x = -\frac{\partial V}{\partial x} \qquad E_y = -\frac{\partial V}{\partial y} \qquad E_z = -\frac{\partial V}{\partial z}$$

The derivatives in the above expressions are called partial derivatives. This means that when the derivative is taken with respect to any one coordinate, any other coordinates that appear in the expression for the potential function are treated as constants.

Since the electrostatic force is a conservative force, the work done by the electrostatic force in moving a charge $q$ from an initial point $A$ to a final point $B$ depends only on the location of the two points and is independent of the path taken between $A$ and $B$. When calculating potential difference, any path between $A$ and $B$ may be chosen to evaluate the integral in Equation 20.3. You should select a path for which the evaluation of the "line integral" will be as convenient as possible.

### Problem-Solving Hints for Electric Potentials

- When working problems involving electric potential, remember that potential is a scalar quantity (rather than a vector quantity like the electric field), so there are no components to worry about. Therefore, when using the superposition principle to evaluate the electric potential at a point due to a system of point charges, you simply take the algebraic sum of the potentials due to each charge. However, you must keep track of signs. The potential created by each positive charge $\left(V = \frac{k_e q}{r}\right)$ is positive, while the potential for each negative charge is negative.

- Just as in mechanics, only changes in electric potential are significant, hence the point where you choose the potential to be zero is arbitrary. When dealing with point charges or a finite-sized charge distribution, we usually define $V = 0$ to be at a point infinitely far from the charges. However, if the charge distribution itself extends to infinity, some other nearby point must be selected as the reference point.

- The electric potential at some point $P$ due to a continuous distribution of charge can be evaluated by dividing the charge distribution into infinitesimal elements of charge $dq$ located at a distance $r$ from the point $P$. You then treat this element as a point charge, so that the potential at $P$ due to the element is $dV = \frac{k_e dq}{r}$. The total potential at $P$ is obtained by integrating $dV$ over the entire charge distribution. In performing the integration for most problems, it is necessary to express $dq$ and $r$ in terms of a single variable. In order to simplify the integration, it is important to give careful consideration of the geometry involved in the problem.

- Another method that can be used to obtain the potential due to a finite continuous charge distribution is to start with the definition of the potential difference given by Equation 20.3. If $\vec{\mathbf{E}}$ is known or can be obtained easily (say from Gauss's law), then the line integral can be evaluated.

- When you know the electric potential at a point, it is possible to obtain the electric field at that point by remembering that the electric field is equal to the negative of the derivative of the potential with respect to some coordinate.

- Note that $V$ is used to denote the value of the electric potential at a point while $\Delta V$ is used to denote the potential difference between two specified points If the electric potential has values $V_A$ and $V_B$ at points A and B, the electric potential difference in moving from point A to point B will be $\Delta V = V_B - V_A$.

In practice, a variety of phrases are used to describe the potential difference between two points, the most common being "voltage." A voltage applied to a device or across a device has the same meaning as the potential difference across the device. For example, if we say that the voltage across a certain capacitor is 12 volts, we mean that the potential difference between the plates of the capacitor is 12 volts.

## Problem-Solving Hints for Capacitance

- When analyzing a series-parallel combination of capacitors to determine the equivalent capacitance, you should make a sequence of circuit diagrams, which show the successive steps in the simplification of the circuit. At each step combine and calculate the equivalent capacitance of those capacitors that are in either simple-series or simple-parallel combination with each other. Continue the sequence of simplifying steps until the final circuit contains a single capacitor, the value of which is the equivalent capacitance of the original circuit. At each step, you know two of the three quantities: $Q$, $\Delta V$, and $C$. You will be able to determine the remaining quantity using the relation $Q = C\Delta V$.

- When calculating capacitance, be careful with your choice of units. To calculate capacitance in farads, make sure that distances are in meters and use the SI value of $\in_0$. When checking consistency of units, remember that the units for electric fields are newtons per coulomb (N/C) or the equivalent volts per meter (V/m).

- When two or more unequal capacitors are connected in series, they carry the same charge, but their potential differences are not the same. The capacitances add as reciprocals, and the equivalent capacitance of the combination is always less than the smallest individual capacitor.

- When two or more capacitors are connected in parallel, the potential differences across them are the same. The charge on each capacitor is proportional to its capacitance; hence, the capacitances add directly to give the equivalent capacitance of the parallel combination.

- A dielectric increases capacitance by the factor $\kappa$ (the dielectric constant) because induced surface charges on the dielectric reduce the electric field inside the material from $E$ to $\dfrac{E}{\kappa}$.

- Be careful about problems in which you may be connecting or disconnecting a battery to a capacitor. It is important to note whether modifications to the capacitor are being made while the capacitor is connected to the battery or after it is disconnected. If the capacitor remains connected to the battery, the voltage across the capacitor necessarily remains the same (equal to the battery voltage), and the charge is proportional to the capacitance, however it may be modified (i.e., by insertion of a dielectric). On the other hand, if you disconnect the capacitor from the battery before making any modifications to the capacitor, then its charge remains the same. In this case, as you vary the capacitance, the voltage across the plates changes in inverse proportion to capacitance, according to $\Delta V = \frac{Q}{C}$.

## REVIEW CHECKLIST

✓ Understand that each point in the vicinity of a charge distribution can be characterized by a scalar quantity called the electric potential, $V$. The values of this potential function over the region (a scalar) are related to the values of the electrostatic field, $\vec{\mathbf{E}}$, over the region (a vector field).

✓ Calculate the electric potential difference between any two points in a uniform electric field, and the electric potential difference between any two points in the vicinity of a group of point charges.

✓ Calculate the electric potential energy associated with a group of point charges.

✓ Obtain an expression for the electric field (a vector quantity) over a region of space if the scalar electric potential function for the region is known.

✓ Calculate the work done by an external force in moving a charge $q$ between any two points in an electric field when (a) an expression giving the field as a function of position is known, or when (b) the charge distribution (either point charges or a continuous distribution of charge) giving rise to the field is known.

✓ Determine the equivalent capacitance of a network of capacitors in series-parallel combination and calculate the final charge on each capacitor and the potential difference across each when a known potential is applied across the combination.

✓ Calculate the capacitance, potential difference, and stored energy of a capacitor that is partially or completely filled with a dielectric.

## ANSWERS TO SELECTED QUESTIONS

**Q20.3** Give a physical explanation showing that the potential energy of a pair of charges with the same sign is positive whereas the potential energy of a pair of charges with opposite signs is negative.

**Answer** You may remember from the chapter on gravitational potential energy that potential energy of a system is defined to be positive when positive work must have been performed by an external agent to change the system from an initial configuration, to which we assign a zero value of potential energy, to a final configuration. For example the system of a flag and the Earth has positive potential energy when the flag is raised if we define zero potential energy as the configuration with the flag at the ground, since positive work must be done by an external force in order to raise it from the ground to the top of the pole.

When assembling like charges from an infinite separation, a configuration for which we have defined the potential energy as having a value of zero, it takes work to move them closer together to some distance $r$; therefore energy is being stored in the system of charges, and the potential energy is positive.

When assembling unlike charges from an infinite separation, the charges tend to accelerate toward each other, and thus energy is released as they approach a separation of distance $r$. Therefore, the potential energy of a pair of unlike charges is negative.

**Q20.11** If the potential difference across a capacitor is doubled, by what factor does the energy stored change?

**Answer** Since $U = \dfrac{C(\Delta V)^2}{2}$, doubling $\Delta V$ will quadruple the stored energy.

**Q20.15** If you were asked to design a capacitor in which small size and large capacitance were required, what factors would be important in your design?

**Answer** You should use a dielectric filled capacitor whose dielectric constant is very large. Furthermore, you should make the dielectric as thin as possible, keeping in mind that dielectric breakdown also be considered.

## SOLUTIONS TO SELECTED PROBLEMS

**P20.1** (a) Calculate the speed of a proton that is accelerated from rest through a potential difference of 120 V.

(b) Calculate the speed of an electron that is accelerated through the same potential difference.

**Solution** **Conceptualize:** Since 120 V is only a modest potential difference, we might expect that the final speed of the particles will be substantially less than the speed of light. We should also expect the speed of the electron to be significantly greater than the proton because, with $m_e << m_p$, and equal force on both particles will result in a much greater acceleration for the electron.

**Categorize:** Conservation of energy of the proton-field system can be applied to this problem to find the final speed from the kinetic energy of the particles. (Review this work-energy theory of motion from Chapter 7 if necessary.)

**Analyze:**

(a) Energy is conserved as the proton moves from high to low potential, which can be defined for this problem as moving from 120 V down to 0 V:

$$K_i + U_i + \Delta E_{\text{mech}} = K_f + U_f: \qquad 0 + qV_i + 0 = \frac{1}{2}mv_p^2 + 0$$

$$\left(1.60\times10^{-19}\ \text{C}\right)(120\ \text{V})\left(\frac{1\ \text{J}}{1\ \text{V}\cdot\text{C}}\right) = \frac{1}{2}\left(1.67\times10^{-27}\ \text{kg}\right)v_p^2$$

$$v_p = 1.52\times10^{5}\ \text{m/s}$$ ◊

(b) The electron will gain speed in moving the other way, from $V_i = 0$ to $V_f = 120$ V:

$$K_i + U_i + \Delta E_{\text{mech}} = K_f + U_f: \qquad 0 + 0 + 0 = \frac{1}{2}mv_e^2 + qV$$

$$0 = \frac{1}{2}\left(9.11\times10^{-31}\ \text{kg}\right)v_e^2 + \left(-1.60\times10^{-19}\ \text{C}\right)(120\ \text{J/C})$$

$$v_e = 6.49\times10^{6}\ \text{m/s}$$ ◊

**Finalize:** Both of these speeds are significantly less than the speed of light as expected, which also means that we were justified in not using the relativistic kinetic energy formula. (For precision to three significant digits, the relativistic formula is only needed if $v$ is greater than about $0.1c$.)

**P20.5** An electron moving parallel to the $x$ axis has an initial speed of $3.70 \times 10^6$ m/s at the origin. Its speed is reduced to $1.40 \times 10^5$ m/s at the point $x = 2.00$ cm. Calculate the potential difference between the origin and that point. Which point is at the higher potential?

**Solution** Use the energy version of the isolated system model to equate the energy of the electron-field system when the electron is at $x = 0$ and $x = 2$ cm.

The unknown will be the difference in potential $V_f - V_i$.

Thus, $K_i + U_i + \Delta E_{\text{mech}} = K_f + U_f$ becomes $\frac{1}{2}mv_i^2 + qV_i + 0 = \frac{1}{2}mv_f^2 + qV_f$

or $\frac{1}{2}m\left(v_i^2 - v_f^2\right) = q\left(V_f - V_i\right)$ so $V_f - V_i = \dfrac{m\left(v_i^2 - v_f^2\right)}{2q}$.

Noting that the electron's charge is negative, and evaluating the potential,

$$V_f - V_i = \frac{\left(9.11 \times 10^{-31}\text{ kg}\right)\left[\left(3.70 \times 10^6\text{ m/s}\right)^2 - \left(1.40 \times 10^5\text{ m/s}\right)^2\right]}{2\left(-1.60 \times 10^{-19}\text{ C}\right)} = -38.9\text{ V}. \quad \Diamond$$

The negative sign means that the 2.00-cm location is lower in potential than the origin. A positive charge would slow in free flight toward the higher voltage, but the negative electron slows as it moves into lower potential. The 2.00-cm distance was unnecessary information for this problem. If the field were uniform, we could find the $x$ component from $\Delta V = -E_x d$.

**P20.11** The three charges in Figure P20.11 are at the vertices of an isosceles triangle. Calculate the electric potential at the midpoint of the base, taking $q = 7.00\ \mu\text{C}$.

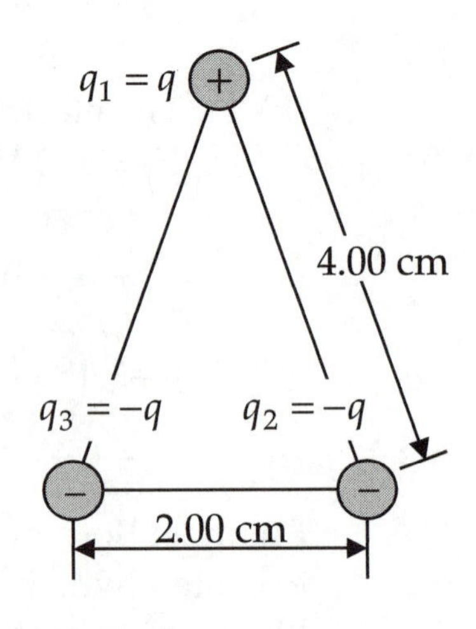


FIG. P20.11

**Solution** Let $q_1 = q$ and $q_2 = q_3 = -q$.

The charges are at distances of

$$r_1 = \sqrt{(0.040\,0\text{ m})^2 - (0.010\,0\text{ m})^2} = 3.87 \times 10^{-2}\text{ m}$$

and $r_2 = r_3 = 0.010\,0$ m.

*continued on next page*

The potential at point $P$ is

$$V_P = \frac{k_e q_1}{r_1} + \frac{k_e q_2}{r_2} + \frac{k_e q_3}{r_3} = k_e(q)\left(\frac{1}{r_1} + \frac{-1}{r_2} + \frac{-1}{r_3}\right)$$

so $V_P = \left(8.99\times10^9\ \text{N}\cdot\text{m}^2/\text{C}^2\right)\left(7.00\times10^{-6}\ \text{C}\right)\left(\frac{1}{0.038\,7\ \text{m}} - \frac{1}{0.010\,0\ \text{m}} - \frac{1}{0.010\,0\ \text{m}}\right)$

and $V_P = -11.0\times10^6\ \text{V}$. ◊

**Related Calculation:** Calculate the electric field vector at the same point due to the three charges. The separate fields of the two negative charges are in opposite directions and add to zero:

$$\vec{\mathbf{E}}_P = \frac{k_e q_1}{r_1^2}\hat{\mathbf{r}}_1 = \frac{\left(8.99\times10^9\ \text{N}\cdot\text{m}^2/\text{C}^2\right)\left(7.00\times10^{-6}\ \text{C}\right)}{(0.040\,0\ \text{m})^2 - (0.010\,0\ \text{m})^2}\ \text{down}$$

$$\vec{\mathbf{E}}_P = \left(42.0\times10^{-6}\ \text{N/C}\right)\left(-\hat{\mathbf{j}}\right)$$

**P20.13** Show that the amount of work required to assemble four identical point charges of magnitude $Q$ at the corners of a square of side $s$ is $\frac{5.41k_eQ^2}{s}$.

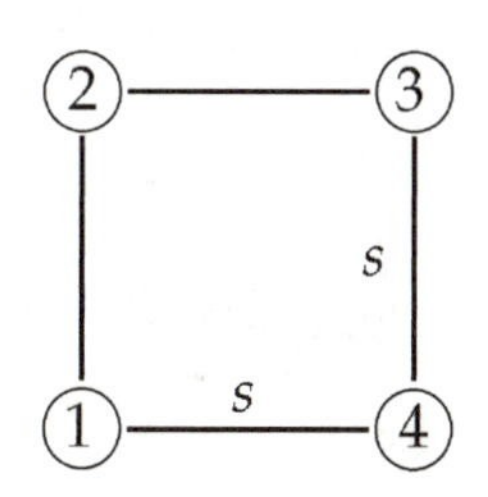


**FIG. P20.13**

**Solution** The work required equals the sum of the potential energies for all pairs of charges. No energy is involved in placing $q_4$ at a given position in empty space.

When $q_3$ is brought from far away and placed close to $q_4$, the system potential energy can be expressed as $q_3V_4$, where $V_4$ is the potential at the position of $q_3$ established by charge $q_4$. When $q_2$ is brought into the system, it interacts with two other charges, so we have two additional terms $q_2V_3$ and $q_2V_4$ in the total potential energy.

Finally, when we bring the fourth charge $q_1$ into the system, it interacts with three other charges, giving us three more energy terms. Thus, the complete expression for the energy is:

$$U = q_1V_2 + q_1V_3 + q_1V_4 + q_2V_3 + q_2V_4 + q_3V_4$$

$$U = \frac{q_1k_eq_2}{r_{12}} + \frac{q_1k_eq_3}{r_{13}} + \frac{q_1k_eq_4}{r_{14}} + \frac{q_2k_eq_3}{r_{23}} + \frac{q_2k_eq_4}{r_{24}} + \frac{q_3k_eq_4}{r_{34}}$$

$$U = \frac{Qk_eQ}{s} + \frac{Qk_eQ}{s\sqrt{2}} + \frac{Qk_eQ}{s} + \frac{Qk_eQ}{s} + \frac{Qk_eQ}{s\sqrt{2}} + \frac{Qk_eQ}{s}$$

*continued on next page*

Evaluating, $U = \frac{k_e Q^2}{s}\left(4 + \frac{2}{\sqrt{2}}\right) = 5.41 k_e \frac{Q^2}{s}$. ◊

Doubling $Q$ makes the work four times larger. Doubling $s$ makes the work two times smaller.

**P20.23** Over a certain region of space, the electric potential is $V = 5x - 3x^2y + 2yz^2$. Find the expressions for the $x$, $y$, and $z$ components of the electric field over this region. What is the magnitude of the field at the point $P$ that has coordinates (1, 0, –2) m?

**Solution** First, we find the $x$, $y$, and $z$ components of the field; then, we evaluate them at point $P$. (We assume that $V$ is given in volts, as a function of distances in meters.)

$$E_x = -\frac{\partial V}{\partial x} = -5 + 6xy \qquad E_y = -\frac{\partial V}{\partial y} = 3x^2 - 2z^2 \qquad E_z = -\frac{\partial V}{\partial z} = -4yz$$ ◊

At point $P$, $E_x = -5 + 6(1.00\text{ m})(0\text{ m}) = -5.00\text{ N/C}$.

At point $P$, $E_y = 3(1.00\text{ m})^2 - 2(-2.00\text{ m})^2 = -5.00\text{ N/C}$.

At point $P$, $E_z = -4(0\text{ m})(-2.00\text{ m}) = 0\text{ N/C}$.

At $P$, the field's magnitude is

$$E = \sqrt{(-5.00\text{ N/C})^2 + (-5.00\text{ N/C})^2 + 0^2} = 7.07\text{ N/C}.$$ ◊

**P20.25** A rod of length $L$ (Fig. P20.25) lies along the $x$ axis with its left end at the origin. It has a nonuniform charge density $\lambda = \alpha x$, where $\alpha$ is a positive constant.

(a) What are the units of $\alpha$?

(b) Calculate the electric potential at $A$.

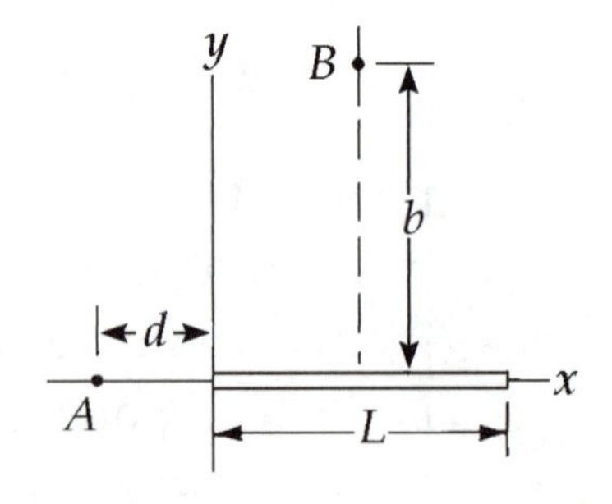


**FIG. P20.25**

**Solution** (a) As a linear charge density, $\lambda$ has units of C/m. So $\alpha = \frac{\lambda}{x}$ must have units of $\text{C/m}^2$. ◊

*continued on next page*

(b) Consider a small segment of the rod at location $x$ and of length $dx$. The amount of charge on it is $\lambda dx = (\alpha x)dx$. Its distance from $A$ is $d+x$, so its contribution to the electric potential at $A$ is

$$dV = k_e \frac{dq}{r} = k_e \frac{\alpha x dx}{d+x}.$$

We must integrate all these contributions for the whole rod, from $x=0$ to $x=L$:

$$V = \int_{\text{all } q} dV = \int_0^L \frac{k_e \alpha x}{d+x} dx.$$

To perform the integral, make a change of variables to

$$u = d+x, \quad du = dx, \quad u(\text{at } x=0) = d, \quad \text{and} \quad u(\text{at } x=\text{L}) = d+L:$$
$$V = \int_d^{d+L} \frac{k_e \alpha (u-d)}{u} du = k_e \alpha \int_d^{d+L} du - k_e \alpha d \int_d^{d+L} \left(\frac{1}{u}\right) du.$$

[Keep track of symbols: the unknown is $V$. The values $k_e$, $\alpha$, $d$, and $L$ are known and constant. Also note that $x$ and $u$ are variables, and will not appear in the answer.]

$$V = k_e \alpha u\Big|_d^{d+L} - k_e \alpha d \ln u\Big|_d^{d+L} = k_e \alpha (d+L-d) - k_e \alpha d (\ln(d+L) - \ln d)$$
$$V = k_e \alpha L - k_e \alpha d \ln\left(\frac{d+L}{d}\right)$$ ◊

We have the answer when the unknown is expressed in terms of the $d$, $L$, and $\alpha$ mentioned in the problem and the universal constant $k_e$.

The potential is directly proportional to the coefficient of charge density $\alpha$. Increasing $L$ makes the potential increase. Increasing $d$ makes the potential decrease. The units of both terms on the right-hand side are $\frac{\text{N}\cdot\text{m}^2}{\text{C}^2}\frac{\text{C}}{\text{m}^2}\text{m} = \frac{\text{J}}{\text{C}} = \text{V}$, as required for the left-hand side.

**P20.29** A spherical conductor has a radius of 14.0 cm and charge of 26.0 $\mu$C. Calculate the electric field and the electric potential at

(a) $r = 10.0$ cm,

(b) $r = 20.0$ cm, and

(c) $r = 14.0$ cm from the center.

**Solution** (a) Inside a conductor when charges are not moving, the electric field is zero and the potential is uniform, the same as on the surface.

$$\vec{\mathbf{E}} = 0 \qquad \Diamond$$

$$V = \frac{k_e q}{R} = \frac{(8.99 \times 10^9 \text{ N} \cdot \text{m}^2/\text{C}^2)(26.0 \times 10^{-6} \text{ C})}{0.140 \text{ m}} = 1.67 \times 10^6 \text{ V} \qquad \Diamond$$

(b) The sphere behaves like a point charge at its center when you stand outside.

$$\vec{\mathbf{E}} = \frac{k_e q}{r^2}\hat{\mathbf{r}} = \frac{(8.99 \times 10^9 \text{ N} \cdot \text{m}^2/\text{C}^2)(26.0 \times 10^{-6} \text{ C})}{(0.200 \text{ m})^2}\hat{\mathbf{r}} = (5.84 \times 10^6 \text{ N/C})\hat{\mathbf{r}} \qquad \Diamond$$

$$V = \frac{k_e q}{r} = 1.17 \times 10^6 \text{ V} \qquad \Diamond$$

(c) $$\vec{\mathbf{E}} = \frac{k_e q}{r^2}\hat{\mathbf{r}} = (11.9 \times 10^6 \text{ N/C})\hat{\mathbf{r}} \qquad \Diamond$$

$$V = 1.67 \times 10^6 \text{ V as in part (a)} \qquad \Diamond$$

**P20.35** An air-filled capacitor consists of two parallel plates, each with an area of 7.60 $\text{cm}^2$, separated by a distance of 1.80 mm. A 20.0-V potential difference is applied to these plates. Calculate

(a) the electric field between the plates,

(b) the surface charge density,

(c) the capacitance, and

(d) the charge on each plate.

*continued on next page*

**Solution** (a) The potential difference between two points in a uniform electric field is $\Delta V = Ed$, so:

$$E = \frac{\Delta V}{d} = \frac{20.0 \text{ V}}{1.80 \times 10^{-3} \text{ m}} = 1.11 \times 10^{4} \text{ V/m}. \qquad \Diamond$$

(b) The electric field between capacitor plates is $E = \dfrac{\sigma}{\epsilon_0}$, so $\sigma = \epsilon_0 E$:

$$\sigma = \left(8.85 \times 10^{-12} \text{ C}^2/\text{N} \cdot \text{m}^2\right)\left(1.11 \times 10^{4} \text{ V/m}\right) = 9.83 \times 10^{-8} \text{ C/m}^2$$

$$\sigma = 98.3 \text{ nC/m}^2. \qquad \Diamond$$

(c) For a parallel-plate capacitor, $C = \dfrac{\epsilon_0 A}{d}$:

$$C = \frac{\left(8.85 \times 10^{-12} \text{ C}^2/\text{N} \cdot \text{m}^2\right)\left(7.60 \times 10^{-4} \text{ m}^2\right)}{1.80 \times 10^{-3} \text{ m}} = 3.74 \times 10^{-12} \text{ F} = 3.74 \text{ pF}. \qquad \Diamond$$

(d) The charge on each plate is $Q = C\Delta V$:

$$Q = \left(3.74 \times 10^{-12} \text{ F}\right)(20.0 \text{ V}) = 7.47 \times 10^{-11} \text{ C} = 74.7 \text{ pC}. \qquad \Diamond$$

**P20.41** Four capacitors are connected as shown in Figure P20.41.

(a) Find the equivalent capacitance between points $a$ and $b$.

(b) Calculate the charge on each capacitor, taking $\Delta V_{ab} = 15.0$ V.

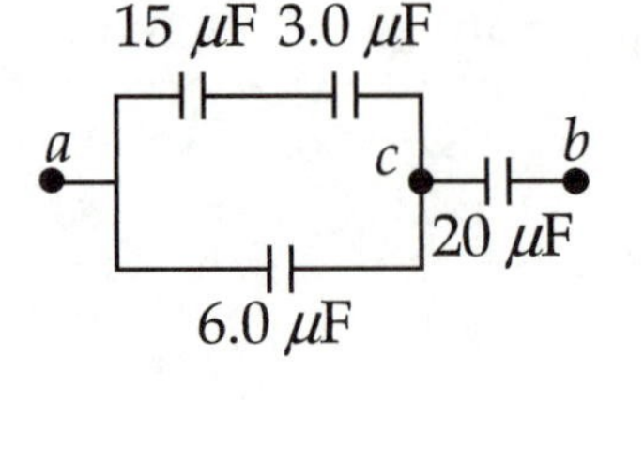


**FIG. P20.41**

**Solution** (a) We simplify the circuit of Figure P20.41 in three steps as shown in Figures (a), (b), and (c).

First, the 15.0 $\mu$F and 30.0 $\mu$F in series are equivalent to

$$\frac{1}{(1/15.0\ \mu\text{F})+(1/30.0\ \mu\text{F})} = 2.50\ \mu\text{F}.$$

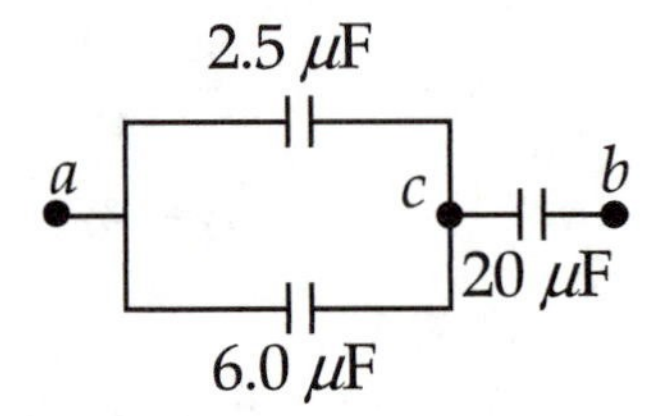


**FIG. P20.41(a)**

Next, 2.50 $\mu$F combines in parallel with 6.00 $\mu$F, creating an equivalent capacitance of 8.50 $\mu$F.

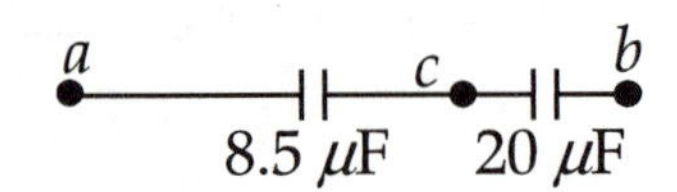


**FIG. P20.41(b)**

At last, 8.5 $\mu$F and 20.0 $\mu$F are in series, equivalent to

$$\frac{1}{(1/8.50\ \mu\text{F})+(1/20.0\ \mu\text{F})} = 5.96\ \mu\text{F}. \qquad \Diamond$$

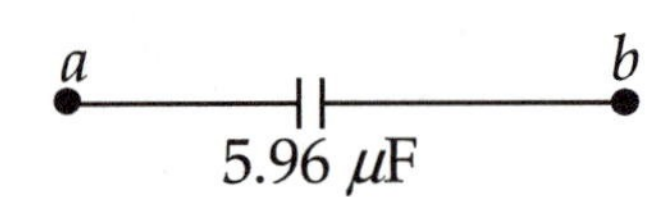


**FIG. P20.41(c)**

(b) We find the charge on and the voltage across each capacitor by working backwards through solution figures (c)–(a), alternately applying $Q = C\Delta V$ and $\Delta V = \frac{Q}{C}$ to every capacitor, real or equivalent. For the 5.96-$\mu$F capacitor, we have

$$Q = C\Delta V = (5.96\ \mu\text{F})(15.0\ \text{V}) = 89.5\ \mu\text{C}.$$

Thus, if $a$ is higher in potential than $b$, just 89.5 $\mu$C flows between the wires and the plates to charge the capacitors in each picture. In (b) we have, for the 8.5-$\mu$F capacitor,

*continued on next page*

$$\Delta V_{ac} = \frac{Q}{C} = \frac{89.5\ \mu\text{C}}{8.5\ \mu\text{F}} = 10.5\ \text{V}.$$

and for the 20.0 $\mu$F in (b), (a), and the original circuit, we have
$Q_{20} = 89.5\ \mu\text{C}$ ◊

$$\Delta V_{cb} = \frac{Q}{C} = \frac{89.5\ \mu\text{C}}{20.0\ \mu\text{F}} = 4.47\ \text{V}.$$

Next, (a) is equivalent to (b), so $\Delta V_{cb} = 4.47\ \text{V}$ and $\Delta V_{ac} = 10.5\ \text{V}$

For the 2.50 $\mu$F, $\Delta V = 10.5\ \text{V}$ and $Q = C\Delta V = (2.50\ \mu\text{F})(10.5\ \text{V}) = 26.3\ \mu\text{C}$

For the 6.00 $\mu$F, $\Delta V = 10.5\ \text{V}$ and $Q_6 = C\Delta V = (6.00\ \mu\text{F})(10.5\ \text{V}) = 63.2\ \mu\text{C}$ ◊

Now, 26.3 $\mu$C having flowed in the upper parallel branch in (a), back in the original circuit we have $Q_{15} = 26.3\ \mu\text{C}$ and $Q_3 = 26.3\ \mu\text{C}$ ◊

**Related Calculation:** An exam problem might also ask for the voltage across each:

$$\Delta V_{15} = \frac{Q}{C} = \frac{26.3\ \mu\text{C}}{15.0\ \mu\text{F}} = 1.75\ \text{V} \text{ and } \Delta V_3 = \frac{Q}{C} = \frac{26.3\ \mu\text{C}}{3.00\ \mu\text{F}} = 8.77\ \text{V}.$$

**P20.43** Consider the circuit shown in Figure P20.43, where $C_1 = 6.00\ \mu\text{F}$, $C_2 = 3.00\ \mu\text{F}$, and $\Delta V = 20.0\ \text{V}$. Capacitor $C_1$ is first charged by the closing of switch $S_1$. Switch $S_1$ is then opened, and the charged capacitor is connected to the uncharged capacitor by the closing of $S_2$. Calculate the initial charge acquired by $C_1$ and the final charge on each capacitor.

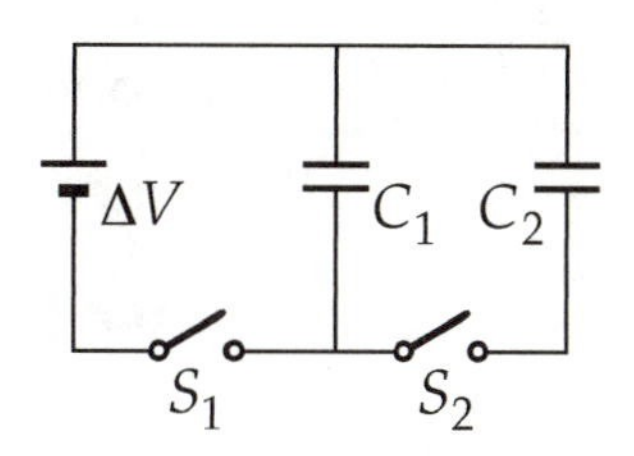


**FIG. P20.43**

**Solution** When $S_1$ is closed, the charge on $C_1$ will be
$Q_1 = C_1\Delta V_1 = (6.00\ \mu\text{F})(20.0\ \text{V}) = 120\ \mu\text{C}.$ ◊

When $S_1$ is opened and $S_2$ is closed, the total charge will remain constant and be shared by the two capacitors: $Q_1' = 120\ \mu\text{C} - Q_2'$. The potential differences across the two capacitors will be equal.

$$\Delta V' = \frac{Q_1'}{C_1} = \frac{Q_2'}{C_2} \quad \text{or} \quad \frac{120\ \mu\text{C} - Q_2'}{6.00\ \mu\text{F}} = \frac{Q_2'}{3.00\ \mu\text{F}}$$

and $Q_2' = 40.0\ \mu\text{C}$ $\qquad Q_1' = 120\ \mu\text{C} - 40.0\ \mu\text{C} = 80.0\ \mu\text{C}$ ◊

**P20.51** A parallel-plate capacitor has a charge $Q$ and plates of area $A$. What force acts on one plate to attract it toward the other plate? Because the electric field between the plates is $E = \dfrac{Q}{A\epsilon_0}$, you might think that the force is $F = QE = \dfrac{Q^2}{A\epsilon_0}$. This conclusion is wrong because the field $E$ includes contributions from both plates, and the field created by the positive plate cannot exert any force on the positive plate. Show that the force exerted on each plate is actually $F = \dfrac{Q^2}{2\epsilon_0 A}$. (*Suggestion*: Let $C = \dfrac{\epsilon_0 A}{x}$ for an arbitrary plate separation $x$; then require that the work done in separating the two charged plates be $W = \int F dx$.) The force exerted by one charged plate on another is sometimes used in a machine shop to hold a workpiece stationary.

**Solution** The electric field in the space between the plates is $E = \dfrac{\sigma}{\epsilon_0} = \dfrac{Q}{A\epsilon_0}$.

You might think that the force on one plate is $F = QE = \dfrac{Q^2}{A\epsilon_0}$

but this is two times too large, because neither plate exerts a force on itself.

The force on one plate is exerted by the other, through its electric field:

$$E = \frac{\sigma}{2\epsilon_0} = \frac{Q}{2A\epsilon_0}.$$

The force on each plate is:

$$F = (Q_{\text{self}})(E_{\text{other}}) = \frac{Q^2}{2A\epsilon_0}.$$

To prove this, we follow the hint, and calculate that the work done in separating the plates, which equals the potential energy stored in the charged capacitor:

$$U = \frac{1}{2}\frac{Q^2}{C} = \int F dx.$$

From the fundamental theorem of calculus, $dU = F dx$

and

$$F = \frac{d}{dx}U = \frac{d}{dx}\left(\frac{Q^2}{2C}\right) = \frac{1}{2}\frac{d}{dx}\left(\frac{Q^2}{A\epsilon_0/x}\right).$$

Solving,

$$F = \frac{1}{2}\frac{d}{dx}\left(\frac{Q^2 x}{A\epsilon_0}\right) = \frac{1}{2}\left(\frac{Q^2}{A\epsilon_0}\right).$$

◊

**P20.63** Calculate the work that must be done to charge a spherical shell of radius $R$ to a total charge $Q$.

**Solution** When the potential of the shell is $V$ due to a charge $q$, the work required to add an additional increment of charge $dq$ is

$dW = Vdq$ where $V = \frac{k_e q}{R}$

$dW = \left(\frac{k_e q}{R}\right)dq$ and $W = \frac{k_e}{R}\int_0^Q q\,dq$; therefore, $W = \left(\frac{k_e}{R}\right)\left(\frac{Q^2}{2}\right)$. ◊

Doubling $Q$ makes the work four times larger. Doubling $R$ makes the work two times smaller.

**P20.73** A parallel-plate capacitor is constructed using a dielectric material whose dielectric constant is 3.00 and whose dielectric strength is $2.00 \times 10^8$ V/m. The desired capacitance is 0.250 $\mu$F, and the capacitor must withstand a maximum potential difference of 4 000 V. Find the minimum area of the capacitor plates.

**Solution** **Conceptualize:** The dielectric strength given is high compared to those of the materials listed in Table 20.1. It is in a sense appropriate for a capacitor with a fairly high breakdown voltage. Then we might expect a typical plate area on the order of 0.1 $m^2$, as for a microfarad capacitor off the shelf at the electronics store.

**Categorize:** In $C = \frac{\kappa \in_0 A}{d}$ we have two unknowns, $A$ and $d$. We can solve because we have a second relationship, $\Delta V_{max} = E_{max} d$. This is a design problem. We can solve most compactly by treating the two as simultaneous equations, combining them by substitution.

**Analyze:** $E_{max} = 2.00 \times 10^8 \text{ V/m} = \frac{\Delta V_{max}}{d}$ so $d = \frac{\Delta V_{max}}{E_{max}}$.

For $C = \frac{\kappa \in_0 A}{d} = 0.250 \times 10^{-6}$ F and $\kappa = 3.00$,

$$A = \frac{Cd}{\kappa \in_0} = \frac{C\Delta V_{max}}{\kappa \in_0 E_{max}} = \frac{(0.250 \times 10^{-6} \text{ F})(4\,000 \text{ V})}{(3.00)(8.85 \times 10^{-12} \text{ F/m})(2.00 \times 10^8 \text{ V/m})} = 0.188 \text{ m}^2. \quad ◊$$

*continued on next page*

**Finalize:** To see how practical our design is, we can evaluate the thickness of the dielectric: $d = \frac{\Delta V_{max}}{E_{max}} = \frac{4\,000 \text{ V}}{2 \times 10^{8} \text{ V/m}} = 2 \times 10^{-5} \text{ m}$. This is thinner than the thinnest garbage bag, but it is hundreds of thousands of atoms thick, so we imagine that it can be manufactured as a coating on a plate if not as a separate film. If $d$ is made somewhat larger on average for manufacturing convenience, then $A$ can be made larger to compensate. It is in this sense that $0.188 \text{ m}^2$ is the minimum area.

# Chapter 21

# Current and Direct Current Circuits

## NOTES FROM SELECTED CHAPTER SECTIONS

### Section 21.1 Electric Current

The direction of conventional current is designated as the direction of motion of positive charge. *In an ordinary metal conductor, the direction of current will be opposite the direction of flow of electrons (which are the charge carriers in this case).*

### Section 21.2 Resistance and Ohm's Law

For ohmic materials, the ratio of the current density to the electric field (that causes the current) is equal to a constant $\sigma$, the conductivity of the material. The reciprocal of the **conductivity** is called the **resistivity**, $\rho$. Each ohmic material has a characteristic resistivity, which depends only on the properties of the specific material and is a function of temperature. Ohmic materials obey Ohm's law: $R$ is constant in $\Delta V = IR$. Resistance has the SI units volts per ampere, called ohms ($\Omega$). Thus, if a potential difference of 1 V across a conductor produces a current of 1 A, the resistance of the conductor is 1 $\Omega$.

### Section 21.4 A Structural Model for Electrical Conduction

In the classical model of electrical conduction in a metal, electrons are treated like molecules in a gas and, in the absence of an electric field, have a zero average velocity.

Under the influence of an electric field, the electrons move along a direction opposite the direction of the applied field with a drift velocity, which is proportional to the average time between collisions with atoms of the metal. The current is proportional to the magnitude of the drift velocity and to the number of free electrons per unit volume.

## Section 21.6 Sources of EMF

The source of emf (for example a battery or generator) maintains the constant current in a closed circuit. The emf, $\mathcal{E}$, of the source is the work done per unit charge in increasing the electric potential of the circulating charges. *One joule of work is required to move one coulomb of charge between two points that differ in potential by one volt.*

The terminal voltage ($\Delta V$ between the positive and negative terminals of a battery or other source) is equal to the emf of the battery when the current in the circuit is zero. In this case the terminal voltage is called the open circuit voltage. When the battery is delivering current to a circuit, the terminal voltage (closed circuit voltage) is less than the emf due to the internal resistance of the battery.

## Section 21.7 Resistors in Series and in Parallel

**For a group of resistors connected in parallel:**

- The reciprocal of the equivalent resistance of the group is the algebraic sum of the reciprocals of the individual resistors. See Equation 21.29.

- The potential differences across the individual resistors have the same value.

- The total current associated with the parallel group is the sum of the currents in the individual resistors.

**For a group of resistors connected in series:**

- The equivalent resistance of the series combination is the algebraic sum of the individual resistances. See Equation 21.27.

- The current is the same for each resistor in the group.

- The total potential difference across the group of resistors equals the sum of the potential differences across the individual resistors.

## Section 21.8 Kirchhoff's Rules and Simple DC Circuits

- **Junction Rule (conservation of charge)**: With currents entering a junction counted as positive and currents leaving the junction counted as negative, the total current accumulating at any junction must be equal to zero. A junction is defined to be any point in the circuit where the current can split. *To generate independent equations, the junction rule may be used a number of times which is one fewer than the number of junction points in the circuit.*

- **Loop Rule (conservation of energy)**: The algebraic sum of the potential differences around any closed loop in a circuit must be zero. *Each application of the loop rule should result in an equation that contains a new circuit element (resistor or battery), or a new current.*

Overall, in order to solve a circuit problem using Kirchhoff's Rules, you must have as many independent equations as the total number of unknown quantities (currents, resistances, or emfs) in the circuit.

## Section 21.9 *RC* Circuits

Consider a battery, capacitor and resistor in series as shown in the figure at the right. When the switch is moved from the "open" to the "closed" position (the capacitor will begin charging) the charge on the capacitor increases with time and the current decreases exponentially. Also, during one time constant, the charging current decreases from its initial maximum value of $I_0 = \frac{\mathcal{E}}{R}$ to 37 percent of $I_0$. Note that $Q = C\mathcal{E}$ is the charge that would be accumulated on the capacitor if time, *t*, is allowed to go to infinity.

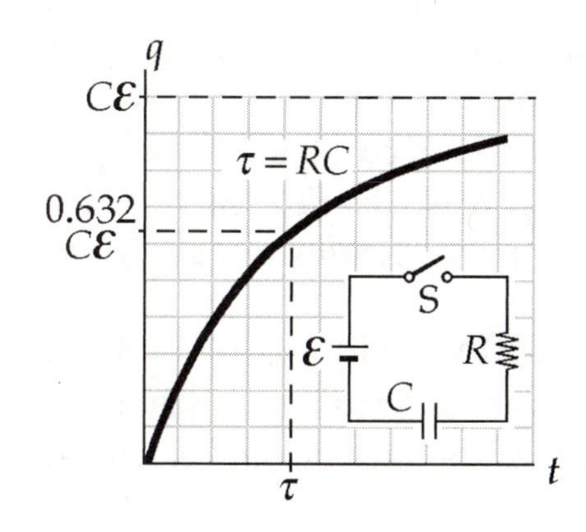


Charging capacitor

When the switch in a circuit containing an initially charged capacitor and a resistor is moved from the "open" to the "closed" position (the capacitor will begin discharging) the charge on the capacitor, current in the circuit, and the voltage across the resistor will all decrease exponentially with time.

In the process of charging or discharging a capacitor, the potential difference between the plates changes as charges are transferred from one plate of the capacitor to the other. The transfer of charge produces a current in the circuit. *The battery does work on the charges to increase their electrostatic potential energy as they move between the plates and the connecting wires. The charges do not move across the gap between the plates of the capacitor.*

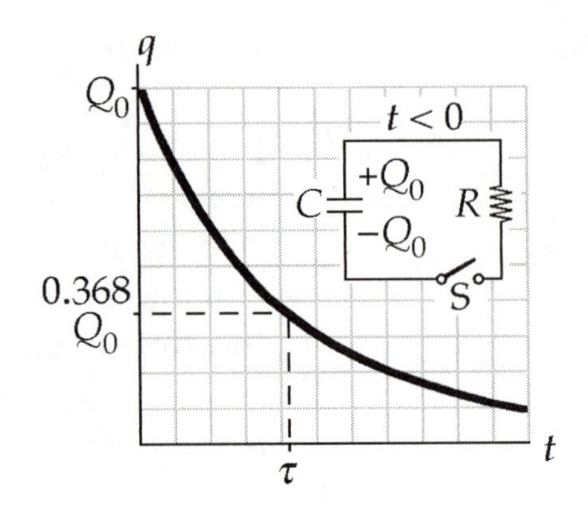


Discharging capacitor

## EQUATIONS AND CONCEPTS

**Electric current,** $I$, is defined as the rate at which charge moves through a cross section of the conductor. The direction of the current is opposite the direction of the flow of electrons in the circuit. The SI unit of current is the **ampere** (A). *Under the action of an electric field, electric charges will move through gases, liquids, and solid conductors.*

$$I \equiv \lim_{t \to 0} \frac{\Delta Q}{\Delta t} = \frac{dQ}{dt} \qquad (21.2)$$

$$1\text{ A} = 1\text{ C/s} \qquad (21.3)$$

The **drift velocity,** $v_d$ (velocity of the charge carriers in a conductor) is actually an average value of the individual velocities.

$$I_{ave} = \frac{\Delta Q}{\Delta t} = nqv_d A \qquad (21.4)$$

$n$ = the number of mobile charge carriers per unit volume of conductor

The **current density,** $J$, in a conductor is defined as the current per unit cross-sectional area of the conductor.

$$J \equiv \frac{I}{A} = nqv_d \qquad (21.5)$$

The **definition of resistance** is expressed in Equation 21.6. According to Ohm's law, $R$ is understood to be independent of $\Delta V$ for ohmic materials, which have a linear current-voltage relationship over a wide range of applied voltages.

$$R \equiv \frac{\Delta V}{I} \qquad (21.6)$$

The **resistance** of a given conductor of uniform cross section depends on the length, cross-sectional area, and a characteristic property of the material of which the conductor is made. The parameter, $\rho$, is the **resistivity** of the material of which the conductor is made. The resistivity is the inverse of the **conductivity** and has units of ohm-meters. The unit of resistance is the ohm (Ω). *The value of the resistivity of a given material depends on the electronic structure of the material and on the temperature.*

$$R = \rho \frac{\ell}{A} \qquad (21.8)$$

$$1\ \Omega = 1\ \text{V/A}$$

The **resistivity**, and therefore the resistance, of a conductor varies with temperature. Over a limited range of temperatures, this variation is approximately linear. *Semiconductors (for example, carbon) are characterized by a negative temperature coefficient of resistivity and in these materials the resistance decreases as the temperature increases.*

$$\rho = \rho_0\left[1 + \alpha(T - T_0)\right] \qquad (21.10)$$

$$R = R_0\left[1 + \alpha(T - T_0)\right] \qquad (21.12)$$

$\alpha$= temperature coefficient of resistivity

$T_0$ is a reference temperature (usually taken to be 20.0°C)

**Power** will be supplied to a resistor or other current-carrying devices when a potential difference is maintained between the terminals of the circuit element. The quantities can be related in an equation called **Joule's law** and the SI unit of power is the watt (W). When the device is resistive, the power delivered can be expressed in alternative forms. *Watt is a unit of power; kilowatt-hour is a unit of energy.*

$$\mathcal{P} = I\Delta V \qquad (21.20)$$

$$\mathcal{P} = I^2 R = \frac{(\Delta V)^2}{R} \qquad (21.21)$$

$$1\ \text{W} = 1\ \text{J/s}$$

$$1\text{kWh} = 3.6 \times 10^6\ \text{J} \qquad (21.22)$$

The **terminal voltage**, $\Delta V$, of a battery will be less than the emf when the battery is providing a current to an external circuit. This is due to **internal resistance** of the battery.

$$\Delta V = \mathcal{E} - Ir \qquad (21.23)$$

The **current**, $I$, delivered by a battery in a simple **dc** circuit, depends on the value of the emf of the source, $\mathcal{E}$, the total load resistance in the circuit, $R$; and the internal resistance of the source, $r$.

$$\mathcal{E} = IR + Ir \qquad (21.24)$$

Many circuits, which contain several resistors, can be reduced to an equivalent single-loop circuit by successive step-by-step combinations of groups of resistors in series and parallel.

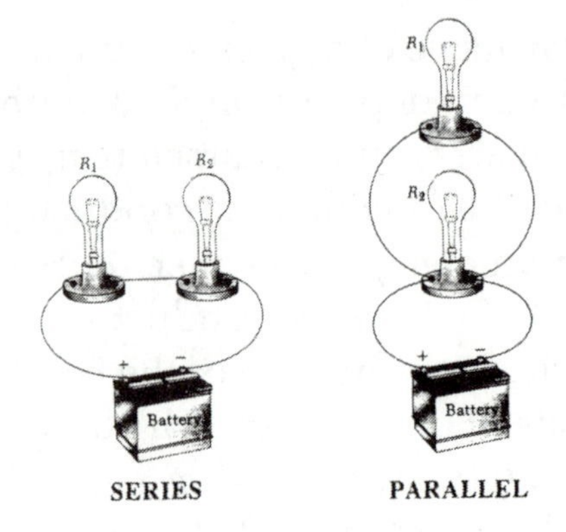


*Resistors in series are connected so that pairs of adjacent resistors share only one common circuit point; there is a common current through each resistor in the group.*

**A series combination of resistors** has total or equivalent resistance equal to the sum of the resistances of the individual resistors.

$$R_{eq} = R_1 + R_2 + R_3 + \ldots \qquad (21.27)$$

(Series combination)

*Resistors in parallel are connected so that each resistor in the parallel group has two circuit points in common with each of the other resistors; there is a common potential difference across each resistor in the group.*

A **parallel combination of resistors** has a total or equivalent resistance, which is the inverse of the sum of the inverses of the individual resistances.

$$\frac{1}{R_{eq}} = \frac{1}{R_1} + \frac{1}{R_2} + \frac{1}{R_3} + \ldots \qquad (21.29)$$

(Parallel combination)

In the most general case successive reduction of a group of resistors to a single equivalent resistor is not possible and you must solve a true multiloop circuit by use of Kirchhoff's rules. Review the procedure suggested in the next section to apply Kirchhoff's rules.

When a battery is used to **charge a capacitor** in series with a resistor, a quantity, $\tau = RC$ called the time constant of the circuit, is used to describe the manner in which the charge on the capacitor varies with time. The charge on the capacitor increases from zero to 63% of its maximum value in a time interval equal to one time constant. Also, during one time constant, the charging current decreases from its initial maximum value of $I_0 = \dfrac{\mathcal{E}}{R}$ to 37% of $I_0$. See figure on left below.

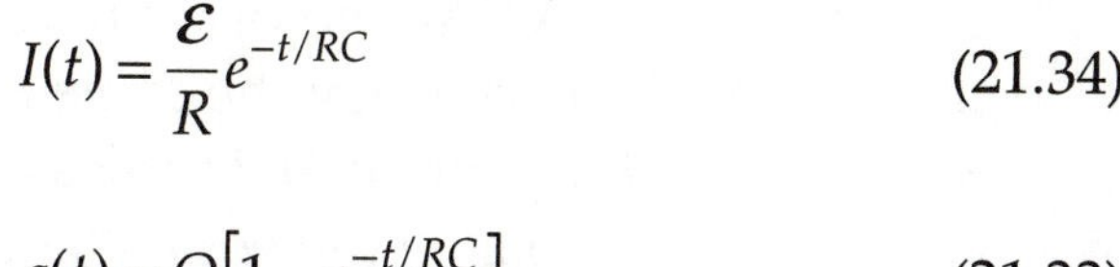

$$I(t) = \frac{\mathcal{E}}{R} e^{-t/RC} \qquad (21.34)$$

$$q(t) = Q\left[1 - e^{-t/RC}\right] \qquad (21.33)$$

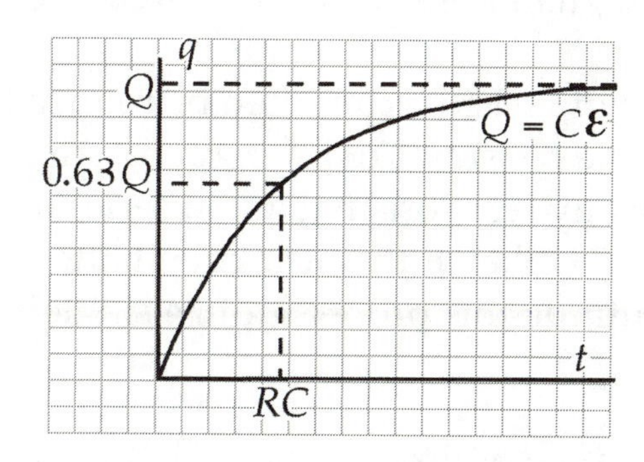


Charging capacitor

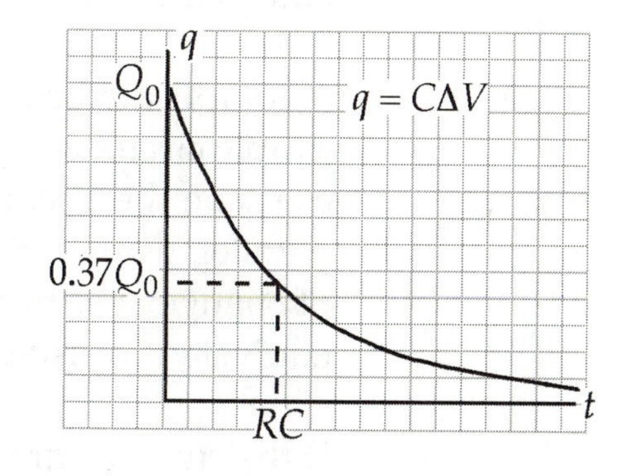


Discharging capacitor

When a **capacitor is discharged** through a resistor, the charge and current decrease exponentially with time. *The negative sign indicates the direction of the current, which is opposite the direction during the charging process.* See figure on right above.

$$q(t) = Qe^{-t/RC} \qquad (21.36)$$

$$I(t) = -I_0 e^{-t/RC} \qquad (21.37)$$

## SUGGESTIONS, SKILLS, AND STRATEGIES

### A Problem-Solving Strategy for Resistors

- Be careful with your choice of units. To calculate the resistance of a device in ohms, make sure that distances are in meters and use the SI value of $\rho$.

- When two or more resistors are in series (connected end-to-end) they each carry the same current, but the potential differences across them are not equal (unless the resistors are equal). The resistors add directly to give the equivalent resistance of the series combination. The equivalent resistance will always be greater than any of the individual resistances.

- When two or more resistors are connected in parallel, the potential differences across them are the same. Since the current is inversely proportional to the resistance, the currents through them are not equal (unless the resistances are equal). The equivalent resistance of a parallel combination of resistors is found through reciprocal addition, and the equivalent resistor is always **less** than the smallest individual resistor.

- A complicated circuit consisting of resistors can often be reduced to a simple circuit containing only one resistor. To do so, repeatedly examine the circuit and replace any resistors that are in series or in parallel, using Eqs. 21.27 and 21.29. Sketch the new circuit after each set of changes has been made. Continue this process until a single equivalent resistance is found.

- If the current through, or the potential difference across, a resistor in the complicated circuit is to be identified, start with the final equivalent circuit found in the last step, and gradually work your way back through the circuits, using $V = IR$ at each step to find either the voltage drop or the current for each resistor in the group.

## A Strategy for Using Kirchhoff's Rules

- First, draw the circuit diagram and assign labels and symbols to all the known and unknown quantities. You must assign directions to the currents in each part of the circuit. Do not be alarmed if your assumption for a current's direction is wrong; the resulting value will be negative, but its magnitude will be correct. Although the assignment of current directions is arbitrary, you must stick with your guess throughout the problem as you apply Kirchhoff's rules. A resulting current with a negative value has a direction opposite that which you assumed.

- Apply the junction rule to any junction in the circuit. The junction rule may be applied as many times as a new current (one not used in a previous application) appears in the resulting equation. In general, the number of times the junction rule can be used is one fewer than the number of junction points in the circuit.

- Now apply Kirchhoff's loop rule to as many loops in the circuit as are needed to solve for the unknowns. Remember you must have as many equations as there are unknowns ($I$'s, $R$'s, and $\mathcal{E}$'s). In order to apply this rule, you must correctly identify the change in potential as you cross each element in traversing the closed loop. Watch out for signs!

  The following figure illustrates convenient rules, which you may use to determine the increase or decrease in potential as the current in a loop crosses a resistor or seat of emf.

$$\Delta V = V_b - V_a = -IR$$

$$\Delta V = V_b - V_a = IR$$

$$\Delta V = V_b - V_a = \mathcal{E}$$

$$\Delta V = V_b - V_a = -\mathcal{E}$$

**Notice that the potential decreases (changes by $-IR$) when the resistor is traversed in the direction of the current. There is an increase in potential of $+IR$ if the direction of travel is opposite the direction of current.** If a seat of emf is traversed in the direction of the emf (from – to + on the battery), the potential increases by $\mathcal{E}$. If the direction of travel is from + to –, the potential decreases by $\mathcal{E}$ (changes by $-\mathcal{E}$).

- Finally, you must solve the equations simultaneously for the unknown quantities. Be careful in your algebraic steps, and check your numerical answers for consistency.

As an illustration of the use of Kirchhoff's rules, consider a three-loop circuit, which has the general form shown in the following figure on the left. In this illustration, the actual circuit elements, *R*s and $\mathcal{E}$'s are not shown but assumed known. There are six possible different values of *I* in the circuit; therefore you will need six independent equations to solve for the six values of *I*. There are four junction points in the circuit (at points *a*, *d*, *f*, and *h*). The first rule applied at any three of these points will yield three equations. The circuit can be thought of as a group of three "blocks" or meshes as shown in the following figure on the right. Kirchhoff's second law, when applied to each of these loops ($abcda$, $ahfga$, and $defhd$), will yield three additional equations. You can then solve the total of six equations simultaneously for the six values of $I_1$, $I_2$, $I_3$, $I_4$, $I_5$, and $I_6$. You can, of course, expect that the sum of the changes in potential difference around any other closed loop in the circuit will be zero (for example, $abcdefga$ or $ahfedcba$); however **the equations found by applying Kirchhoff's second rule to these additional loops will not be** independent of the six equations found previously.

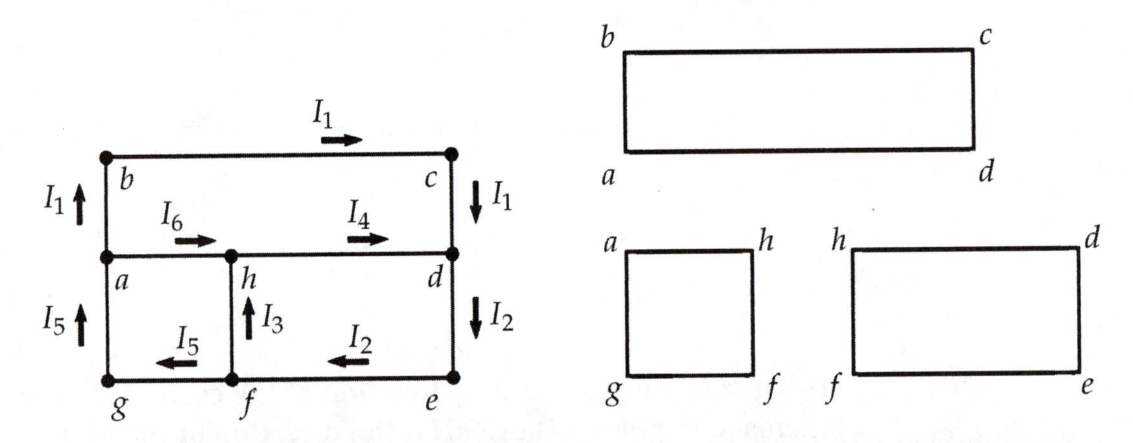


## REVIEW CHECKLIST

✓ Define electric current in terms of the rate of charge flow; and define the ampere as the unit of currents. Calculate electron drift velocity, and the quantity of charge passing a point in a given time interval in a specified current-carrying conductor.

✓ Make calculations of the variation of resistance with temperature, which involves the concept of the temperature coefficient of resistivity.

✓ Determine the power delivered to a resistor.

✓ Calculate the equivalent resistance of a group of resistors in parallel, series, or series-parallel combination.

✓ Apply Kirchhoff's rules to solve multiloop circuits; that is, find the currents and the potential difference between any two points.

✓ Calculate the charging (discharging) current $I(t)$ and the accumulated (residual) charge $q(t)$ during charging (and discharging) of a capacitor in an $RC$ circuit.

## ANSWERS TO SELECTED QUESTIONS

**Q21.3** Two wires A and B of circular cross-section are made of the same metal and have equal lengths, but the resistance of wire A is three times greater than that of wire B. What is the ratio of their cross-sectional areas? How do their radii compare?

**Answer** Since $R = \dfrac{\rho\ell}{\text{Area}}$, the ratio of resistances is given by $\dfrac{R_A}{R_B} = \dfrac{\text{Area}_B}{\text{Area}_A}$.

Hence, the ratio of their areas is three to one; that is, the area of wire B is three times that of wire A. From the ratio of the areas, we can calculate that the radius of wire B is $\sqrt{3}$ times the radius of wire A.

**Q21.5** Use the atomic theory of matter to explain why the resistance of a material should increase as its temperature increases.

**Answer** As the temperature increases, the amplitude of atomic vibrations increases. This makes it more likely that the drifting electrons will be scattered by atomic vibrations, and makes it more difficult for charges to participate in organized motion inside the conductor.

**Q21.7** If charges flow very slowly through a metal, why does it not require several hours for a light to come on when you throw a switch?

**Answer** Individual electrons move with a small average velocity through the conductor, but as soon as the voltage is applied, electrons all along the conductor start to move. Actually the current does not flow "immediately", but is limited by the speed of light.

**Q21.13** Why is it possible for a bird to sit on a high-voltage wire without being electrocuted?

**Answer** The resistance of the short segment of wire between the bird's feet is so small that the potential difference $\Delta V = IR$ between the feet is negligible. In order for the bird to be electrocuted, a potential difference is required. There is negligible potential difference between the bird's feet.

It could also be said like this: A small amount of current does flow through the bird's feet, according to the rule of parallel resistors. However, since the resistance of the bird's feet is much higher than that of the wire, the amount of current that flows through the bird is still not enough to harm it.

**Q21.15** Referring to Figure Q21.15, describe what happens to the lightbulb after the switch is closed. Assume that the capacitor has a large capacitance and is initially uncharged, and assume that the lightbulb illuminates when connected directly across the battery terminals.

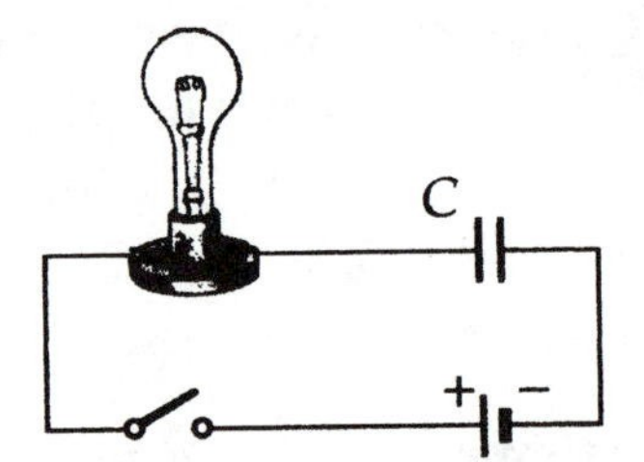


**FIG. Q21.15**

**Answer** The bulb will light up for an instant as the capacitor is being charged and there is a current in the circuit. As soon as the capacitor is fully charged, the current in the circuit will drop to zero, and the bulb will cease to glow.

## SOLUTIONS TO SELECTED PROBLEMS

**P21.3** Suppose that the current in a conductor decreases exponentially with time according to the equation $I(t) = I_0 e^{-t/\tau}$ where $I_0$ is the initial current (at $t = 0$), and $\tau$ is a constant having dimensions of time. Consider a fixed observation point within the conductor.

(a) How much charge passes this point between $t = 0$ and $t = \tau$?

(b) How much charge passes this point between $t = 0$ and $t = 10\tau$?

(c) How much charge passes this point between $t = 0$ and $t = \infty$?

*continued on next page*

**Solution** From $I = \dfrac{dQ}{dt}$, we have $$dQ = Idt\,.$$

From this, we derive a general integral $$Q = \int dQ = \int Idt\,.$$

In all three cases, define an end-time, $T$. $$Q = \int_0^T I_0 e^{-t/\tau} dt\,.$$

Integrating from time $t = 0$ to time $t = T$, $$Q = \int_0^T (-I_0\tau)e^{-t/\tau}\left(-\frac{dt}{\tau}\right).$$

Setting $Q = 0$ at $t = 0$, $$Q = -I_0\tau\left(e^{-T/\tau} - e^0\right) = I_0\tau\left(1 - e^{-T/\tau}\right).$$

(a) If $T = \tau$, $$Q = I_0\tau\left(1 - e^{-1}\right) = 0.632\,1 I_0\tau\,. \quad \lozenge$$

(b) If $T = 10\tau$, $$Q = I_0\tau\left(1 - e^{-10}\right) = 0.999\,95 I_0\tau\,. \quad \lozenge$$

(c) If $T = \infty$, $$Q = I_0\tau\left(1 - e^{-\infty}\right) = I_0\tau\,. \quad \lozenge$$

**P21.7** A 0.900-V potential difference is maintained across a 1.50-m length of tungsten wire that has a cross-sectional area of $0.600\ \text{mm}^2$. What is the current in the wire?

**Solution** From the definition of resistance $I = \dfrac{\Delta V}{R}$ where $R = \dfrac{\rho\ell}{A}$.

Therefore, $$I = \frac{\Delta VA}{\rho\ell} = \frac{(0.900\ \text{V})\left(6.00\times10^{-7}\ \text{m}^2\right)}{\left(5.6\times10^{-8}\ \Omega\cdot\text{m}\right)(1.50\ \text{m})} = 6.43\ \text{A}\,. \quad \lozenge$$

**P21.9** An aluminum wire with a diameter of 0.100 mm has a uniform electric field of 0.200 V/m imposed along its entire length. The temperature of the wire is 50.0°C. Assume one free electron per atom.

(a) Use the information in Table 21.1 and determine the resistivity.

(b) What is the current density in the wire?

(c) What is the total current in the wire?

*continued on next page*

(d) What is the drift speed of the conduction electrons?

(e) What potential difference must exist between the ends of a 2.00-m length of the wire to produce the stated electric field?

**Solution** The resistivity is found from $\rho = \rho_0[1+\alpha(T-T_0)]$.

(a) $\rho = (2.82\times10^{-8}\ \Omega\cdot\text{m})[1+(3.90\times10^{-3}\ {}^\circ\text{C}^{-1})(30.0\ {}^\circ\text{C})] = 3.15\times10^{-8}\ \Omega\cdot\text{m}$ ◊

(b) $J = \sigma E = \dfrac{E}{\rho} = \left(\dfrac{0.200\ \text{V/m}}{3.15\times10^{-8}\ \Omega\cdot\text{m}}\right)\left(\dfrac{1\ \Omega\cdot\text{A}}{\text{V}}\right) = 6.35\times10^{6}\ \text{A/m}^2$ ◊

(c) $J = \dfrac{I}{A} = \dfrac{I}{\pi r^2}$

$I = J\pi r^2 = (6.35\times10^{6}\ \text{A/m}^2)\left[\pi(5.00\times10^{-5}\ \text{m})^2\right] = 49.9\ \text{mA}$ ◊

(d) The mass density gives the number-density of free electrons; we assume that each atom donates one free electron:

$$n = \left(\frac{2.70\times10^{3}\ \text{kg}}{\text{m}^3}\right)\left(\frac{1\ \text{mol}}{26.98\ \text{g}}\right)\left(\frac{10^3\ \text{g}}{\text{kg}}\right)\left(\frac{6.02\times10^{23}\ \text{free e}^-}{1\ \text{mol}}\right)$$
$$= 6.02\times10^{28}\ \text{e}^-/\text{m}^3.$$

Now $J = nqv_d$ gives:

$$v_d = \frac{J}{nq} = \frac{(6.35\times10^{6}\ \text{A/m}^2)}{(6.02\times10^{28}\ \text{e}^-/\text{m}^3)(-1.60\times10^{-19}\ \text{C/e}^-)} = -6.59\times10^{-4}\ \text{m/s}.$$ ◊

The sign indicates that the electrons drift opposite to the field and current.

(e) $\Delta V = E\ell = (0.200\ \text{V/m})(2.00\ \text{m}) = 0.400\ \text{V}$ ◊

**P21.13** If the magnitude of the drift velocity of free electrons in a copper wire is $7.84 \times 10^{-4}$ m/s, what is the electric field in the conductor?

**Solution** **Conceptualize:** For electrostatic cases, we learned that the electric field inside a conductor is always zero. On the other hand, if there is a current, a non-zero electric field must be maintained by a battery or other source to make the charges flow. Therefore, we might expect the electric field to be small, but definitely **not** zero.

**Categorize:** The drift velocity of the electrons can be used to find the current density, which can be used with Ohm's law to find the electric field inside the conductor.

**Analyze:** We first need the electron density in copper, which from Example 21.1 is $n = 8.49 \times 10^{28}\ \text{e}^-/\text{m}^3$. The current density in this wire is then

$$J = nqv_d = \left(8.49 \times 10^{28}\ \text{e}^-/\text{m}^3\right)\left(1.60 \times 10^{-19}\ \text{C}/\text{e}^-\right)\left(7.84 \times 10^{-4}\ \text{m/s}\right)$$
$$J = 1.06 \times 10^{7}\ \text{A/m}^2$$

Ohm's law can be stated as

$$J = \sigma E = \frac{E}{\rho} \quad \text{where } \rho = 1.7 \times 10^{-8}\ \Omega \cdot \text{m for copper,}$$

so then, $E = \rho J = \left(1.70 \times 10^{-8}\ \Omega \cdot \text{m}\right)\left(1.06 \times 10^{7}\ \text{A/m}^2\right) = 0.181\ \text{V/m}$. ◊

**Finalize:** This electric field is certainly smaller than typical static values outside charged objects. The direction of the electric field must be along the length of the conductor; otherwise the electrons would be forced to leave the wire! When the current is switched on, excess charges arrange themselves on the surface of the wire to create an electric field that "steers" the free electrons to flow along the length of the wire from low to high potential (opposite the direction of a positive test charge). It is also interesting to note that when the electric field is being established it travels at the speed of light; but the drift velocity of the electrons is literally at a "snail's pace"!

**P21.17** Suppose that a voltage surge produces 140 V for a moment. By what percentage does the power output of a 120-V, 100-W lightbulb increase? Assume that its resistance does not change.

**Solution** **Conceptualize:** The voltage increases by about 20%, but since $\mathscr{P} = \dfrac{(\Delta V)^2}{R}$, the power will increase as the square of the voltage:

$$\frac{P_f}{P_i} = \frac{(\Delta V_f)^2/R}{(\Delta V_i)^2/R} = \frac{(140\text{ V})^2}{(120\text{ V})^2} = 1.361 \text{ or a } 36.1\% \text{ increase.}$$

**Categorize:** We have already found an answer to this problem by reasoning in terms of ratios, but we can also calculate the power explicitly for the bulb and compare with the original power by using Ohm's law and the equation for electrical power. To find the power, we must first find the resistance of the bulb, which should remain relatively constant during the power surge (we can check the validity of this assumption later).

**Analyze:** $\mathscr{P} = \dfrac{(\Delta V)^2}{R}$, we find that $\quad R = \dfrac{(\Delta V_i)^2}{\mathscr{P}} = \dfrac{(120\text{ V})^2}{100\text{ W}} = 144\ \Omega$.

The final current is $\quad I_f = \dfrac{\Delta V_f}{R} = \dfrac{140\text{ V}}{144\ \Omega} = 0.972\text{ A}$.

The power during surge is $\quad \mathscr{P} = \dfrac{(\Delta V_f)^2}{R} = \dfrac{(140\text{ V})^2}{144\ \Omega} = 136\text{ W}$.

So the percentage increase is $\quad \dfrac{136\text{ W} - 100\text{ W}}{100\text{ W}} = 0.361 = 36.1\%$. ◊

**Finalize:** Our result tells is that this 100-W lightbulb momentarily acts like a 136-W lightbulb, which explains why it would suddenly get brighter. Some electronic devices (like computers) are sensitive to voltage surges like this, which is the reason that **surge protectors** are recommended to protect these devices from being damaged.

In solving this problem, we assumed that the resistance of the bulb did not change during the voltage surge, but we should check this assumption. Let us assume that the filament is made of tungsten and that its resistance will change linearly with temperature according to Equation 21.12. Let us further assume that the increased voltage lasts for a time long enough so that the filament comes to a new equilibrium temperature. The temperature change can be estimated from the power surge according to Stefan's law (Equation 17.35), assuming that all the

*continued on next page*

power loss is due to radiation. By this law, $T \propto \sqrt[4]{\mathcal{P}}$ so that a 36% change in power should correspond to only about 8% increase in temperature. A typical operating temperature of a white lightbulb is about 3000 °C, so $\Delta T \approx 0.08(3\ 273\ \text{K}) = 260°\text{C}$. Then the increased resistance would be roughly

$$R = R_0\left[1+\alpha(T-T_0)\right] = (144\ \Omega)\left[1+\left(4.5\times10^{-3}\right)(260)\right] \approx 310\ \Omega$$

It appears that the resistance could double from $144\ \Omega$. On the other hand, if the voltage surge lasts only a very short time, the 136 W we calculated originally accurately describes the conversion of electrically transmitted energy into internal energy in the filament.

**P21.19** A certain toaster has a heating element made of Nichrome resistance wire. When the toaster is first connected to a 120-V source (and the wire is at a temperature of 20.0°C) the initial current is 1.80 A. The current begins to decrease as the resistive element warms up. When the toaster reaches its final operating temperature, the current drops to 1.53 A.

(a) Find the power delivered to the toaster when it is at its operating temperature.

(b) What is the final temperature of the heating element?

**Solution** **Conceptualize:** Most toasters are rated at about 1 000 W (usually stamped on the bottom of the unit), so we might expect this one to have a similar power rating. The temperature of the heating element should be hot enough to toast bread but low enough that the nickel-chromium alloy element does not melt. (The melting point of nickel is 1 455°C, and chromium melts at 1 907°C.)

**Categorize:** The power can be calculated directly by multiplying the current and the voltage. The temperature can be found from the linear resistivity equation for Nichrome, with $\alpha = 0.4\times10^{-3}\ °\text{C}^{-1}$ from Table 21.1.

**Analyze:**

(a) $\mathcal{P} = \Delta V I = (120\ \text{V})(1.53\ \text{A}) = 184\ \text{W}$ ◊

(b) The resistance at 20.0°C is $R_0 = \dfrac{\Delta V}{I} = \dfrac{120\ \text{V}}{1.80\ \text{A}} = 66.7\ \Omega$.

At operating temperature, $R = \dfrac{120\ \text{V}}{1.53\ \text{A}} = 78.4\ \Omega$.

*continued on next page*

Neglecting thermal expansion,

we have $$R = \frac{\rho \ell}{A} = \frac{\rho_0 [1+\alpha(T-T_0)]\ell}{A} = R_0[1+\alpha(T-T_0)]$$

$$T = T_0 + \frac{R/R_0 - 1}{\alpha} = 20.0°\text{C} + \frac{78.4\ \Omega/66.7\ \Omega - 1}{0.4\times10^{-3}\ °\text{C}^{-1}} = 461°\text{C}. \quad \Diamond$$

**Finalize:** Although this toaster appears to use significantly less power than most, the temperature seems high enough to toast a piece of bread in a reasonable amount of time. In fact, the temperature of a typical 1 000-W toaster would not be vastly higher because Stefan's radiation law (Equation 17.35) tells us that (assuming all power is lost through radiation) $T \propto \sqrt[4]{\mathcal{P}}$, so that the temperature might be about 850°C. In either case, the operating temperature is well below the melting point of the heating element.

**P21.23** An electric car is designed to run off a bank of 12.0-V batteries with total energy storage of $2.00\times10^7$ J.

(a) If the electric motor draws 8.00 kW, what is the current delivered to the motor?

(b) If the electric motor draws 8.00 kW as the car moves at a steady speed of 20.0 m/s, how far will the car travel before it is "out of juice"?

**Solution** (a) Since $\mathcal{P} = \Delta VI$

$$I = \frac{\mathcal{P}}{\Delta V} = \frac{8.00\times10^3\ \text{W}}{12.0\ \text{V}} = 667\ \text{A}. \quad \Diamond$$

(b) The time the car runs is

$$\Delta t = \frac{U}{\mathcal{P}} = \left(\frac{2.00\times10^7\ \text{J}}{8.00\times10^3\ \text{W}}\right)\left(\frac{1\ \text{W}\cdot\text{s}}{\text{J}}\right) = 2.50\times10^3\ \text{s}.$$

So it moves a distance of

$$\Delta x = v\Delta t = (20.0\ \text{m/s})(2.50\times10^3\ \text{s}) = 50.0\ \text{km}. \quad \Diamond$$

**P21.25** A battery has an emf of 15.0 V. The terminal voltage of the battery is 11.6 V when it is delivering 20.0 W of power to an external load resistor $R$.

(a) What is the value of $R$?

(b) What is the internal resistance of the battery?

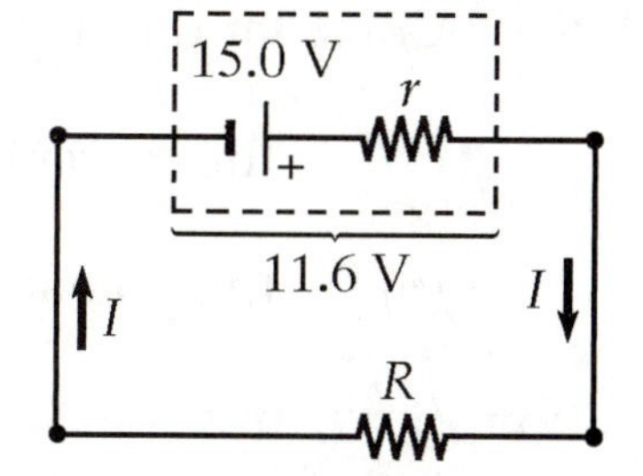


**FIG. P21.25**

**Solution** **Conceptualize:** The internal resistance of a battery usually is less than $1\ \Omega$, with physically larger batteries having less resistance due to the larger anode and cathode areas. The voltage of this battery drops significantly (23%), when the load resistance is added, so a sizable amount of current must be drawn from the battery. If we assume that the internal resistance is about $1\ \Omega$, then the current must be about 3 A to give the 3.4-V drop across the battery's internal resistance. If this is true, then the load resistance must be about $R \approx \dfrac{12\text{ V}}{3\text{ A}} = 4\ \Omega$.

**Categorize:** We can find $R$ exactly by using Joule's law for the power delivered to the load resistor when the voltage is 11.6 V. Then we can find the internal resistance of the battery by summing the electric potential differences around the circuit.

**Analyze:**

(a) Combining Joule's law, $\mathcal{P} = \Delta V I$, and the definition of resistance, $\Delta V = IR$, gives

$$R = \frac{(\Delta V)^2}{\mathcal{P}} = \frac{(11.6\text{ V})^2}{20.0\text{ W}} = 6.73\ \Omega. \qquad \Diamond$$

(b) The electromotive force if the battery must equal the voltage drops across the resistances: $\mathcal{E} = IR + Ir$, where $I = \dfrac{\Delta V}{R}$.

$$r = \frac{\mathcal{E} - IR}{I} = \frac{(\mathcal{E} - \Delta V)R}{\Delta V} = \frac{(15.0\text{ V} - 11.6\text{ V})(6.73\ \Omega)}{11.6\text{ V}} = 1.97\ \Omega \qquad \Diamond$$

**Finalize:** The resistance of the battery is larger than $1\ \Omega$, but it is reasonable for an old battery or for a battery consisting of several small electric cells in series. The load resistance agrees reasonably well with our prediction, despite the fact that the battery's internal resistance was about twice as large as we assumed. Note that in our initial guess we did not consider the power of the load resistance; however, there is not sufficient information to accurately solve this problem without this data.

**P21.29** Consider the circuit shown in Figure 21.29. Find

(a) the current in the 20.0-Ω resistor and

(b) the potential difference between points $a$ and $b$.

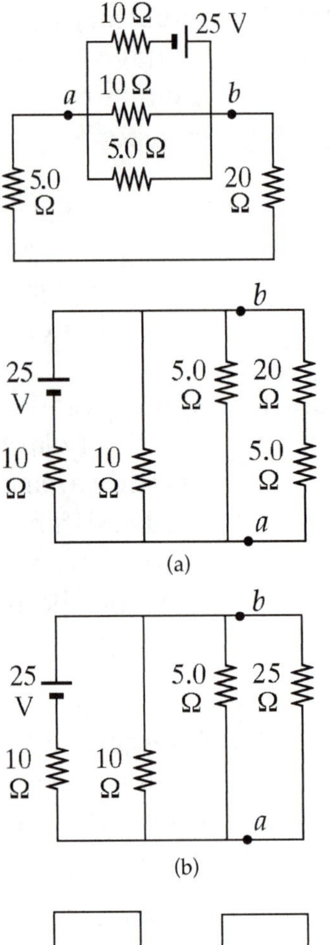


**Solution** If we turn the given diagram on its side, we find that it is the same as figure (a). The 20.0-Ω and 5.00-Ω resistors are in series, so the first reduction is as shown in (b). In addition, since the 10.0-Ω, 5.00-Ω, and 25.0-Ω resistors are then in parallel, we can solve for their equivalent resistance as:

$$R_{eq} = \frac{1}{(1/10.0\ \Omega)+(1/5.00\ \Omega)+(1/25.0\ \Omega)} = 2.94\ \Omega .$$

This is shown in figure (c), which in turn reduces to the circuit shown in (d).

Next, we work backwards through the diagrams, applying $I = \dfrac{\Delta V}{R}$ and $\Delta V = IR$ alternately to each resistor, real or equivalent. The 12.94-Ω resistor is connected across 25.0-V, so the current through the voltage source in every diagram is

$$I = \frac{\Delta V}{R} = \frac{25.0\ \text{V}}{12.94\ \Omega} = 1.93\ \text{A} .$$

In figure (c), this 1.93 A goes through the 2.94-Ω equivalent resistor to give a voltage drop of:

$$\Delta V = IR = (1.93\ \text{A})(2.94\ \Omega) = 5.68\ \text{V} .$$

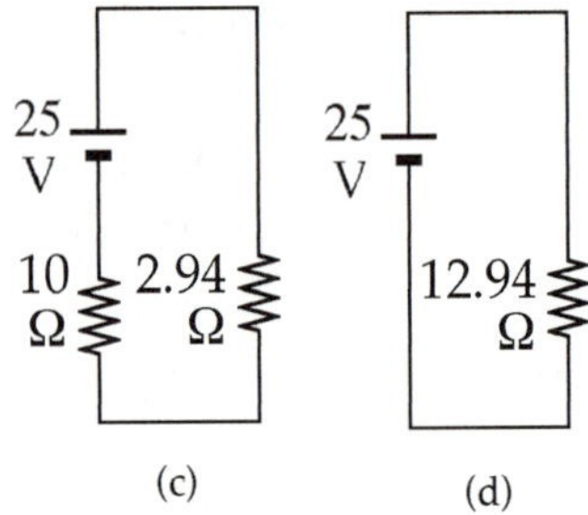


**FIG. P21.29**

From figure (b), we see that this voltage drop is the same across $\Delta V_{ab}$, the 10-Ω resistor , and the 5.00-Ω resistor.

(b) Therefore, $V_{ab} = 5.68\ \text{V}$. ◊

(a) Since the current through the 20-Ω resistor is also the current through the 25.0-Ω line $ab$,

$$I = \frac{\Delta V_{ab}}{R_{ab}} = \frac{5.68\ \text{V}}{25.0\ \Omega} = 0.227\ \text{A} .$$ ◊

**P21.35** Determine the current in each branch of the circuit shown in Figure P21.35.

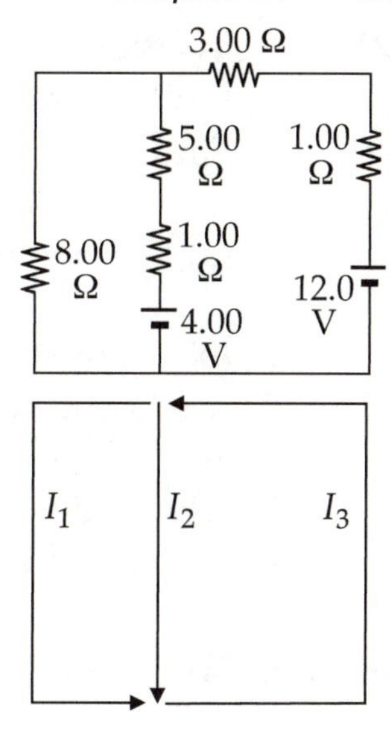


**FIG. P21.35**

**Solution** First, we arbitrarily define the initial current directions and names, as shown in the second diagram. The current rule then says that

$$I_3 - I_1 - I_2 = 0 \qquad (1)$$

By the voltage rule, clockwise around the left-hand loop,

$$+I_1(8.00\ \Omega) - I_2(5.00\ \Omega) - I_2(1.00\ \Omega) - 4.00\ \text{V} = 0 \qquad (2)$$

Clockwise around the right-hand loop, combining the 1 and 5 Ω resistors and the 1 and 3 Ω resistors,

$$4.00\ \text{V} + I_2(6.00\ \Omega) + I_3(4.00\ \Omega) - 12.0\ \text{V} = 0 \qquad (3)$$

We substitute $(I_1 + I_2)$ for $I_3$, and reduce our three equations to:

$$(8.00\ \Omega)I_1 - (6.00\ \Omega)I_2 - 4.00\ \text{V} = 0$$

and $$4.00\ \text{V} + (6.00\ \Omega)I_2 + (4.00\ \Omega)(I_1 + I_2) - 12.0\ \text{V} = 0.$$

Solving the top equation for $I_2$, $\quad I_2 = \dfrac{(8.00\ \Omega)I_1 - 4.00\ \text{V}}{6.00\ \Omega}$

and rearranging the bottom equation, $\quad I_1 + \dfrac{10.0\ \Omega}{4.00\ \Omega}I_2 = \dfrac{8.00\ \text{V}}{4.00\ \Omega}.$

We can substitute for $I_2$, $\quad I_1 + 3.33I_1 - 1.67\ \text{V}/\Omega = 2.00\ \text{V}/\Omega$

and solve for $I_1$, down the 8-Ω resistor: $I_1 = 0.846\ \text{V}/\Omega = 0.846\ \text{A}$. ◊

If we were solving the three simultaneous equations by determinants or by calculator matrix inversion, we would have to do much more work to find $I_2$ and $I_3$. Our method of solving by substitution is more generally useful, and makes it easy to work out the remaining answers after the first, just as when the cat has kittens. We substitute again, putting the value for $I_1$ into equations we already solved for the other unknowns.

Thus, $I_2 = \dfrac{(8.00\ \Omega)(0.846\ \text{A}) - 4.00\ \text{V}}{6.00\ \Omega} = 0.462\ \text{A}$ down the middle branch ◊

$I_3 = 0.846\ \text{A} + 0.462\ \text{A} = 1.31\ \text{A}$ up in the right-hand branch. ◊

**P21.41** Consider a series *RC* circuit (see Figure 21.25) for which $R = 1.00\ \text{M}\Omega$, $C = 5.00\ \mu\text{F}$, and $\mathcal{E} = 30.0\ \text{V}$. Find

(a) the time constant of the circuit and

(b) the maximum charge on the capacitor after the switch is closed.

(c) the current in the resistor 10.0 s after the switch is closed.

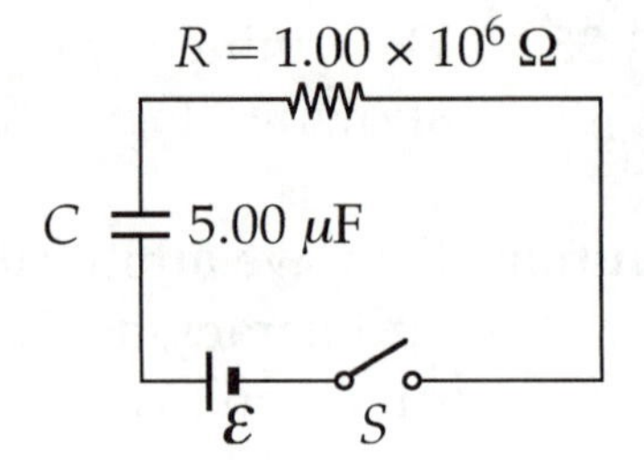


**FIG. P21.41**

**Solution** (a) $\tau = RC = (1.00\times10^{6}\ \Omega)(5.00\times10^{-6}\ \text{F}) = 5.00\ \Omega\cdot\text{F} = 5.00\ \text{s}$ ◊

(b) After a long time, the capacitor is "charged to thirty volts", separating charges of

$$Q = C\Delta V = (5.00\times10^{-6}\ \text{F})(30.0\ \text{V}) = 150\ \mu\text{C}$$ ◊

(c) $I = I_0 e^{-t/\tau}$ where $I_0 = \dfrac{\mathcal{E}}{R}$ and $\tau = RC = 5.00\ \text{s}$

$$I = \frac{\mathcal{E}}{R}e^{-t/\tau} = \left(\frac{30.0\ \text{V}}{1.00\times10^{6}\ \Omega}\right)e^{-10.0\ \text{s}/5.00\ \text{s}}$$

$I = 4.06\times10^{-6}\ \text{A} = 4.06\ \mu\text{A}$ ◊

**P21.45** The circuit in Figure P21.45 has been connected for a long time.

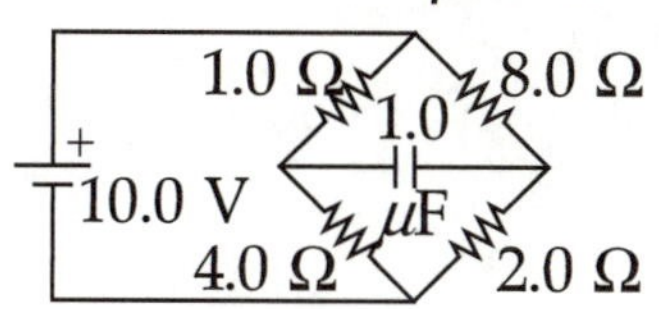


FIG. P21.45

(a) What is the voltage across the capacitor?

(b) If the battery is disconnected, how long does it take the capacitor to discharge to one tenth of its initial voltage?

**Solution** (a) After a long time the capacitor branch will carry negligible current. The current flow is as shown in figure (a). To find the voltage at point $a$, we first find the current, using the voltage rule:

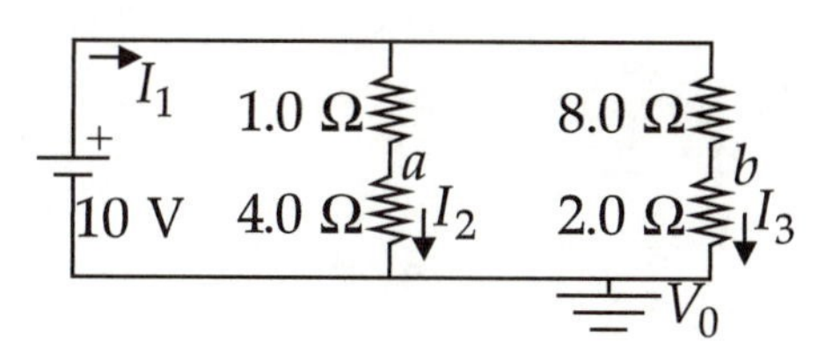


FIG. P21.45(a)

$$10.0\text{ V} - (1.00\ \Omega)I_2 - (4.00\ \Omega)I_2 = 0$$

$$I_2 = 2.00\text{ A}$$

$$V_a - V_0 = (4.00\ \Omega)I_3 = 8.00\text{ V}$$

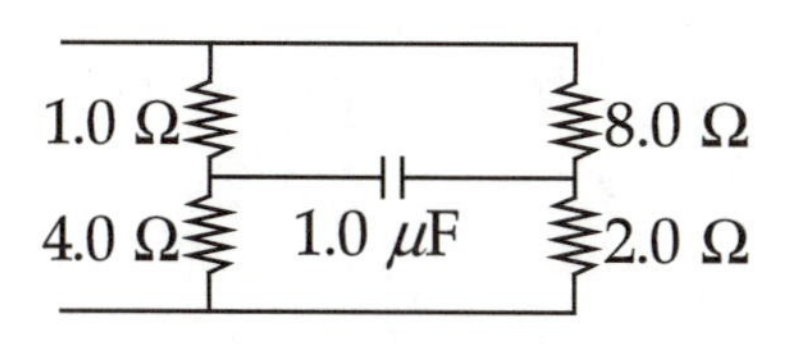


FIG. P21.45(b)

Similarly,

$$10.0\text{ V} - (8.00\ \Omega)I_3 - (2.00\ \Omega)I_3 = 0$$

$$I_3 = 1.00\text{ A}.$$

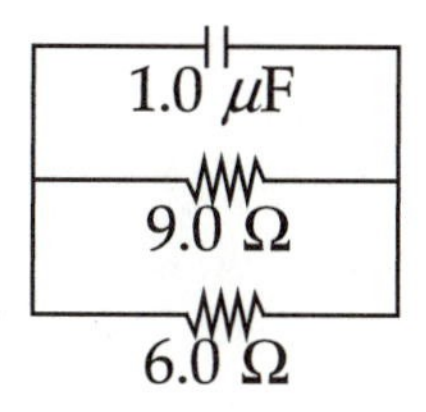


FIG. P21.45(c)

At point $b$,

$$V_b - V_0 = (2.00\ \Omega)I_3 = 2.00\text{ V}.$$

Thus, the voltage across the capacitor is

$$V_a - V_b = 8.00\text{ V} - 2.00\text{ V} = 6.00\text{ V}. \quad \Diamond$$

1.0 μF

$$\frac{1}{\frac{1}{9}+\frac{1}{6}} = 3.60\ \Omega$$

FIG. P21.45(d)

(b) We suppose the battery is pulled out leaving an open circuit. We are left with figure (b), which can be reduced to equivalent circuits (c) and (d).

From (d), we can see that the capacitor discharges through a 3.60-Ω equivalent resistance.

According to $q = Qe^{-t/RC}$

we calculate that $qC = QCe^{-t/RC}$

and $\Delta V = \Delta V_i e^{-t/RC}$.

*continued on next page*

Solving,

$$\frac{1}{10}\Delta V_i = \Delta V_i e^{-t/(3.60\ \Omega)(1.00\ \mu\text{F})}$$

$$e^{-t/3.60\ \mu\text{s}} = 0.100$$

$$\frac{-t}{3.60\ \mu\text{s}} = \ln(0.100) = -2.30$$

$$\frac{t}{3.60\ \mu\text{s}} = 2.30$$

$$t = (2.30)(3.60\ \mu\text{s}) = 8.29\ \mu\text{s}$$

◊

# Chapter 22

# Magnetic Forces and Magnetic Fields

## NOTES FROM SELECTED CHAPTER SECTIONS

### Section 22.2 The Magnetic Field

Particles with charge $q$, moving with speed $\vec{\mathbf{v}}$ in a magnetic field $\vec{\mathbf{B}}$, experience a magnetic force $\vec{\mathbf{F}}_B$. Properties of the magnetic force are:

- The magnetic force is proportional to the charge $q$ and speed $v$ of the particle.
- The magnitude and direction of the magnetic force depend on the angle between the velocity vector of the particle and the magnetic field vector.
- When a charged particle moves in a direction parallel to the magnetic field vector, the magnetic force $\vec{\mathbf{F}}_B$ on the charge is zero.
- The magnetic force acts in a direction perpendicular to both $\vec{\mathbf{v}}$ and $\vec{\mathbf{B}}$; that is, $\vec{\mathbf{F}}_B$ is perpendicular to the plane formed by $\vec{\mathbf{v}}$ and $\vec{\mathbf{B}}$.
- The magnetic force on a positive charge is in the direction opposite to the force on a negative charge moving in the same direction.
- If the velocity vector makes an angle $\theta$ with the magnetic field, the magnitude of the magnetic force is proportional to sin $\theta$.

There are several important differences between electric and magnetic forces:

- The electric force is always along the direction of the electric field, whereas the magnetic force is perpendicular to the magnetic field.
- The electric force acts on a charged particle independent of the particle's velocity, whereas the magnetic force acts on a charged particle only when the particle is in motion.
- The electric force does work in displacing a charged particle, whereas the magnetic force associated with a steady magnetic field does no work when a particle is displaced.

## Section 22.7 The Biot-Savart Law

The Biot-Savart law says that if a wire carries a steady current $I$, the magnetic field $d\mathbf{B}$ at a point $P$ associated with an element $d\vec{\mathbf{s}}$ has the following properties:

- The vector $d\vec{\mathbf{B}}$ is perpendicular both to $d\vec{\mathbf{s}}$ (which is in the direction of the current) and to the unit vector $\hat{\mathbf{r}}$ (directed from the current element to the point $P$).
- The magnitude of $d\vec{\mathbf{B}}$ is inversely proportional to $r^2$, where $r$ is the distance from the element to the point $P$.
- The magnitude of $d\vec{\mathbf{B}}$ is proportional to the current and to the length of the element, $d\vec{\mathbf{s}}$.
- The magnitude of $d\vec{\mathbf{B}}$ is proportional to $\sin\theta$, where $\theta$ is the angle between the vectors $d\vec{\mathbf{s}}$ and $\hat{\mathbf{r}}$.

## Section 22.8 The Magnetic Force Between Two Parallel Conductors

Parallel conductors carrying currents in the same direction attract each other, whereas parallel conductors carrying currents in opposite directions repel each other.

The force between two parallel wires each carrying a current is used to define the ampere as follows:

If two long, parallel wires 1 m apart in a vacuum carry the same current and the force per unit length on each wire is $2\times10^{-7}$ N/m, then the current is defined to be 1 A.

## Section 22.9 Ampère's Law

The direction of the magnetic field due to a current in a conductor is given by the right-hand rule:

If the wire is grasped in the right hand with the thumb in the direction of the current, the fingers will wrap (or curl) in the direction of $\vec{\mathbf{B}}$.

*Ampère's law is valid only for steady currents and is useful only in those cases where the current configuration has a high degree of symmetry.*

## EQUATIONS AND CONCEPTS

The **magnetic force** exerted on a positive electric charge moving in a magnetic field can be expressed as a vector cross product.

$$\vec{\mathbf{F}}_B = q\vec{\mathbf{v}} \times \vec{\mathbf{B}} \tag{22.1}$$

The **magnitude of the magnetic force** will be a maximum when the charge moves along a direction perpendicular to the direction of the magnetic field.

$$F_B = |q| vB\sin\theta \tag{22.2}$$

The **magnetic field $\vec{\mathbf{B}}$** at some point in space is defined in terms of the magnetic force exerted on a moving positive electric charge at that point. The SI unit of the magnetic field is the tesla (T) or weber per square meter $\left(\text{Wb/m}^2\right)$. The cgs unit of magnetic field is the gauss (G).

$$B \equiv \frac{F_B}{|q| v\sin\theta}$$

$$1\ \text{T} = 1\ \frac{\text{N}}{\text{C}\cdot\text{m/s}}$$

$$1\ \text{T} = 10^4\ \text{G}$$

The **direction of the magnetic force on a moving charge** can be determined by applying the right-hand-rule. Two versions for application of the rule are shown in the figure below. The direction shown for $\vec{\mathbf{F}}$ is the direction of force on a positive charge; if the charge is negative the direction of the force will be reversed.

**Version 1**

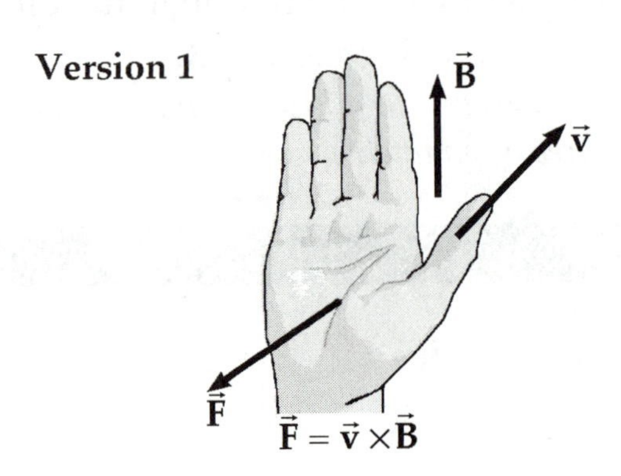


**Version 2**

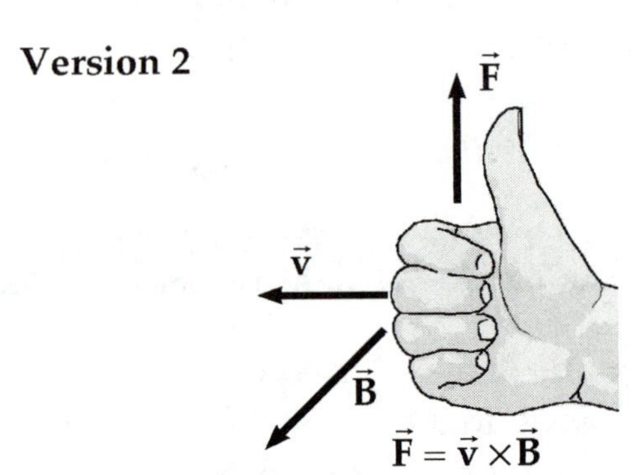


| | Hold your open right hand with your thumb pointing in the direction of $\vec{\mathbf{v}}$ (*the first named vector quantity*) and your fingers pointing in the direction of $\vec{\mathbf{B}}$ (*the second named vector quantity*). By necessity, your fingers and thumb cannot stretch farther than 180°, so if you do this, your hand will be oriented properly. The force vector $\vec{\mathbf{F}}$ now is directed out of the palm of your hand. | Orient your hand so that your fingers point in the direction of $\vec{\mathbf{v}}$ (*the first named vector quantity*) and then curl your fingers to point in the direction of $\vec{\mathbf{B}}$ (*the second named vector quantity*). Note that since your fingers cannot bend farther than 180°, you may have to flip your hand upside down to do this. Your thumb now points in the direction of $\vec{\mathbf{F}}$, where $\vec{\mathbf{F}}$ is perpendicular to both $\vec{\mathbf{v}}$ and $\vec{\mathbf{B}}$. |
|---|---|---|

**Motion** of a charged particle entering a uniform magnetic field with the velocity vector initially perpendicular to the field has the following characteristics:

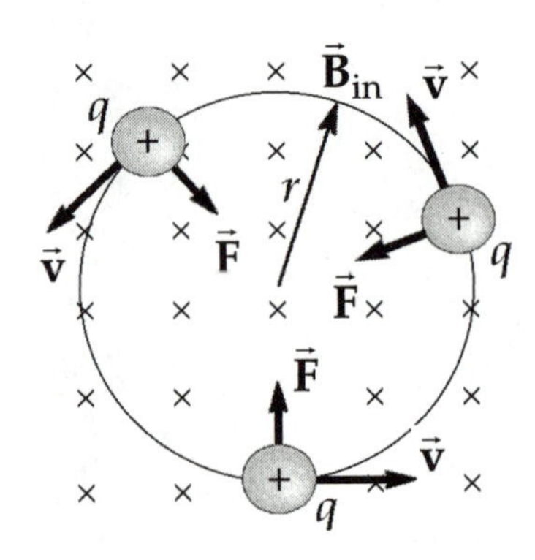

- The **path** of the particle will be circular and in a plane perpendicular to the direction of the field. The direction of rotation of the particle will be as determined by the right-hand rule.

$$r = \frac{mv}{qB} \quad (22.3)$$

- The **radius** of the circular path will be proportional to the linear momentum of the charged particle.

$$\omega = \frac{qB}{m} \quad (22.4)$$

- The **angular frequency** (or cyclotron frequency) of the particle will be proportional to the ratio of charge to mass.

$$T = \frac{2\pi m}{qB} \quad (22.5)$$

*Note that the frequency, and hence the period of rotation do not depend on the radius of the path.*

If the **initial velocity is not perpendicular** to the field, the particle will move in a helical path whose axis is parallel to the field. The "pitch" of the helix will depend on the component of velocity parallel to the field.

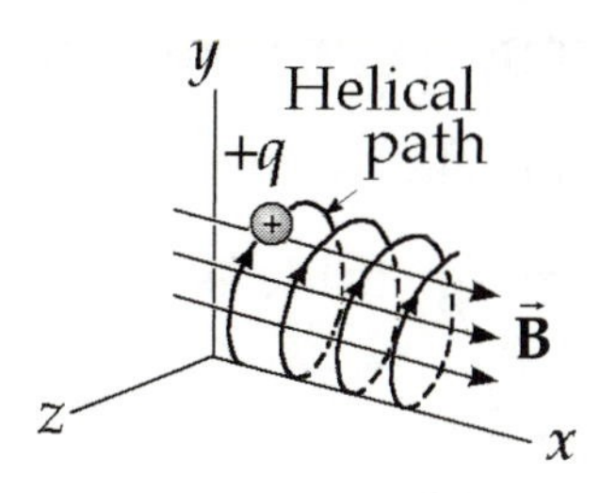


**Force on a current-carrying conductor** in a magnetic field.

$$\vec{\mathbf{F}}_B = I\vec{\ell} \times \vec{\mathbf{B}} \quad (22.10)$$

**For a straight conductor**, the magnitude of the force depends on the angle between the direction of the current and the direction of the field. The force is maximum when the conductor is perpendicular to the direction of the field.

$$F_B = BI\ell \sin\theta$$

$$F_{B,\,\max} = BI\ell$$

The **direction of the force is determined** by the right-hand rule. In this case, your thumb must point in the direction of the current. *Use the right-hand rule to confirm that the direction of* $\vec{\mathbf{B}}$ *in the figure at right is into the plane of the page.*

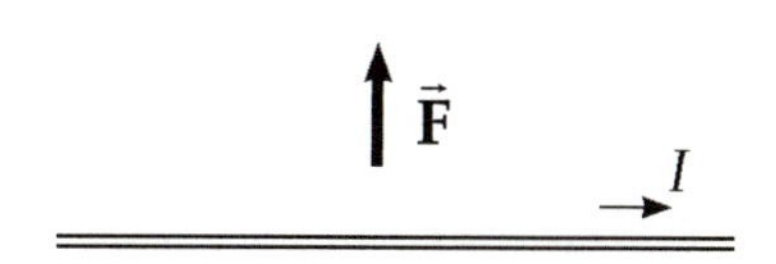

**For a wire of arbitrary shape** the magnetic force is found by integrating over the length of the wire. In these equations the direction of $\ell$ and $ds$ is that of the current. *The magnetic force on a curved current-carrying wire in a uniform magnetic field is equal to that of a straight wire connecting the end points and carrying the same current. Also, the net magnetic force on a closed current loop in a uniform field is zero.*

$$\vec{\mathbf{F}}_B = I\int_a^b d\vec{\mathbf{s}} \times \vec{\mathbf{B}} \quad (22.12)$$

**Applications of the motion of charged particles in a magnetic field** include:

- **Velocity Selector**—When a beam of charged particles is directed into a region where uniform electric and magnetic fields are perpendicular to each other and to the initial direction of the particle beam, only those particles with a velocity $v = \frac{E}{B}$ will emerge undeflected.

$$v = \frac{E}{B} \quad (22.7)$$

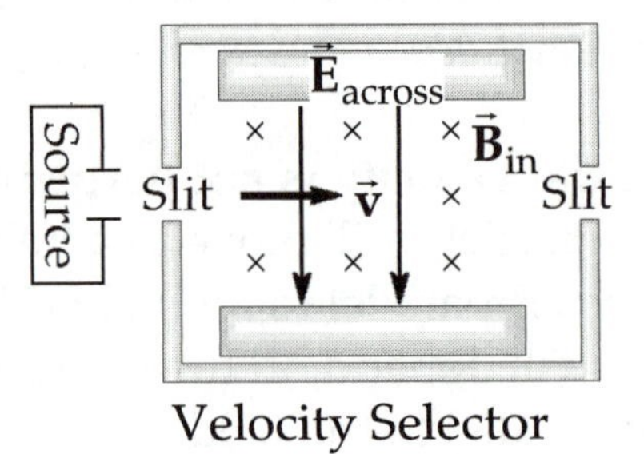

Velocity Selector

- **Mass Spectrometer**—If an ion beam, after passing through a velocity selector, is directed perpendicularly into a second uniform magnetic field, the ratio of charge to mass for the isotopic species can be determined by measuring the radius of curvature of the beam in the second field.

$$\frac{m}{q} = \frac{rB_0B}{E} \quad (22.8)$$

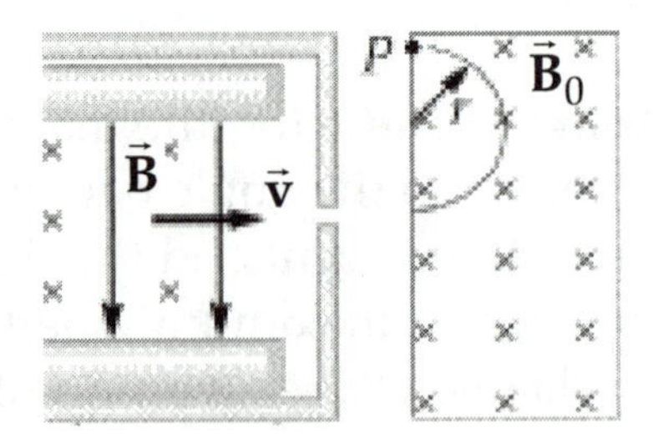

- **Cyclotron**—The maximum kinetic energy acquired by an ion in a cyclotron depends on the radius of the "dees" and the intensity of the magnetic field. This relationship holds until the ion reaches relativistic energies (~ 20 MeV). *This is true for ions of proton mass or greater, but is not true for electrons.*

$$K = \frac{1}{2}mv^2 = \frac{q^2B^2R^2}{2m} \quad (22.9)$$

A **net torque** is exerted on a conducting loop carrying a current when placed in an external magnetic field. The magnitude of vector $\vec{\mathbf{A}}$ is numerically equal to the area of the loop and is directed perpendicular to the area of the loop. *When the fingers of the right hand are curled around the loop in the direction of the current, the thumb points in the direction of* $\vec{\mathbf{A}}$.

$$\vec{\tau} = I\vec{\mathbf{A}} \times \vec{\mathbf{B}} \quad (22.14)$$

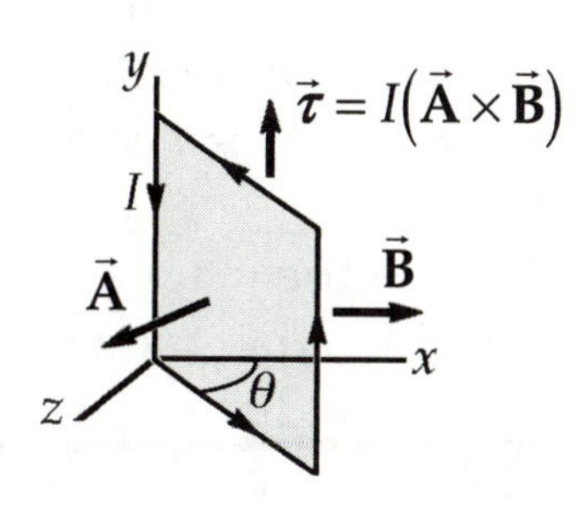


The torque on a current loop can also be expressed in terms of $\vec{\mu}$, the **magnetic dipole moment** of the loop.

$$\vec{\tau} = \vec{\mu} \times \vec{\mathbf{B}} \quad (22.16)$$

$$\vec{\mu} = I\vec{\mathbf{A}} \quad (22.15)$$

The **magnitude of the torque** will depend on the angle between the direction of the magnetic field and the direction of the normal (or perpendicular) to the plane of the loop and will be maximum when the magnetic field is parallel to the plane of the loop.

$$\tau = IAB\sin\theta$$

$$\tau_{\max} = IAB \quad (22.13)$$

The **direction of rotation of the loop** is in the direction of decreasing values of $\tau$ (i.e. vector $\vec{\mathbf{A}}$ rotates toward the direction of the magnetic field). When $\vec{\mathbf{A}}$ becomes parallel to $\vec{\mathbf{B}}$ the torque on the loop will be zero. The loop shown in the figure will rotate counterclockwise as seen from above. *The direction of the vector torque* $\vec{\tau}$ *is indicated by the thumb of the right hand when the fingers curl* $\vec{\mathbf{A}}$ *into the direction of* $\vec{\mathbf{B}}$.

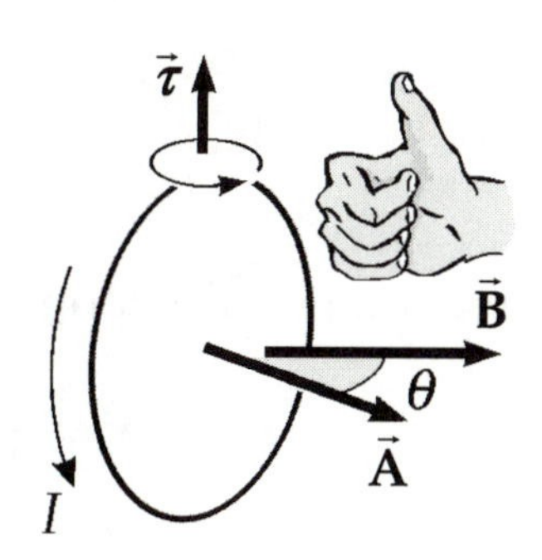

The **Biot–Savart law** gives the magnetic field at a point in space due to an element of conductor $d\vec{\mathbf{s}}$, which carries a current $I$ and is at a distance $r$ away from the point. *The vector $d\vec{\mathbf{B}}$ is perpendicular to both $d\vec{\mathbf{s}}$ (which points in the direction of the current) and to the unit vector $\hat{\mathbf{r}}$ (which points from the current element toward the point where the field is to be determined).* At point $P$ as shown in the figure the direction of $d\vec{\mathbf{B}}$ is out of the page. The **total magnetic field** at a point in the vicinity of a current of finite length is found by integrating the Biot–Savart law expression over the entire current distribution. *Remember, the integrand is a vector quantity.*

$$d\vec{\mathbf{B}} = k_m \frac{I\, d\vec{\mathbf{s}} \times \hat{\mathbf{r}}}{r^2} \tag{22.17}$$

$$\mu_0 = 4\pi \times 10^{-7}\ \mathrm{T \cdot m/A} \tag{22.19}$$

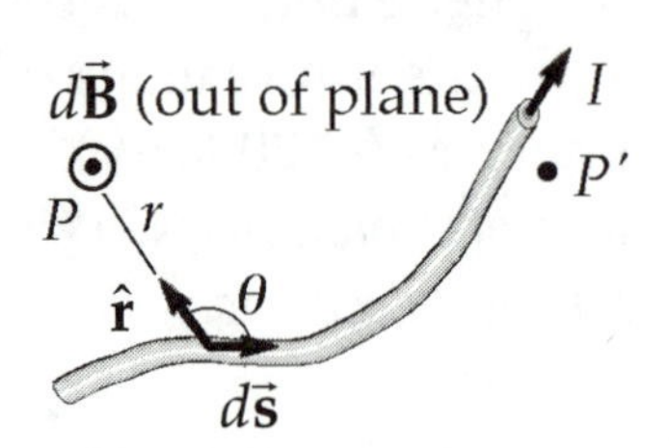


The **magnitude of the magnetic field** due to several important geometric arrangements of a current-carrying conductor can be calculated by use of the Biot-Savart law. Some examples are:

$B$ at a distance $r$ from a **long straight conductor,** carrying a current $I$.

$$B = \frac{\mu_0 I}{2\pi r} \tag{22.21}$$

$B$ at the **center of an arc** of radius $R$ **that** subtends an angle $\theta$ (in radians) at the center of the arc.

$$B = \frac{\mu_0 I}{4\pi R}\theta$$

$B_x$ on the axis of a **circular loop** of radius $R$ and at a distance $x$ from the plane of the loop.

$$B_x = \frac{\mu_0 I R^2}{2\left(x^2 + R^2\right)^{3/2}} \tag{22.23}$$

$$B \approx \frac{\mu_0 I R^2}{2x^3} \quad (\text{for } x >> R) \tag{22.25}$$

$B$ at the **center of a circular loop**.

$$B = \frac{\mu_0 I}{2R} \tag{22.24}$$

The **magnetic force per unit length** between very long parallel conductors depends on the distance $a$ between the conductors and the magnitudes of the two currents. Parallel conductors carrying currents in the same direction attract each other, and parallel conductors carrying currents in opposite directions repel each other. *The forces on the two conductors will be equal in magnitude regardless of the relative magnitude of the two currents.*

$$\frac{F}{\ell} = \frac{\mu_0}{2\pi}\frac{I_1 I_2}{a} \tag{22.27}$$

**Ampère's law**: The total current passing through a surface bounded by the closed path. This technique is most useful in calculating the magnetic field due to currents that have a high degree of symmetry. Two examples are:

$$\oint \vec{\mathbf{B}} \cdot d\vec{\mathbf{s}} = \mu_0 I \tag{22.29}$$

$B$ **inside a toroid** having $N$ turns and at a distance $r$ from the center of the toroid.

$$B = \frac{\mu_0 N I}{2\pi r} \tag{22.31}$$

$B$ near the center of a **solenoid** of $n$ turns per unit length.

$$B = \mu_0 n I \tag{22.32}$$

The **direction of the magnetic field produced by a current** can be determined by using the right-hand rule.

**For a long straight conductor**

Hold the conductor in the right hand with the thumb pointing in the direction of the conventional current. The fingers will then wrap around the wire in the direction of the magnetic field lines. *The magnetic field is tangent to the circular field lines at every point in the region around the conductor.*

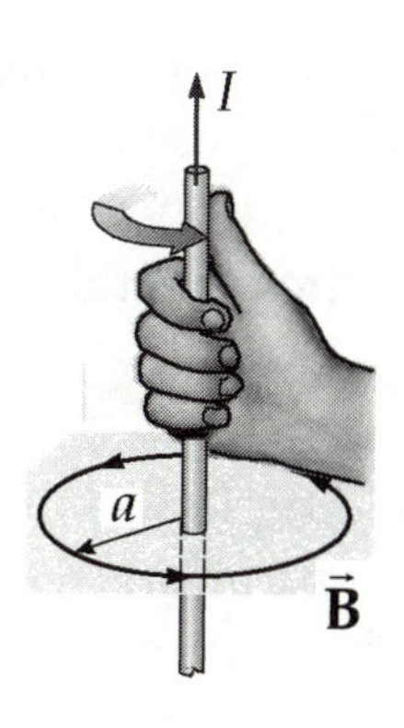


**For a current loop**

The magnetic field is perpendicular to the plane of the loop and directed in the sense given by the right-hand rule.

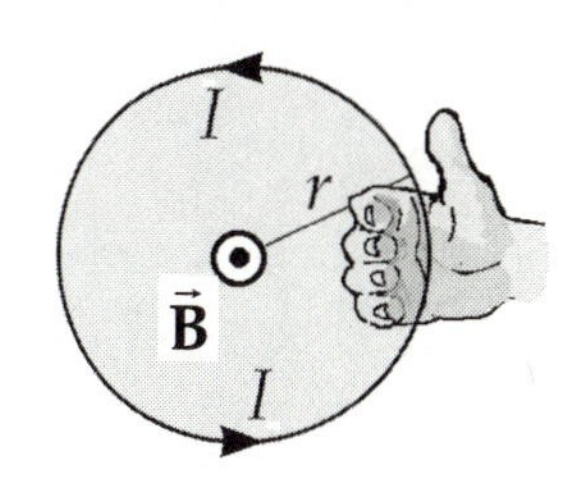

**Within a solenoid**

The magnetic field is parallel to the axis of the solenoid and pointing in a sense determined by applying the right-hand rule to one of the coils.

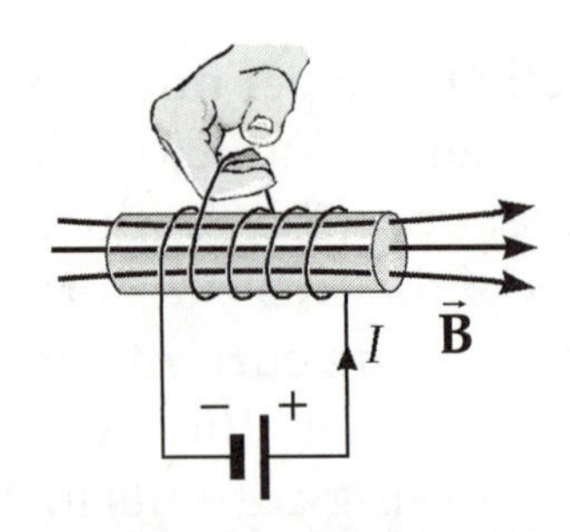


## REVIEW CHECKLIST

✓ Use the defining equation for a magnetic field $\vec{\mathbf{B}}$ to determine the magnitude and direction of the magnetic force exerted on an electric charge moving in a region where there is a magnetic field. You should understand clearly the important differences between the forces exerted on electric charges by electric fields and those forces exerted on moving electric charges by magnetic fields.

✓ Calculate the magnitude and direction of the magnetic force on a current-carrying conductor when placed in an external magnetic field.

✓ Determine the magnitude and direction of the torque exerted on a closed current loop in an external magnetic field. You should understand how to designate the direction of the area vector corresponding to a given current loop, and to incorporate the magnetic moment of the loop into the calculation of the torque on the loop.

✓ Use the Biot-Savart law to calculate the magnetic field at a specified point in the vicinity of a current element, and by integration find the total magnetic field due to a number of important geometric arrangements. Your use of the Biot-Savart law must include a clear understanding of the direction of the magnetic field contribution relative to the direction of the current element, which produces it and the direction of the vector, which locates the point at which the field is to be calculated.

✓ Use Ampère's law to calculate the magnetic field due to steady current configurations, which have a sufficiently high degree of symmetry such as a long straight conductor, a long solenoid, and a toroidal coil.

## ANSWERS TO SELECTED QUESTIONS

**Q22.1** Two charged particles are projected into a magnetic field perpendicular to their velocities. If the charges are deflected in opposite directions, what can you say about them?

**Answer** We know the magnetic field is constant, and the velocity vectors are the same, but one force is the negative of the other. From $\vec{\mathbf{F}}_B = q\left(\vec{\mathbf{v}} \times \vec{\mathbf{B}}\right)$, we can conclude that the only thing that could cause the force to be of opposite sign is if the charges were of opposite sign.

**Q22.5** Is it possible to orient a current loop in a uniform magnetic field such that the loop does not tend to rotate? Explain.

**Answer** Yes. If the magnetic field is perpendicular to the plane of the loop, the forces on opposite sides will be equal and opposite, but will produce no net torque.

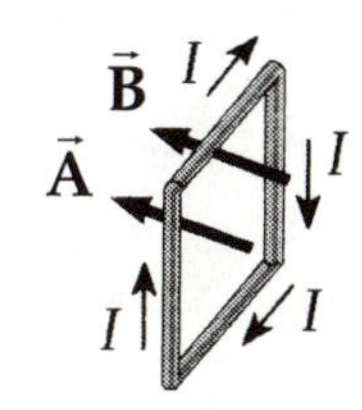


**FIG. Q22.5**

**Q22.9** Explain why two parallel wires carrying currents in opposite directions repel each other.

**Answer** The figure at the right will help you understand this result. The magnetic field due to wire 2 at the position of wire 1 is directed out of the paper. Hence, the magnetic force on wire 1, given by $I_1\vec{\ell}_1 \times \vec{\mathbf{B}}_2$, must be directed to the left since $\vec{\ell}_1 \times \vec{\mathbf{B}}_2$ is directed to the left. Likewise, you can show that the magnetic force on wire 2 due to the field of wire 1 is directed towards the right.

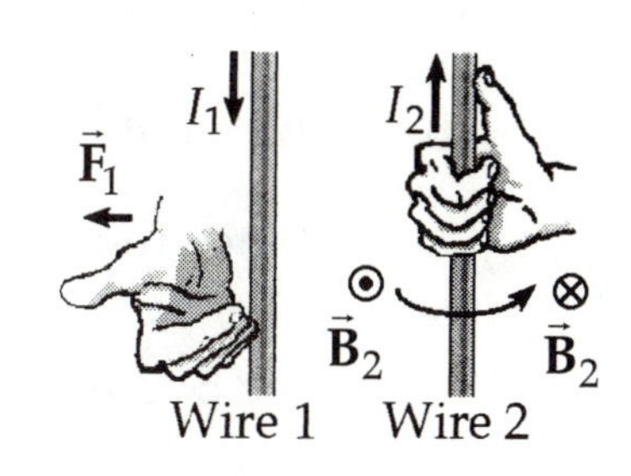


**FIG. Q22.9**

**Q22.11** A hollow copper tube carries a current along its length. Why is $\vec{\mathbf{B}} = 0$ inside the tube? Is $\vec{\mathbf{B}}$ nonzero outside the tube?

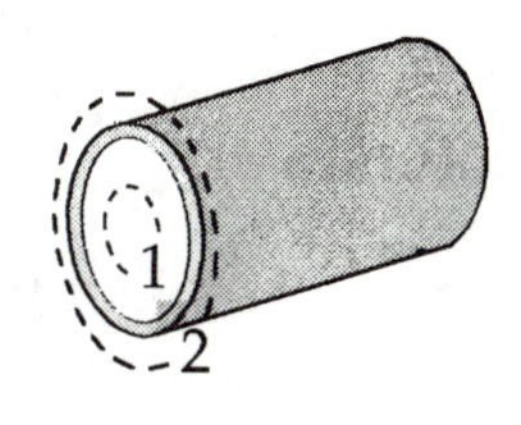


**FIG. Q22.11**

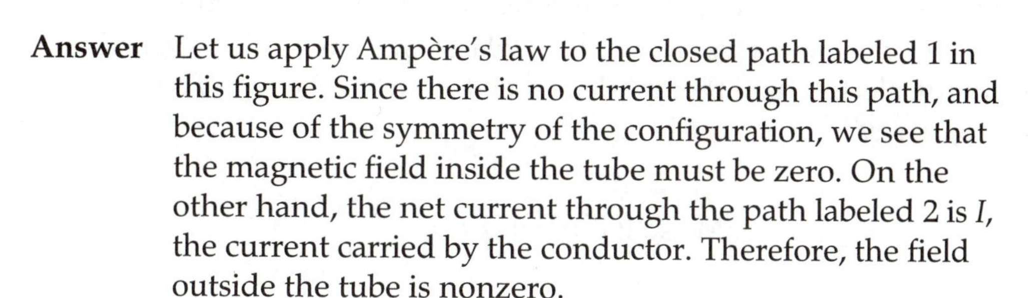

**Answer** Let us apply Ampère's law to the closed path labeled 1 in this figure. Since there is no current through this path, and because of the symmetry of the configuration, we see that the magnetic field inside the tube must be zero. On the other hand, the net current through the path labeled 2 is $I$, the current carried by the conductor. Therefore, the field outside the tube is nonzero.

## SOLUTIONS TO SELECTED PROBLEMS

**P22.1** Determine the initial direction of the deflection of charged particles as they enter the magnetic fields as shown in Figure P22.1.

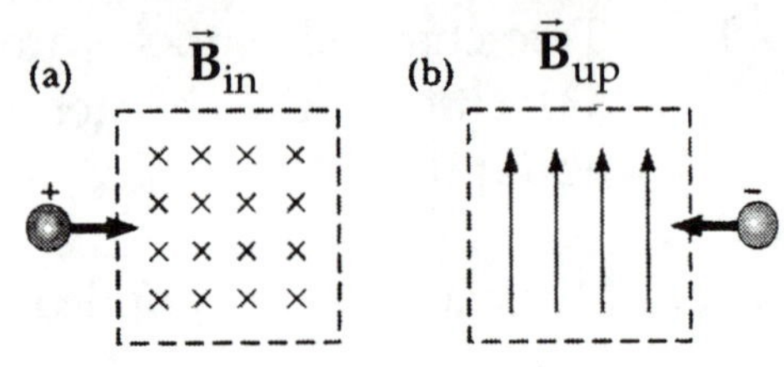


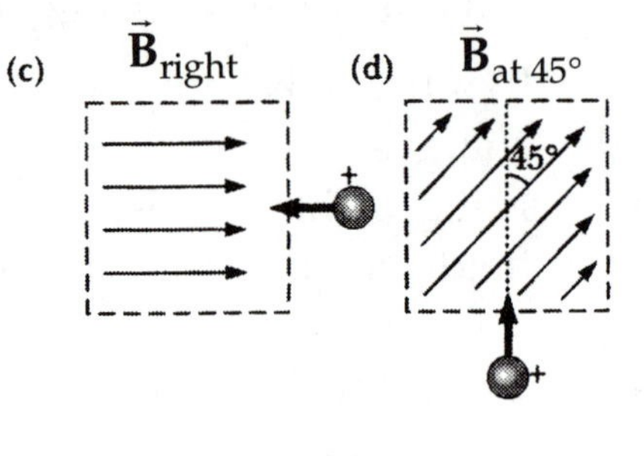


FIG. P22.1

**Solution**

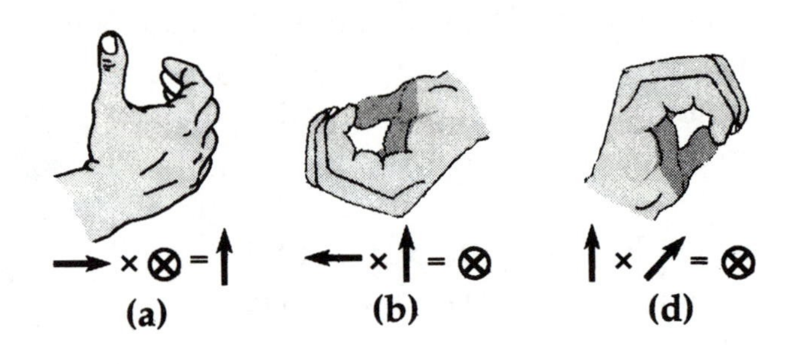


(a) By solution figure (a), $\vec{\mathbf{v}} \times \vec{\mathbf{B}}$ is (right) × (away) = up. ◊

(b) By solution figure (b), $\vec{\mathbf{v}} \times \vec{\mathbf{B}}$ is (left) × (up) = away.

Since the charge is negative, $q\vec{\mathbf{v}} \times \vec{\mathbf{B}}$ is toward you. ◊

(c) $\vec{\mathbf{v}} \times \vec{\mathbf{B}}$ is zero since the angle between $\vec{\mathbf{v}}$ and $\vec{\mathbf{B}}$ is 180° and $\sin 180° = 0$.

There is no deflection. ◊

(d) $\vec{\mathbf{v}} \times \vec{\mathbf{B}}$ is (up) × (up and right), or away from you. ◊

**P22.9** A cosmic-ray proton in interstellar space has an energy of 10.0 MeV and executes a circular orbit having a radius equal to that of Mercury's orbit around the Sun $\left(5.80 \times 10^{10} \text{ m}\right)$. What is the magnetic field in that region of space?

**Solution** Think of the proton as having accelerated through a potential difference $\Delta V = 10^7 \text{ V}$. We use the energy version of the isolated system model, applied to the particle and the electric field that made it speed up from the rest. The proton's kinetic energy is

$$E = \frac{1}{2}mv^2 = e\Delta V \text{ so its speed is } v = \sqrt{\frac{2e\Delta V}{m}}.$$

*continued on next page*

Now we use the particle in a magnetic field model and the particle in uniform circular motion model.

$\sum F = ma$ becomes $\dfrac{mv^2}{R} = evB\sin 90°$.

So $$B = \frac{mv}{eR} = \frac{m}{eR}\sqrt{\frac{2e\Delta V}{m}} = \frac{1}{R}\sqrt{\frac{2m\Delta V}{e}}$$

and $$B = \frac{1}{5.80 \times 10^{10}\ \text{m}}\sqrt{\frac{2\left(1.67 \times 10^{-27}\ \text{kg}\right)\left(10^{7}\ \text{V}\right)}{1.60 \times 10^{-19}\ \text{C}}} = 7.88 \times 10^{-12}\ \text{T}.$$ ◊

**P22.13** The picture tube in a television uses magnetic deflection coils rather than electric deflection plates. Suppose an electron beam is accelerated through a 50.0-kV potential difference and then through a region of uniform magnetic field 1.00 cm wide. The screen is located 10.0 cm from the center of the coils and is 50.0 cm wide. When the field is turned off, the electron beam hits the center of the screen. What field magnitude is necessary to deflect the beam to the side of the screen? Ignore relativistic corrections.

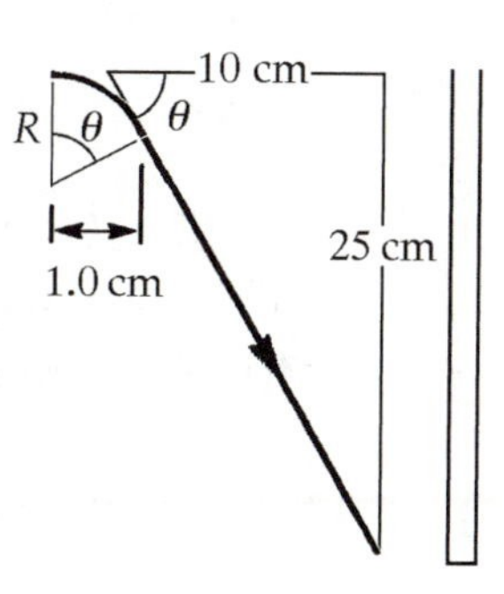


**FIG. P22.13**

**Solution** The beam is deflected by the angle $\theta = \tan^{-1}\left(\dfrac{25.0\ \text{cm}}{10.0\ \text{cm}}\right) = 68.2°$

The two angles $\theta$ shown are equal because their sides are perpendicular, right side to right side and left side to left side. The radius of curvature of the electrons in the field is

$$R = \frac{1.00\ \text{cm}}{\sin 68.2°} = 1.077\ \text{cm}.$$

Now $\dfrac{1}{2}mv^2 = q\Delta V$ so $v = \sqrt{\dfrac{2q\Delta V}{m}} = 1.33 \times 10^{8}\ \text{m/s}$

$\sum \vec{\mathbf{F}} = m\vec{\mathbf{a}}$ becomes $\dfrac{mv^2}{R} = |q|vB\sin 90°$

$$B = \frac{mv}{|q|R} = \frac{\left(9.11 \times 10^{-31}\ \text{kg}\right)\left(1.33 \times 10^{8}\ \text{m/s}\right)}{\left(1.60 \times 10^{-19}\ \text{C}\right)\left(1.077 \times 10^{-2}\ \text{m}\right)} = 70.1\ \text{mT}.$$ ◊

**P22.17** *A nonuniform magnetic field exerts a net force on a magnetic dipole.* A strong magnet is placed under a horizontal conducting ring of radius $r$ that carries current $I$ as shown in Figure P22.17. If the magnetic field $\vec{\mathbf{B}}$ makes an angle $\theta$ with the vertical at the ring's location, what are the magnetic and direction of the resultant force on the ring?

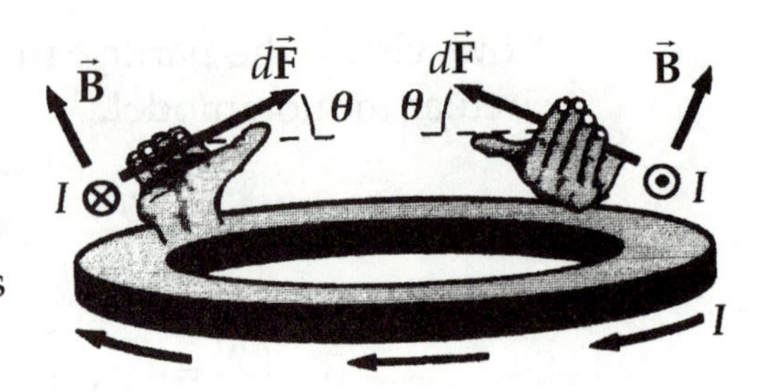


**FIG. P22.17**

**Solution** The magnetic force on each bit of ring is inward and upward, at an angle $\theta$ above the radial line, according to:

$$\left|d\vec{\mathbf{F}}\right| = I\left|d\vec{\mathbf{s}} \times \vec{\mathbf{B}}\right| = IdsB.$$

The radially inward components tend to squeeze the ring, but cancel out as forces. The upward components $IdsB\sin\theta$ all add to

$$\vec{\mathbf{F}} = I(2\pi r)B\sin\theta \text{ up}. \qquad \Diamond$$

The magnetic moment of the ring is down. This problem is a model for the force on a dipole in a nonuniform magnetic field, or for the force that one magnet exerts on another magnet.

**P22.21** A rectangular coil consists of $N = 100$ closely wrapped turns and has dimensions $a = 0.400$ m and $b = 0.300$ m. The coil is hinged along the $y$ axis, and its plane makes an angle $\theta = 30.0°$ with the $x$ axis (Fig. P22.21). What is the magnitude of the torque exerted on the coil by a uniform magnetic field $B = 0.800$ T directed along the $x$ axis when the current is $I = 1.20$ A in the direction shown? What is the expected direction of rotation of the coil?

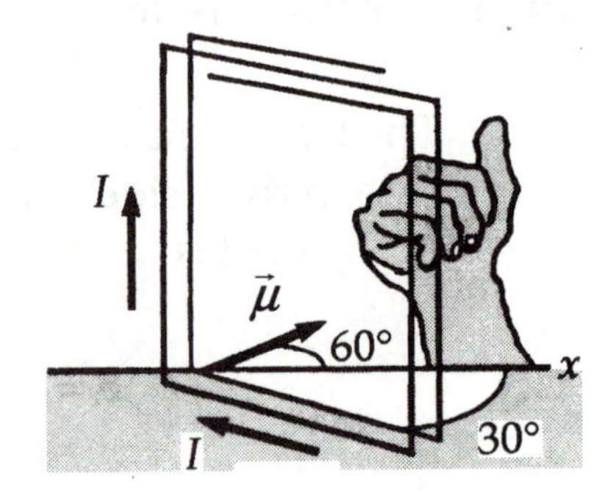


**Solution** The magnetic moment of the coil is $\mu = NIA$, perpendicular to its plane and making a 60° angle with the $x$ axis as shown to the right. The torque on the dipole is then

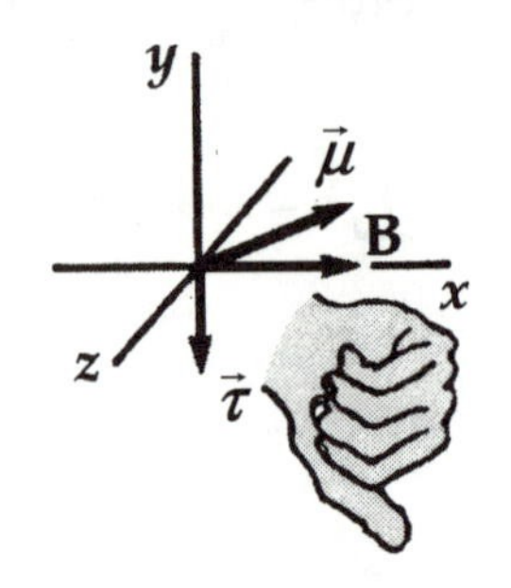


$$\vec{\tau} = \vec{\mu} \times \vec{\mathbf{B}} = NBAI\sin\phi \text{ down}$$

**FIG. P22.21**

*continued on next page*

having a magnitude $\tau = NBAI\sin\phi$

$$\tau = (100)(0.800 \text{ T})\left(0.400 \times 0.300 \text{ m}^2\right)(1.20 \text{ A})(\sin 60.0°) = 9.98 \text{ N}\cdot\text{m} \qquad \Diamond$$

Note that $\phi$ is the angle between the magnetic moment and the $\vec{\mathbf{B}}$ field. We model the coil as a rigid body under a net torque; the coil will rotate so as to align the magnetic moment with the $\vec{\mathbf{B}}$ field. Looking down along the $y$ axis, we will see the coil rotate in the clockwise direction. $\Diamond$

**P22.25** Determine the magnetic field at a point $P$ located a distance $x$ from the corner of an infinitely long wire bent at a right angle, as shown in Figure P22.25. The wire carries a steady current $I$.

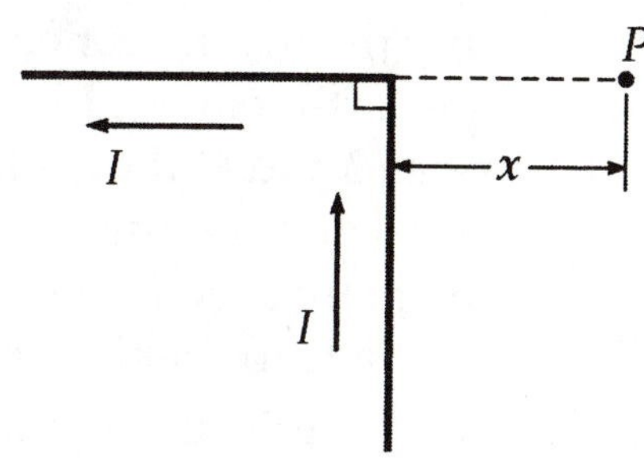


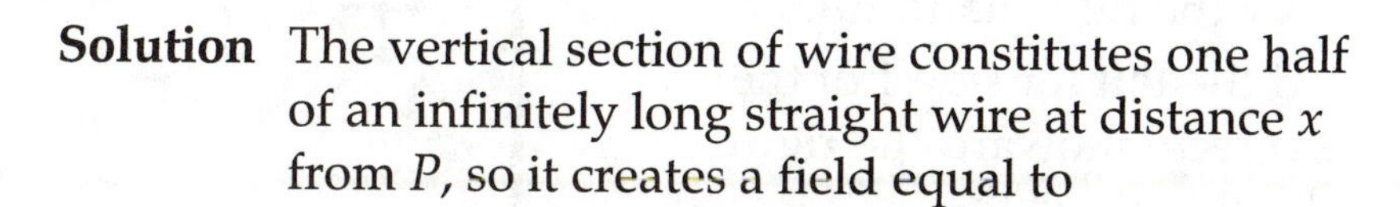

**Solution** The vertical section of wire constitutes one half of an infinitely long straight wire at distance $x$ from $P$, so it creates a field equal to

$$B = \frac{1}{2}\left(\frac{\mu_0 I}{2\pi x}\right)$$

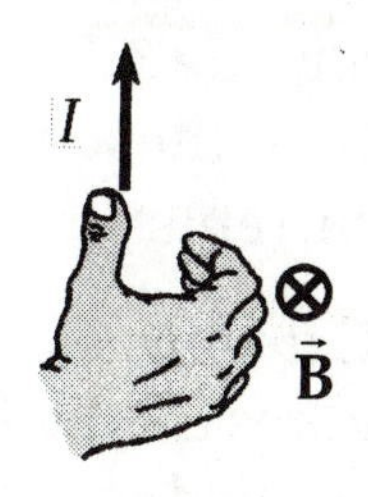


FIG. P22.25

Hold your right hand with extended thumb in the direction of the current; the field is away from you, into the paper. For each bit of the horizontal section of wire $d\vec{\mathbf{s}}$ is to the left and $\hat{\mathbf{r}}$ is to the right, so $d\vec{\mathbf{s}} \times \hat{\mathbf{r}} = 0$. The horizontal current produces zero field at $P$.

Thus, $\vec{\mathbf{B}} = \dfrac{\mu_0 I}{4\pi x}$ into the paper. $\Diamond$

**P22.35** In Figure P22.35, the current in the long, straight wire is $I_1 = 5.00$ A and the wire lies in the plane of the rectangular loop, which carries the current $I_2 = 10.0$ A. The dimensions are $c = 0.100$ m, $a = 0.150$ m, and $\ell = 0.450$ m. Find the magnitude and direction of the net force exerted on the loop by the magnetic field created by the wire.

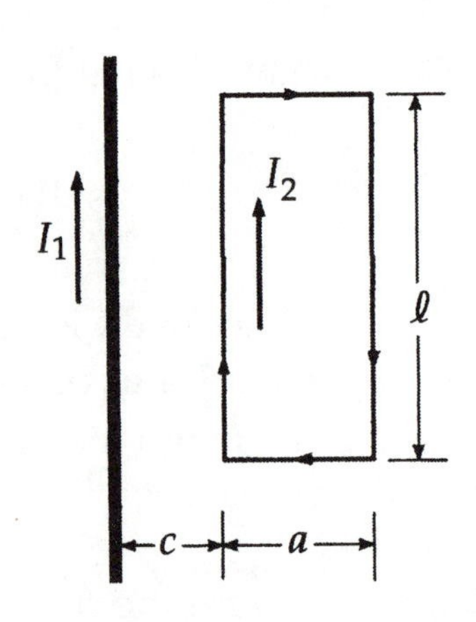


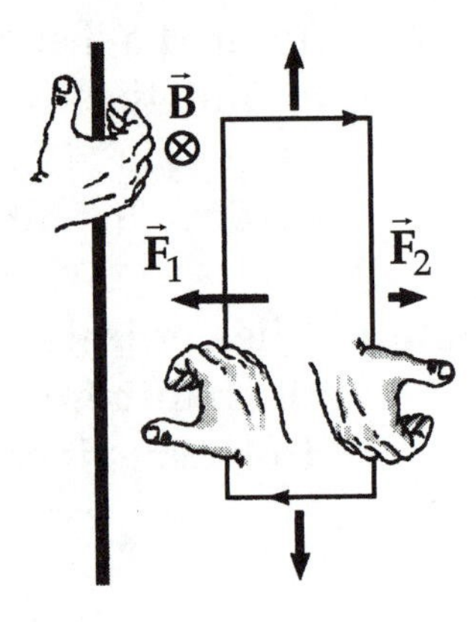


FIG. P22.35

**Solution** **Conceptualize:** Even though there are forces in opposite directions on the loop, we must remember that the magnetic field is stronger near the wire than it is farther away. By symmetry the forces exerted on sides 2 and 4 (the horizontal segments of length $a$) are equal and opposite, and therefore cancel. The magnetic field in the plane of the loop is directed into the page to the right of $I_1$. By the right-hand rule, $\vec{F} = I\vec{\ell} \times \vec{B}$ is directed toward the **left** for side 1 of the loop and a smaller force is directed toward the **right** for side 3. Therefore, we should expect the net force to be to the left, possibly in the $\mu$N range for the currents and distances given.

**Categorize:** The magnetic force between two parallel wires can be found from Equation 22.27, which can be applied to sides 1 and 3 of the loop to find the net force resulting from these opposing force vectors.

**Analyze:** $\vec{F} = \vec{F}_1 + \vec{F}_2 = \dfrac{\mu_0 I_1 I_2 \ell}{2\pi}\left(\dfrac{1}{c+a} - \dfrac{1}{c}\right)\hat{i} = \dfrac{\mu_0 I_1 I_2 \ell}{2\pi}\left(\dfrac{-a}{c(c+a)}\right)\hat{i}$

$$\vec{F} = \frac{\left(4\pi \times 10^{-7}\ \text{N/A}^2\right)(5.00\ \text{A})(10.0\ \text{A})(0.450\ \text{m})}{2\pi}\left(\frac{-0.150\ \text{m}}{(0.100\ \text{m})(0.250\ \text{m})}\right)\hat{i}$$

$\vec{F} = \left(-2.70 \times 10^{-5}\,\hat{i}\right)$ N or $\vec{F} = 2.70 \times 10^{-5}$ N toward the left ◊

**Finalize:** The net force is to the left and in the $\mu$N range as we expected. The symbolic representation of the net force on the loop shows that the net force would be zero if either current disappeared, if either dimension of the loop became very small ($a \to 0$ or $\ell \to 0$), or if the magnetic field were uniform ($c \to \infty$).

**P22.37** Four long, parallel conductors carry equal currents of $I = 5.00\text{ A}$. Figure P22.37 is an end view of the conductors. The current direction is into the page at points $A$ and $B$ (indicated by the crosses) and out of the page at $C$ and $D$ (indicated by the dots). Calculate the magnitude and direction of the magnetic field at point $P$, located at the center of the square of edge length 0.200 m.

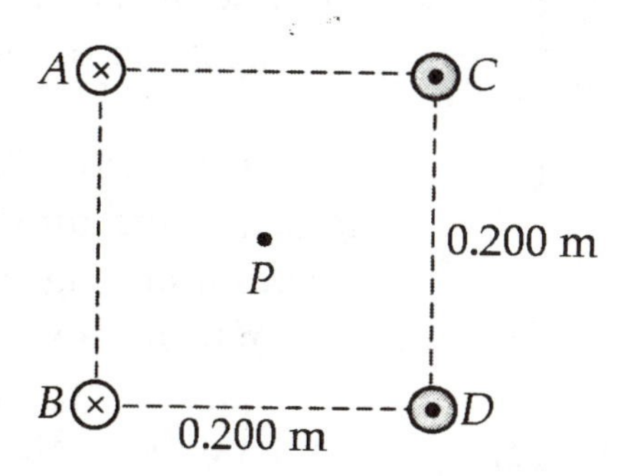


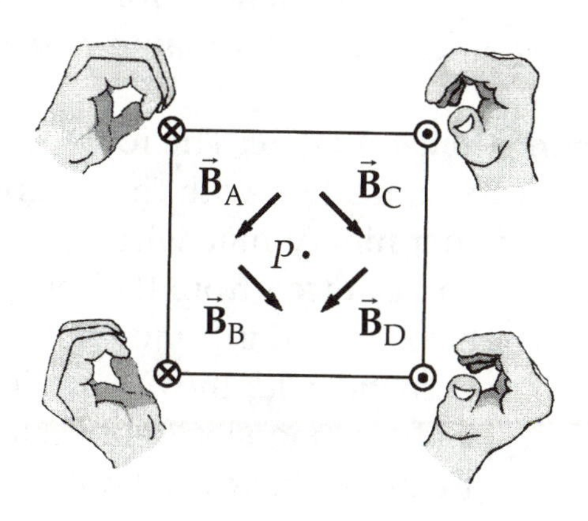


**FIG. P22.37**

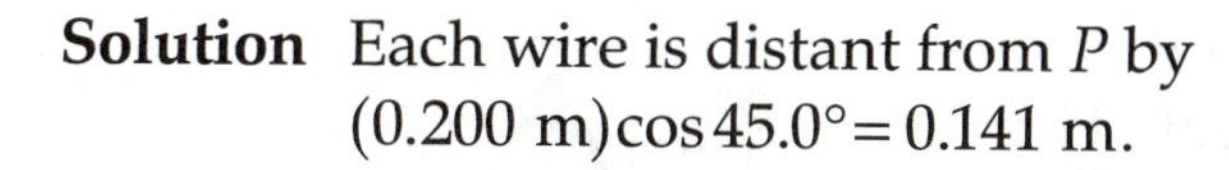

**Solution** Each wire is distant from $P$ by $(0.200\text{ m})\cos 45.0° = 0.141\text{ m}$.

Each wire produces a field at $P$ of equal magnitude:

$$B = \frac{\mu_0 I}{2\pi a}$$

$$= \frac{\left(2.00\times10^{-7}\text{ T}\cdot\text{m/A}\right)(5.00\text{ A})}{0.141\text{ m}}$$

$$= 7.07\ \mu\text{T}.$$

Carrying currents away from you, the left-hand wires produce fields at $P$ of $7.07\ \mu\text{T}$, in the following directions:

$A$: to the bottom and left, at 225°
$B$: to the bottom and right, at 315°.

Carrying currents toward you, the wires to the right also produce fields at $P$ of $7.07\ \mu\text{T}$, in the following directions:

$C$: downward and to the right, at 315°
$D$: downward and to the left, at 225°.

The total field is then $4(7.07\ \mu\text{T})\sin 45.0° = 20.0\ \mu\text{T}$ toward the bottom of the page.

◊

**P22.39** A packed bundle of 100 long, straight, insulated wires forms a cylinder of radius $R = 0.500$ cm.

(a) If each wire carries 2.00 A, what are the magnitude and direction of the magnetic force per unit length acting on a wire located 0.200 cm from the center of the bundle?

(b) Would a wire on the outer edge of the bundle experience a force greater or smaller than the value calculated in part (a)?

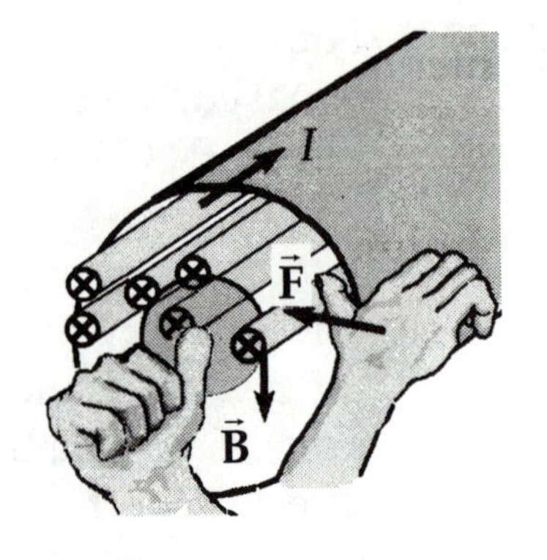


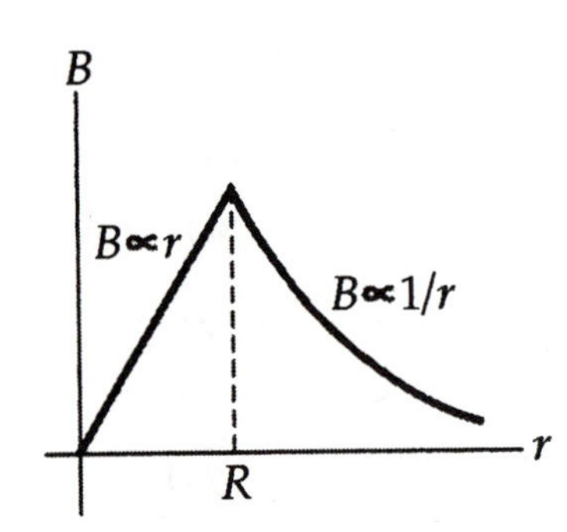


FIG. P22.39

**Solution** **Conceptualize:** The force on one wire comes from its interaction with the magnetic field created **by** the other ninety-nine wires. According to Ampère's law, at a distance $r$ from the center, only the wires enclosed within a radius $r$ contribute to this net magnetic field; the other wires outside the radius produce magnetic field vectors in opposite directions that cancel out at $r$. Therefore, the magnetic field (and also the force on a given wire at radius $r$) will be greater for larger radii within the bundle, and will decrease for distances beyond the radius of the bundle, as shown in the graph to the right. Applying $\vec{\mathbf{F}} = I\vec{\ell} \times \vec{\mathbf{B}}$, the magnetic force on a single wire will be directed toward the center of the bundle, so that all the wires tend to attract each other.

**Categorize:** We model the current carried by the ninety-nine wires as uniformly spread across the whole cylinder. Using Ampère's law, we can find the magnetic field at any radius, so that the magnetic force $\vec{\mathbf{F}} = I\vec{\ell} \times \vec{\mathbf{B}}$ on a single wire can then be calculated.

*continued on next page*

**Analyze:**

(a) Ampère's law is used to derive Equation 22.30, which we can use to find the magnetic field at $r = 0.200$ cm from the center of the cable:

$$B = \frac{\mu_0 I_0 r}{2\pi R^2} = \frac{\left(4\pi\times10^{-7}\ \text{T}\cdot\text{m/A}\right)(99)(2.00\ \text{A})\left(0.200\times10^{-2}\ \text{m}\right)}{2\pi\left(0.500\times10^{-2}\ \text{m}\right)^2} = 3.17\times10^{-3}\ \text{T}.$$

This field points tangent to a circle of radius 0.200 cm and exerts a force $\vec{\mathbf{F}} = I\vec{\ell}\times\vec{\mathbf{B}}$ toward the center of the bundle, on the single hundredth wire:

$$\frac{F}{\ell} = IB\sin\theta = (2.00\ \text{A})\left(3.17\times10^{-3}\ \text{T}\right)(\sin 90°) = 6.34\ \text{mN/m}. \quad \Diamond$$

(b) As shown in the graph in Figure P22.39, the magnetic field increases linearly as a function of $r$ until it reaches a maximum at the outer surface of the cable. Therefore, the force on a single wire at the outer radius $r = 5.00$ cm would be greater than at $r = 2.00$ cm by a factor of $\frac{5}{2}$. $\Diamond$

**Finalize:** We did not estimate the expected magnitude of the force, but 200 amperes is a lot of current. It would be interesting to see if the magnetic force that pulls together the individual wires in the bundle is enough to hold them against their own weight: If we assume that the insulation accounts for about half the volume of the bundle, then a single copper wire in this bundle would have a cross sectional area of about

$$\frac{1}{2}(0.01)\pi(0.500\ \text{cm})^2 = 4\times10^{-7}\ \text{m}^2$$

with a weight per unit length of

$$\rho g A = \left(8\,920\ \text{kg/m}^3\right)(9.8\ \text{N/kg})\left(4\times10^{-7}\ \text{m}^2\right) = 0.03\ \text{N/m}.$$

Therefore, the outer wires experience an inward magnetic force that is about half the magnitude of their own weight. If placed on a table, this bundle of wires would form a loosely held mound without the outer sheathing to hold them together. They would jerk and squirm when you turn the current off or on.

**P22.45** What current is required in the windings of a long solenoid that has 1 000 turns uniformly distributed over a length of 0.400 m, to produce at the center of the solenoid a magnetic field of magnitude $1.00 \times 10^{-4}$ T?

**Solution** The magnetic field at the center of a solenoid is $B = \mu_0 \frac{N}{\ell} I$.

So $$I = \frac{B\ell}{\mu_0 N} = \frac{(1.00 \times 10^{-4}\ \text{T} \cdot \text{A})(0.400\ \text{m})}{(4\pi \times 10^{-7}\ \text{T} \cdot \text{m})(1\,000)}.$$

Caution: If you use your calculator, it may not understand the keystrokes

| 4 | × | π | exp | +/− | 7 |
|---|---|---|---|---|---|

You may need to use

| 4 | exp | +/− | 7 | × | π |
|---|---|---|---|---|---|

Make sure you get $I = 31.8$ mA. ◊

**P22.57** A positive charge $q = 3.20 \times 10^{-19}$ C moves with a velocity $\vec{\mathbf{v}} = (2\hat{\mathbf{i}} + 3\hat{\mathbf{j}} - \hat{\mathbf{k}})$ m/s through a region where both a uniform magnetic field and a uniform electric field exist.

(a) Calculate the total force on the moving charge (in unit-vector notation), taking $\vec{\mathbf{B}} = (2\hat{\mathbf{i}} + 4\hat{\mathbf{j}} + \hat{\mathbf{k}})$ T and $\vec{\mathbf{E}} = (4\hat{\mathbf{i}} - \hat{\mathbf{j}} - 2\hat{\mathbf{k}})$ V/m.

(b) What angle does the force vector make with the positive $x$ axis?

**Solution** The total force is the Lorentz force,

(a)
$$\vec{\mathbf{F}} = q\vec{\mathbf{E}} + q(\vec{\mathbf{v}} \times \vec{\mathbf{B}}) = q(\vec{\mathbf{E}} + \vec{\mathbf{v}} \times \vec{\mathbf{B}})$$
$$\vec{\mathbf{F}} = q\left[(4\hat{\mathbf{i}} - \hat{\mathbf{j}} - 2\hat{\mathbf{k}}) \text{ V/m} + (2\hat{\mathbf{i}} + 3\hat{\mathbf{j}} - \hat{\mathbf{k}}) \text{ m/s} \times (2\hat{\mathbf{i}} + 4\hat{\mathbf{j}} + \hat{\mathbf{k}}) \text{ T}\right]$$
$$\vec{\mathbf{F}} = q\left[(4\hat{\mathbf{i}} - \hat{\mathbf{j}} - 2\hat{\mathbf{k}}) \text{ V/m} + (7\hat{\mathbf{i}} - 4\hat{\mathbf{j}} + 2\hat{\mathbf{k}}) \text{ T} \cdot \text{m/s}\right]$$
$$\vec{\mathbf{F}} = q\left[(11\hat{\mathbf{i}} - 5\hat{\mathbf{j}}) \text{ V/m}\right] = q\left[(11\hat{\mathbf{i}} - 5\hat{\mathbf{j}}) \text{ N/C}\right]$$
$$\vec{\mathbf{F}} = (3.20 \times 10^{-19} \text{ C})\left[(11\hat{\mathbf{i}} - 5\hat{\mathbf{j}}) \text{ N/C}\right] = (3.52\hat{\mathbf{i}} - 1.60\hat{\mathbf{j}}) \times 10^{-18} \text{ N} \quad \Diamond$$

(b) $\vec{\mathbf{F}} \cdot \hat{\mathbf{i}} = F\cos\theta = F_x$:
$$\theta = \cos^{-1}\left(\frac{F_x}{F}\right) = \cos^{-1}\left(\frac{3.52}{3.87}\right) = 24.4° \quad \Diamond$$

**P22.63** A very long, thin strip of metal of width $w$ carries a current $I$ along its length as shown in Figure P22.63. Find the magnetic field at the point $P$ in the diagram. The point $P$ is in the plane of the strip at distance $b$ away from it.

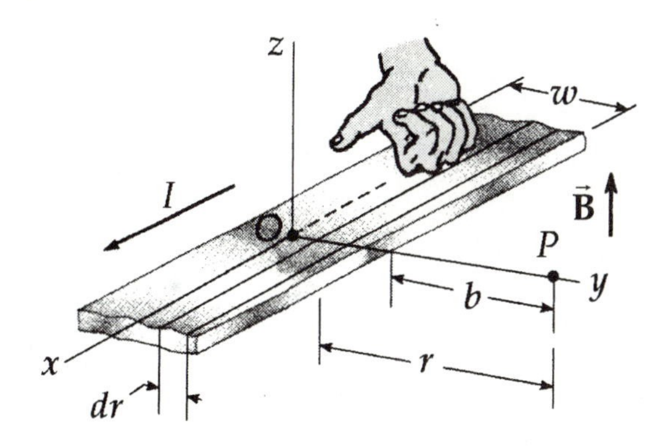


**FIG. P22.63**

**Solution** Consider a long filament of the strip, which has width $dr$ and is a distance $r$ from point $P$. The magnetic field at a distance $r$ from a long conductor is

$$B = \frac{\mu_0 I}{2\pi r}.$$

*continued on next page*

Thus, the field due to the thin filament is $d\vec{\mathbf{B}} = \frac{\mu_0 dI}{2\pi r}\hat{\mathbf{k}}$ where $dI = I\left(\frac{dr}{w}\right)$

so $$\vec{\mathbf{B}} = \int_b^{b+w} \frac{\mu_0}{2\pi r}\left(I\frac{dr}{w}\right)\hat{\mathbf{k}} = \frac{\mu_0 I}{2\pi w}\int_b^{b+w} \frac{dr}{r}\hat{\mathbf{k}} = \frac{\mu_0 I}{2\pi w}\ln\left(1+\frac{w}{b}\right)\hat{\mathbf{k}}.$$ ◊

The units of the right-hand side are $\frac{\text{T}\cdot\text{m}}{\text{A}}\frac{\text{A}}{\text{m}}1 = \text{T}$, in agreement with the units of the left-hand side. The field is directly proportional to the current. For fixed $b$, the field decreases as $w$ increases. As $b$ increases relative to $w$, we use $\ln(1+x) \approx x$ to find that the field approaches $B \approx \frac{\mu_0 I}{2\pi w}\frac{w}{b} = \frac{\mu_0 I}{2\pi b}$, in agreement with that of a long thin wire.

*Chapter* 23

# Faraday's Law and Inductance

## NOTES FROM SELECTED CHAPTER SECTIONS

### Section 23.1 Faraday's Law of Induction

The emf induced in a circuit is equal to the time rate of change of magnetic flux through the circuit.

An emf can be induced in the circuit in several ways:

- The magnitude of the magnetic field can change as a function of time.
- The area of the circuit can change with time.
- The direction of the magnetic field relative to the circuit can change with time.
- Any combination of the above can change.

### Section 23.2 Motional emf

A potential difference, $\Delta V$, will be maintained across a conductor moving in a magnetic field. The potential difference will be zero when the direction of motion of the conductor is parallel or antiparallel to the field direction. If the motion is reversed, the polarity of the potential difference will also be reversed.

### Section 23.3 Lenz's Law

The polarity of the induced emf is such that it tends to produce a current that will create a magnetic flux to **oppose the change in flux** through the circuit.

## Section 23.5 Self-Inductance

The self-induced emf is always proportional to the **time rate of change** of current in the circuit. The inductance of a device (an inductor) depends on its geometry.

## Section 23.6 *RL* Circuits

If a resistor and an inductor are connected in series to a battery, the current in the circuit will reach an equilibrium value $\left(\frac{\varepsilon}{R}\right)$ after a time that is long compared to the time constant of the circuit, $\tau = \frac{L}{R}$.

## Section 23.7 Energy Stored in a Magnetic Field

In an *RL* circuit, the rate at which energy is supplied by the battery equals the sum of the rate at which energy is delivered to the resistor and the rate at which energy is stored in the inductor. *The energy density is proportional to the square of the magnetic field.*

# EQUATIONS AND CONCEPTS

The **total magnetic flux** threading a circuit is the integral of the normal component of the magnetic field over the area bounded by the circuit.

$$\Phi_B = \int \vec{\mathbf{B}} \cdot d\vec{\mathbf{A}} \qquad (23.1)$$

$d\vec{\mathbf{A}}$ is a vector perpendicular to the surface whose magnitude equals the area $dA$

The **magnetic flux through a plane area,** $A$, placed in a uniform magnetic field depends on the angle between the direction of the magnetic field and the direction perpendicular to the surface area. *The maximum flux through the area occurs when the magnetic field is perpendicular to the plane of the surface area.*

$$\Phi_B \equiv B_{\perp} A = BA\cos\theta$$

$$\Phi_{\text{max}} = BA$$

The unit of magnetic flux is the weber, Wb.

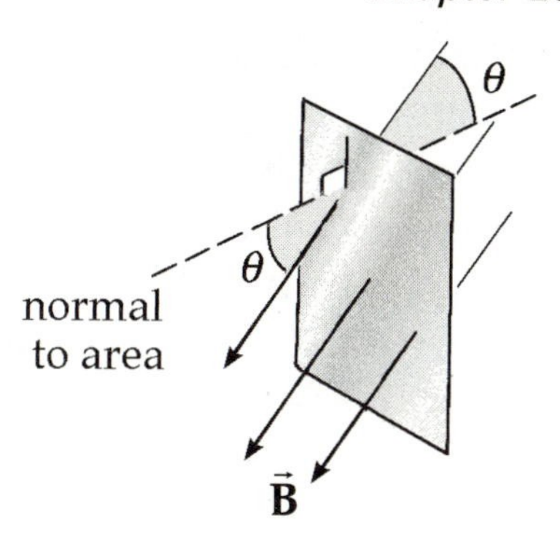


**Faraday's law of induction** states that the emf induced in a circuit is the rate of change of magnetic flux through the circuit. *The minus sign is included to indicate the polarity of the induced emf, which can be found by use of Lenz's law.*

$$\mathcal{E} = -\frac{d\Phi_B}{dt} \tag{23.2}$$

The **induced emf** can be expressed in terms of magnetic field ($\vec{\mathbf{B}}$), surface area ($A$) and the angle $\theta$ between $\vec{\mathbf{B}}$ and the normal to the surface. From Equation 23.4, an emf will be induced in a circuit when any one or more of the following change with time:

$$\mathcal{E} = -\frac{d}{dt}(BA\cos\theta) \tag{23.4}$$

- magnitude of $\vec{\mathbf{B}}$
- area enclosed by the circuit
- angle $\theta$ between $\vec{\mathbf{B}}$ and the normal

**Lenz's law** states that the polarity of the induced emf in a loop (and the direction of the associated current in a closed circuit) produces a current whose magnetic field opposes the change in the flux through the loop. That is, the induced emf produces a current (and the corresponding induced magnetic field) that tends to maintain the original flux through the circuit.

A **motional emf** is induced in a conductor moving through a magnetic field. The induced emf is given by Equation 23.5 when a conductor of length $\ell$ moves with speed $v$ perpendicular to a constant magnetic field. *Recall the significance of the negative sign as required by Lenz's law.*

$$\mathcal{E} = -B\ell v \quad (23.5)$$

An **induced current** will exist in a conductor moving in a magnetic field if it is part of a complete circuit. *Confirm that Lenz's law correctly predicts the direction of the induced current as shown in the figure.*

$$I = \frac{B\ell v}{R} \quad (23.6)$$

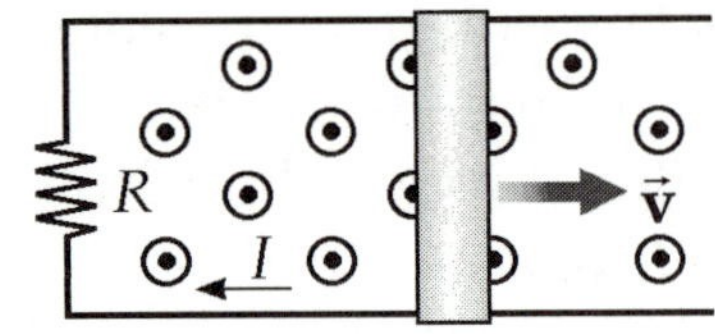


A **sinusoidally varying emf** is produced when a conducting loop of $N$ turns and cross-sectional area, $A$, rotates with a constant angular velocity in a magnetic field. *For a given loop, the maximum value of the induced emf will be proportional to the angular velocity of the loop.*

$$\mathcal{E} = NAB\omega \sin \omega t \quad (23.8)$$

$$\mathcal{E}_{\text{max}} = NAB\omega$$

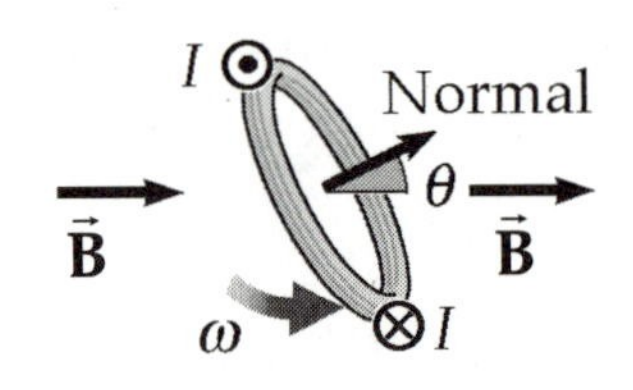


**Faraday's law in a more general form** can be written as the integral of the electric field around a closed path. In this form the electric field is a nonconservative filed that is generated by a changing magnetic field.

$$\oint \vec{\mathbf{E}} \cdot d\mathbf{s} = -\frac{d}{dt}\int \vec{\mathbf{B}} \cdot dA$$

A **self-induced emf** is present in a coil when the current in the coil changes in time. *The inductance, L, is a measure of the opposition of the coil to a change in the current. Recall that resistance, R, is a measure of opposition to current.*

$$\mathcal{E}_L = -L\frac{dI}{dt} \quad (23.10)$$

**Inductance** of a given device (coil, solenoid, toroid, coaxial cable, or other conducting device) can be calculated if the flux and current are known (Eq. 23.11) or as the ratio of the induced emf magnitude divided by the time rate of change of current in the circuit (Eq. 23.12). The SI unit of inductance is the henry (H).

$$L = \frac{N\Phi_B}{I} \quad (23.11)$$

$$L = -\frac{\mathcal{E}_L}{dI/dt} \quad (23.12)$$

$$1\,\text{H} = 1\,\frac{\text{V}\cdot\text{s}}{\text{A}} = 1\,\Omega\cdot\text{s}$$

**Increasing current in an *RL* circuit** (which contains a battery, resistor, and inductor) produces a back emf. If the switch in the series circuit shown below is closed in position 1 at time $t = 0$, current in the circuit will increase in a characteristic fashion toward a maximum value of $\left(\frac{\mathcal{E}}{R}\right)$. This is shown in the graph to the right.

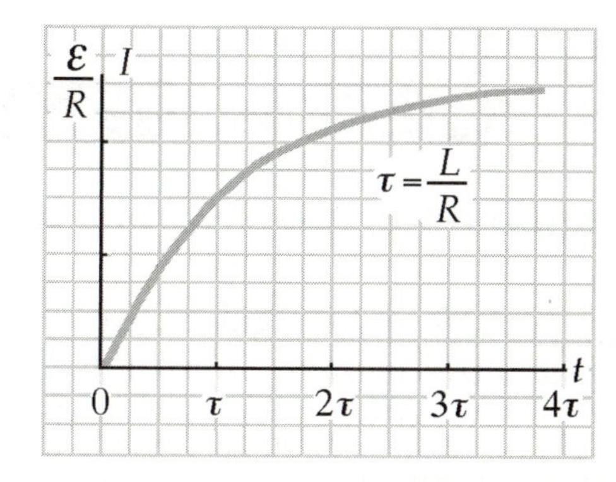


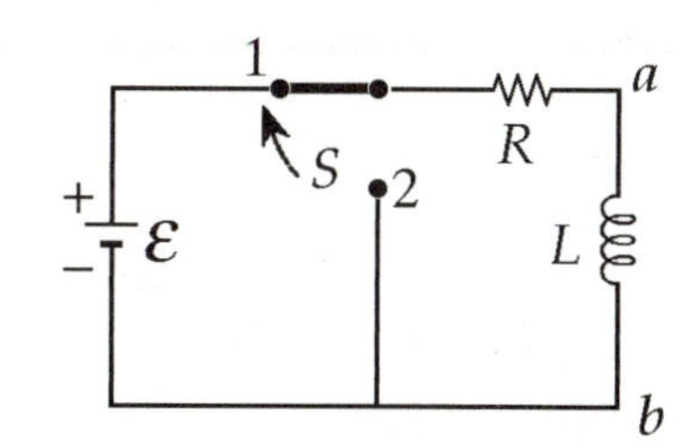


$$I(t) = \frac{\mathcal{E}}{R}\left(1 - e^{-t/\tau}\right) \quad (23.14)$$

Let the switch in the circuit shown above be at position 1 with the current at its maximum value $I_0 = \frac{\mathcal{E}}{R}$. If the switch is thrown to position 2 (as shown below) at $t = 0$, the current will decay exponentially with time. The graph at right shows the manner in which the current decays.

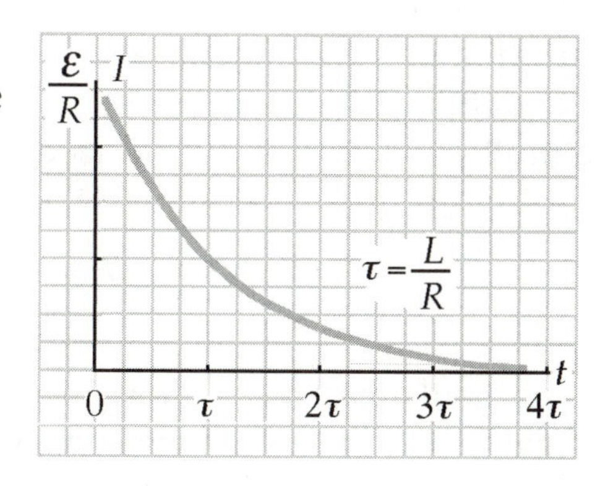


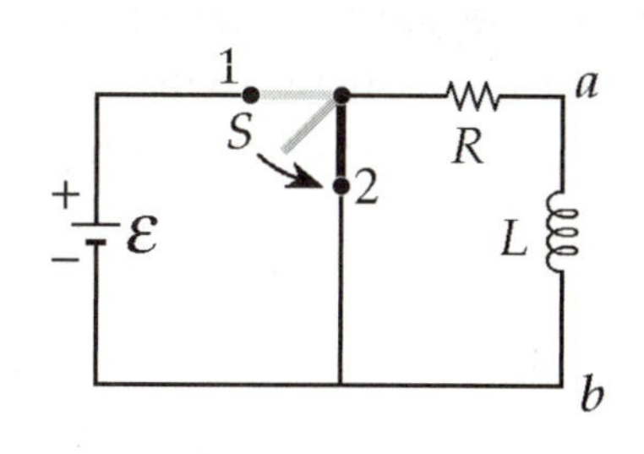


$$I(t) = \frac{\mathcal{E}}{R}e^{-t/\tau} = I_i e^{-t/\tau} \quad (23.18)$$

where $\frac{\mathcal{E}}{R} = I_i$

The **time constant** of the circuit, $\tau$, is the time required for the current to reach 63.2% of its maximum value.

$$\tau = \frac{L}{R} \qquad (23.15)$$

The **energy stored** in the magnetic field of an inductor is proportional to the square of the current in the inductor. The **energy density** $u_B$; that is, energy per unit volume.

$$U_B = \frac{1}{2}LI^2 \text{ (energy)} \qquad (23.20)$$

$$u_B = \frac{B^2}{2\mu_0} \text{ (energy density)} \qquad (23.22)$$

## SUGGESTIONS, SKILLS, AND STRATEGIES

### Instantaneous and Average Induced Emf

It is important to distinguish clearly between the **instantaneous value** of emf induced in a circuit and the **average value** of the emf induced in the circuit over a finite time interval.

To calculate the **average induced emf**, it is often useful to write Equation 23.3 as

$$\mathcal{E}_{av} = -N\left(\frac{d\Phi_B}{dt}\right)_{av} = -N\frac{\Delta\Phi_B}{\Delta t} \qquad \text{or} \qquad \mathcal{E}_{av} = -N\left(\frac{\Phi_{B,f} - \Phi_{B,i}}{\Delta t}\right)$$

where the subscripts $i$ and $f$ refer to the magnetic flux through the circuit at the beginning and end of the time interval $\Delta t$. For a circuit (or coil) in a single plane, $\Phi_B = BA\cos\theta$, where $\theta$ is the angle between the direction of the normal to plane of the circuit (conducting loop) and the direction of the magnetic field.

Equation 23.4 can be used to calculate the **instantaneous value of an induced emf**. For a multiple turn coil, the induced emf is

$$\mathcal{E} = -N\frac{d}{dt}(BA\cos\theta)$$

where in a particular case $B$, $A$, $\theta$, or any combination of those parameters can be time dependent while the others remain constant. The expression resulting from the differentiation is then evaluated using the values of $B$, $A$, and $\theta$ corresponding to the specified value.

## AN EXAMPLE OF LENZ'S LAW

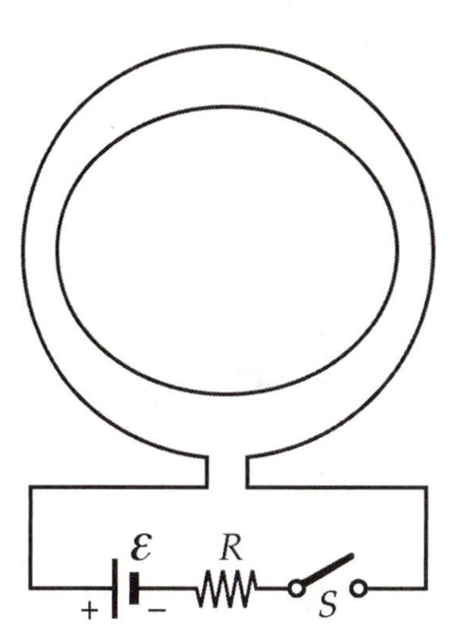


Consider two single-turn, concentric coils **lying in the plane of the paper** as shown in the figure. The outside coil is part of a circuit containing a resistor ($R$), a battery ($\mathcal{E}$) and switch ($S$). The inner coil is not part of the circuit. When the switch is moved from "**open**" to "**closed**" the direction of the induced current in the inner coil can be predicted by Lenz's law.

Consider the following steps:

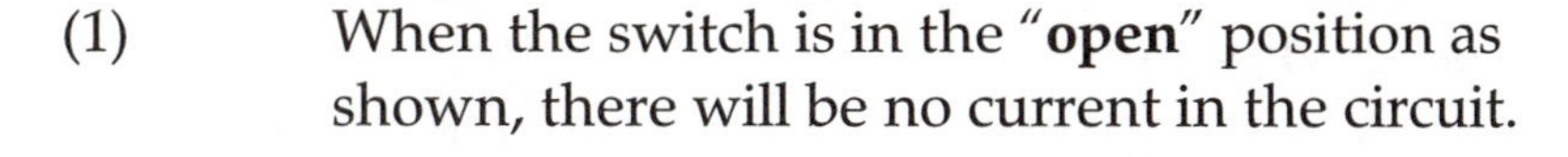

(1) When the switch is in the "**open**" position as shown, there will be no current in the circuit.

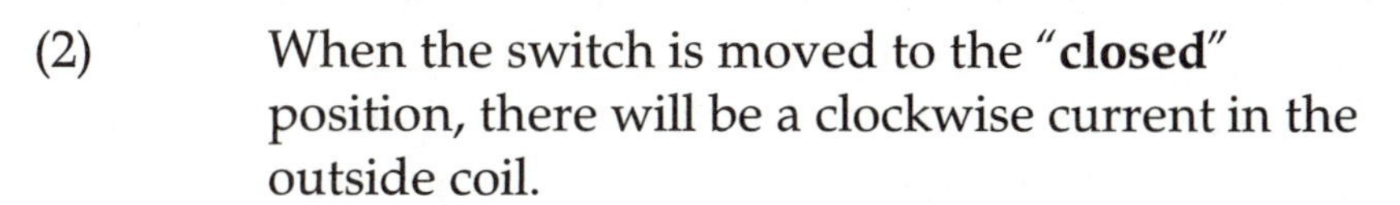

(2) When the switch is moved to the "**closed**" position, there will be a clockwise current in the outside coil.

(3) Magnetic field lines due to current in the outside coil will be directed into the page through the area enclosed by the coil. **Use the right-hand rule (first application) to confirm this**. Magnetic flux into the page will penetrate the entire area enclosed by the outside coil **including the area of the inner coil**.

(4) By Faraday's law, the increasing flux produces an induced emf (and current) in the inner coil.

(5) Lenz's law requires that the induced current have a direction, which will tend to maintain the initial flux condition (**which in this case was zero**). By using the **right-hand rule (second application)**, you should be able to determine that the direction of the induced current in the inner coil must be counterclockwise, contributing to a flux out of the page. Try this!

As a second example you should follow steps similar to those above to predict the direction of the induced current in the inner coil when the switch is moved from "**closed**" to "**open**".

## REVIEW CHECKLIST

✓ Calculate the emf (or current) induced in a circuit when the magnetic flux through the circuit is changing in time. The variation in flux might be due to a change in (a) the area of the circuit, (b) the magnitude of the magnetic field, (c) the direction of the magnetic field, or (d) the orientation/location of the circuit in the magnetic field.

✓ Calculate the emf induced between the ends of a conducting bar as it moves through a region where there is a constant magnetic field (motional emf). Apply Lenz's law to determine the direction of an induced emf or current. You should also understand that Lenz's law is a consequence of the law of conservation of energy.

✓ Calculate the inductance of a device of suitable geometry.

✓ Calculate the magnitude and direction of the self-induced emf in a circuit containing one or more inductive elements when the current changes with time.

✓ Determine instantaneous values of the current in an *RL* circuit while the current is either increasing or decreasing with time.

✓ Calculate the total magnetic energy stored in a magnetic field. You should be able to perform this calculation if (1) you are given the values of the inductance of the device with which the field is associated and the current in the circuit, or (2) given the value of the magnitude of the magnetic field throughout the region of space in which the magnetic field exists. In the latter case, you must integrate the expression for the energy density $u_B$ over an appropriate volume.

## ANSWERS TO SELECTED QUESTIONS

**Q23.5** The bar in Figure Q23.5 moves on rails to the right with a velocity $\vec{\mathbf{v}}$, and the uniform, constant magnetic field is directed out of the page. Why is the induced current clockwise? If the bar were moving to the left, what would be the direction of the induced current?

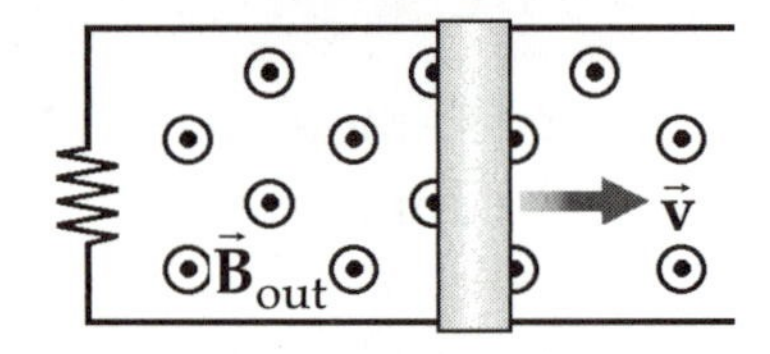


**FIG. Q23.5**

**Answer** The external magnetic field is out of the paper, as the area $A$ enclosed by the loop increases, the external flux increases according to $\Phi_B = BA\cos\theta = BA$.

As predicted by Lenz's law, due to the increase in flux through the loop, free electrons will produce a current to create a magnetic flux inside the loop to oppose the change in flux.

In this case, the magnetic field due to the current must point into the paper in order to oppose the increasing magnetic flux of the external field coming out of the paper. By the right-hand rule (with your thumb pointing in the direction of the current along each wire), the current must be clockwise.

If the bar were moving toward the left, the area would decrease, and the flux would decrease. The flux due to $\vec{\mathbf{B}}$ pointing out of the paper, then, would be decreasing. In order to cancel this change, the current would have to create a magnetic field inside the loop pointing out of the paper. This time, by the right-hand rule, we see that the current must be counter-clockwise.

**Q23.7** In a hydroelectric dam, how is the energy converted that is then transferred out by electrical transmission? That is, how is the energy of motion of the water converted to energy that is transmitted by AC electricity?

**Answer** As the water falls, it gains kinetic energy. It is then forced to pass through a water wheel, transferring some of its energy to the rotor of a large AC electric generator.

The rotor of the generator is supplied with a small amount of DC current, which powers electromagnets in the rotor. Because the rotor is spinning, the electromagnets then create a magnetic flux that changes with time, according to the equation $\Phi_B = BA\cos\omega t$.

Coils of wire that are placed near the rotor then experience an induced emf according to the equation $\mathcal{E} = -N\dfrac{d\Phi_B}{dt}$.

Finally, a small amount of this electricity is used to supply the rotor with its DC current; the rest is sent out over power lines to supply customers with electricity.

In terms of energy, the hydroelectric generating station is a nonisolated system in steady state. It takes in mechanical energy and puts out a precisely equal quantity of energy, nearly all of it by electrical transmission.

**Q23.17** If the current in an inductor is doubled, by what factor does the stored energy change?

**Answer** The energy stored in an inductor carrying a current $I$ is given by $U = \frac{1}{2}LI^2$. Therefore, doubling the current will quadruple the energy stored in the inductor.

## SOLUTIONS TO SELECTED PROBLEMS

**P23.3** A strong electromagnet produces a uniform magnetic field of 1.60 T over a cross-sectional area of 0.200 m$^2$. We place a coil having 200 turns and a total resistance of 20.0 Ω around the electromagnet. We then smoothly reduce the current in the electromagnet until it reaches zero in 20.0 ms. What is the current induced in the coil?

**Solution** **Conceptualize:** A strong magnetic field turned off in a short time (20.0 ms) will produce a large emf, maybe on the order of 1 kV. With only 20.0 Ω of resistance in the coil, the induced current produced by this emf will probably be larger than 10 A but less than 1 000 A.

**Categorize:** According to Faraday's law, if the magnetic field is reduced uniformly, then a constant emf will be produced. The definition of resistance can be applied to find the induced current from the emf.

*continued on next page*

**Analyze:** The induced voltage is

$$\mathcal{E} = -N\frac{d(\vec{\mathbf{B}}\cdot\vec{\mathbf{A}})}{dt} = -N\left(\frac{0 - B_i A\cos\theta}{\Delta t}\right)$$

$$\mathcal{E} = \frac{(+200)(1.60\text{ T})(0.200\text{ m}^2)(\cos 0°)}{20.0\times 10^{-3}\text{ s}}\left(\frac{1\text{ N}\cdot\text{s/C}\cdot\text{m}}{\text{T}}\right)\left(\frac{1\text{ V}\cdot\text{C}}{\text{N}\cdot\text{m}}\right) = 3\,200\text{ V}$$

(Note: The unit conversions come from $\vec{\mathbf{F}} = q\vec{\mathbf{v}}\times\vec{\mathbf{B}}$ and $U = qV$.)

$$I = \frac{\mathcal{E}}{R} = \frac{3\,200\text{ V}}{20.0\ \Omega} = 160\text{ A}$$ ◊

**Finalize:** This is a large current, as we expected. The positive sign means that the current in the coil is in the same direction as the current in the electromagnet.

**P23.13** Figure P23.13 shows a top view of a bar that can slide without friction. The resistor is 6.00 Ω and a 2.50-T magnetic field is directed perpendicularly downward, into the paper. Let $\ell = 1.20$ m.

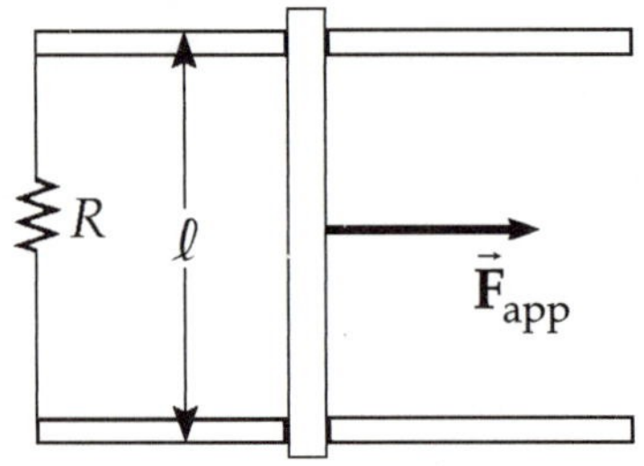

(a) Calculate the applied force required to move the bar to the right at a constant speed of 2.00 m/s.

(b) At what rate is energy delivered to the resistor?

**FIG. P23.13**

**Solution** (a) We use the rigid body in equilibrium model. At constant speed, the net force on the moving bar equals zero,

or $$\left|\vec{\mathbf{F}}_{\text{app}}\right| = I\left|\vec{\ell}\times\vec{\mathbf{B}}\right|$$

where the current in the bar is $$I = \frac{\mathcal{E}}{R}$$

and $$\mathcal{E} = B\ell v.$$

Therefore, $$F_{\text{app}} = \left(\frac{B\ell v}{R}\right)\ell B = \frac{B^2\ell^2 v}{R}$$

$$F_{\text{app}} = \frac{(2.50\text{ T})^2(1.20\text{ m})^2(2.00\text{ m/s})}{6.00\ \Omega} = 3.00\text{ N}$$ ◊

*continued on next page*

(b) $\mathcal{P} = F_{app}v = (3.00\ \text{N})(2.00\ \text{m/s}) = 6.00\ \text{W}$ ◊

In terms of energy, the circuit is a nonisolated system in steady state. The energy, six joules every second, delivered to the circuit by work done by the applied force is delivered to the resistor by electrical transmission. The energy can leave the resistor as energy transferred by heat into the surrounding air. When we studied an electric circuit consisting of a resistor connected across a battery, we did not consider the details of the chemical reaction in the battery and the battery's loss of chemical energy. The circuit in this problem is equally simple electrically. The outside agent pulling the bar through the magnetic field takes the place of the battery. This is the first circuit in which we can account completely for the energy transformations.

**P23.21** A conducting rectangular loop of mass $M$, resistance $R$, and dimensions $w$ by $\ell$ falls from rest into a magnetic field $\vec{\mathbf{B}}$ as shown in Figure P23.21. During the time interval before the top edge of the loop reaches the field, the loop approaches a terminal speed $v_T$.

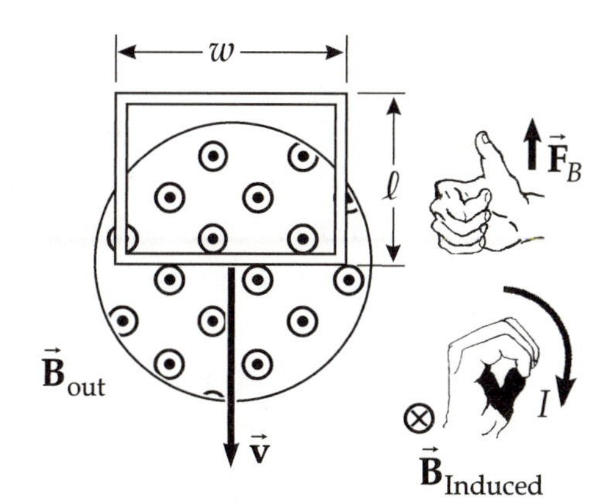


FIG. P23.21

(a)  Show that $v_T = \dfrac{MgR}{B^2w^2}$.

(b) Why is $v_T$ proportional to $R$?

(c) Why is it inversely proportional to $B^2$?

**Solution** Let $y$ represent the average vertical height of the strong-field region above the bottom edge of the loop.

As the loop falls, $y$ increases; the loop encloses increasing flux toward you and has induced in it an emf to produce a current to make its own magnetic field away from you. This current is to the left in the bottom side of the loop, and feels an upward force in the external field.

(a) Symbolically, the flux is $\Phi = BA\cos\theta = Bwy\cos 0°$.

The emf is $\mathcal{E} = -N\dfrac{d}{dt}(Bwy) = -Bw\dfrac{dy}{dt} = -Bwv$.

The magnitude of the current is $I = \dfrac{|\mathcal{E}|}{R} = \dfrac{Bwv}{R}$

*continued on next page*

and the force is $$\vec{\mathbf{F}}_B = I\vec{\mathbf{w}} \times \vec{\mathbf{B}} = \left(\frac{Bwv}{R}\right) wB\sin 90° = \frac{B^2w^2v}{R}.$$

When the loop is moving at terminal speed, we may model it as a rigid body in equilibrium:

$\sum F_y = 0$ becomes $$\frac{+B^2w^2v_T}{R} - Mg = 0.$$

Thus, $$v_T = \frac{MgR}{B^2w^2}.$$ ◊

(b) The emf is directly proportional to $v$, but the current is inversely proportional to $R$. A large $R$ means a small current at a given speed, so the loop must travel faster to get $F_B$ = weight. ◊

(c) At a given speed, the current is directly proportional to the magnetic field. But the force is proportional to the product of the current and the field. For a small $B$, the speed must increase to compensate for both the small $B$ and also the current, so $v_T \propto B^{-2}$. ◊

**P23.23** A coil of area 0.100 $m^2$ is rotating at 60.0 rev/s with the axis of rotation perpendicular to a 0.200-T magnetic field.

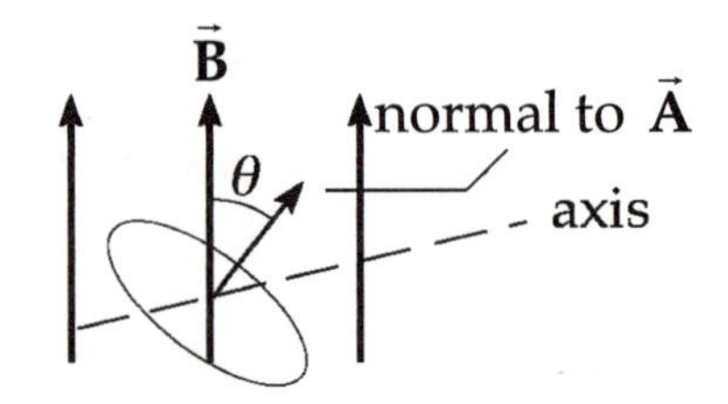


**FIG. P23.23**

(a) If the coil has 1 000 turns, what is the maximum emf generated in it?

(b) What is the orientation of the coil with respect to the magnetic field when the maximum induced voltage occurs?

**Solution** For a coil rotating in a magnetic field, $B$ and $A$ are constant in

$$\mathcal{E} = -N\frac{d}{dt}BA\cos\theta = -NBA\frac{d}{dt}\cos\omega t = +NBA\omega\sin\omega t.$$

(a) The maximum value of $\sin\theta$ is 1

so $\mathcal{E}_{\max} = NBA\omega$.

*continued on next page*

With these data,

$$\mathcal{E}_{\max} = (1\,000)(0.200 \text{ T})\left(0.100 \text{ m}^2\right)(60.0 \text{ rev/s})(2\pi \text{ rad/rev})$$

$$\mathcal{E}_{\max} = 7\,540 \text{ V}. \qquad \lozenge$$

(b) $|\mathcal{E}|$ approaches $\mathcal{E}_{\max}$ when $|\sin\theta|$ approaches 1 or $\theta = \pm\frac{\pi}{2}$. Therefore, at maximum emf, the plane of the coil is parallel to the field.

**P23.25** A magnetic field directed into the page changes with time according to $B = \left(0.030\,0t^2 + 1.40\right)$ T, where $t$ is in seconds. The field has a circular cross-section of radius $R = 2.50$ cm (Fig. P23.25). What are the magnitude and direction of the electric field at point $P_1$ when $t = 3.00$ s and $r_1 = 0.020\,0$ m?

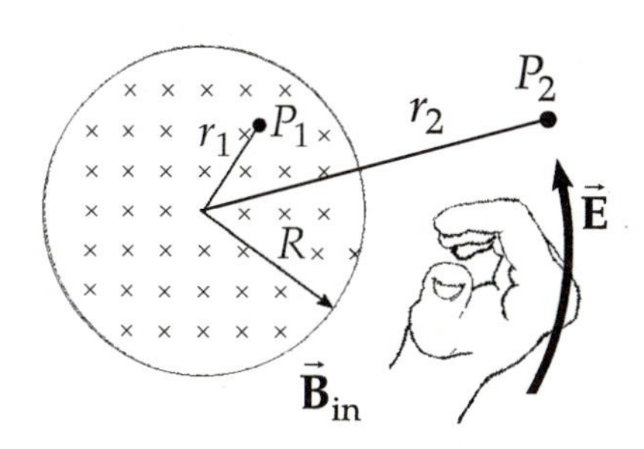


**FIG. P23.25**

**Solution** $\oint \vec{\mathbf{E}} \cdot d\vec{\mathbf{s}} = -\frac{d\Phi_B}{dt}$

Consider a circular integration path of radius $r_1$

$$E(2\pi r_1) = -\frac{d}{dt}(BA) = -A\left(\frac{dB}{dt}\right)$$

$$|E| = \frac{A}{2\pi r_1}\frac{d}{dt}\left(0.030\,0t^2 + 1.40\right) \text{ T} = \frac{\pi r_1^2}{2\pi r_1}(0.060\,0t) = \frac{r_1}{2}(0.060\,0t)$$

At $t = 3.00$ s, $\quad E = \frac{1}{2}(0.020\,0 \text{ m})(0.060\,0 \text{ T/sec})(3.00 \text{ sec}) = 1.80 \times 10^{-3} \text{ N/C}$. $\quad \lozenge$

If there were a circle of wire with radius $r_1$, it would enclose increasing magnetic flux due to a magnetic field away from you. It would carry counterclockwise current to make its own magnetic field toward you, to oppose the change. Even without the wire and current, the counterclockwise electric field that would cause the current is lurking. At point $P_1$, it is upward and to the left, perpendicular to $r_1$. $\quad \lozenge$

**P23.29** A 10.0-mH inductor carries a current $I = I_{\max}\sin\omega t$, with $I_{\max} = 5.00$ A and $\frac{\omega}{2\pi} = 60.0$ Hz. What is the self-induced emf as a function of time?

*continued on next page*

**Solution** $\mathcal{E} = -L\frac{dI}{dt} = -L\frac{d}{dt}(I_{\text{max}} \sin\omega t)$

$$\mathcal{E} = -L\omega I_{\text{max}} \cos\omega t = -(0.010\,0 \text{ H})\left(120\pi \text{ s}^{-1}\right)(5.00 \text{ A})\cos(120\pi t)$$

$$\mathcal{E} = -(18.8 \text{ V})\cos(377t) \qquad \Diamond$$

The answer is not a single number, but a function of time, like a graph on an oscilloscope screen. It can be described as a sinusoidally oscillating voltage with amplitude 18.8 V, angular frequency 377 rad/s and with initial phase angle such that it starts from –18.8 V at $t = 0$.

**P23.33** A 12.0-V battery is connected into a series circuit containing a 10.0-Ω resistor and a 2.00-H inductor. In what time interval will the current reach

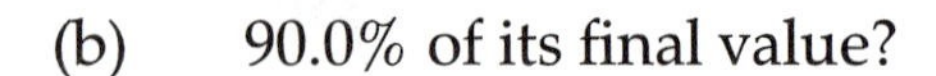


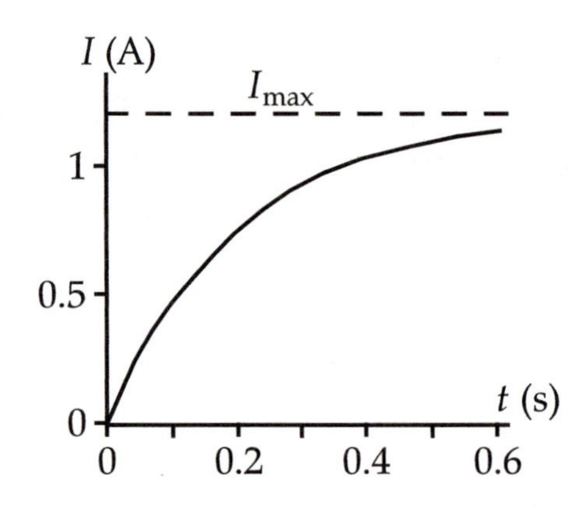


**FIG. P23.33**

**Solution** **Conceptualize:** The time constant for this circuit is $\tau = \frac{L}{R} = 0.2$ s, which means that in 0.2 s, the current will reach $1 - \frac{1}{e} = 63\%$ of its final value, as shown in the graph above. We can see from this graph that the time to reach 50% of $I_{\text{max}}$ should be slightly less than the time constant, perhaps about 0.15 s, and the time to reach $0.9I_{\text{max}}$ should be about $2.5\tau = 0.5$ s.

**Categorize:** The precise times can be found from the equation that describes the rising current in the above graph, and gives the current as a function of time for a known emf, resistance, and time constant.

**Analyze:** At time $t$ after connecting the circuit, $I(t) = \frac{\mathcal{E}\left(1 - e^{-t/\tau}\right)}{R}$

where, after a long time, the current $I_{\text{max}} = \frac{\mathcal{E}\left(1 - e^{-\infty}\right)}{R} = \frac{\mathcal{E}}{R}.$

*continued on next page*

(a) At 50% of this maximum value, $I(t) = 0.500 I_{max}$.

So $0.500\dfrac{\mathcal{E}}{R} = \dfrac{\mathcal{E}\left(1 - e^{-t/0.200\text{ s}}\right)}{R}$.

This then yields $0.500 = 1 - e^{-t/0.200\text{ s}}$.

Isolating the constants on the right, $\ln\left(e^{-t/0.200\text{ s}}\right) = \ln(0.500)$

and solving for $t$, $-\dfrac{t}{0.200\text{ s}} = -0.693$.

Thus, to reach 50% of $I_{max}$ takes: $t = 0.139\text{ s}$. ◊

(b) Similarly, to reach 90% of $I_{max}$, $0.900 = 1 - e^{-t/\tau}$

so $t = -\tau\ln(1 - 0.900)$ or $t = -(0.200\text{ s})\ln(0.100) = 0.461\text{ s}$. ◊

**Finalize:** The calculated times agree reasonably well with our predictions. We must be careful to avoid confusing the equation for the rising current with the similar equation for the falling current. Checking our answers against predictions is a safe way to prevent such mistakes.

**P23.41** A 140-mH inductor and a 4.90-Ω resistor are connected with a switch to a 6.00-V battery as shown in Figure P23.41.

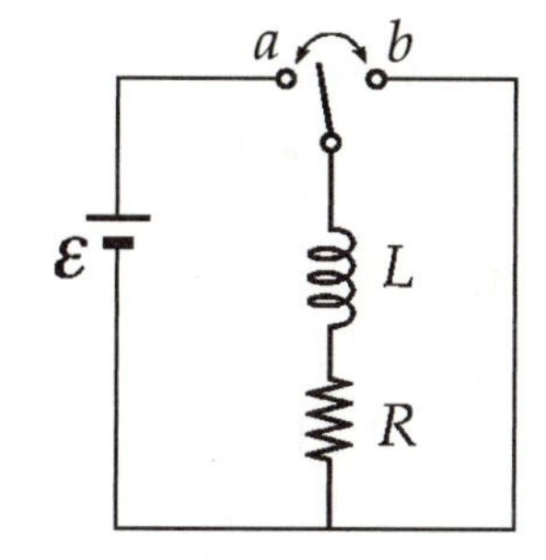


**FIG. P23.41**

(a) If the switch is thrown to the left (connecting the battery) how much time elapses before the current reaches 220 mA?

(b) What is the current in the inductor 10.0 s after the switch is closed?

(c) Now the switch is quickly thrown from *a* to *b*. How much time elapses before the current falls to 160 mA?

*continued on next page*

**Solution** The general equation for increasing current in an $LR$ circuit is obtained by combining Equations 23.14 and 23.15:

(a) $$I = \frac{\mathcal{E}\left(1 - e^{-Rt/L}\right)}{R}$$

or $$0.220 \text{ A} = \frac{6.00 \text{ V}}{4.90\ \Omega}\left(1 - e^{-(4.90\ \Omega)t/0.140 \text{ H}}\right)$$

$$0.180 = 1 - e^{-\left(35.0 \text{ s}^{-1}\right)t} \quad \text{or} \quad e^{\left(35.0 \text{ s}^{-1}\right)t} = 1.22.$$

Thus, $$t = \frac{\ln 1.22}{35.0 \text{ s}^{-1}} = 5.66 \text{ ms}.$$ ◊

(b) Again referring to the general equation,

$$I = \frac{6.00 \text{ V}}{4.90\ \Omega}\left(1 - e^{\left(-35.0 \text{ s}^{-1}\right)(10.0 \text{ s})}\right) = 1.22 \text{ A}.$$ ◊

(c) Now falling current is described by $I = I_i e^{-Rt/L}$

$$0.160 \text{ A} = (1.22 \text{ A})e^{(-4.90\ \Omega)t/0.140 \text{ H}}$$

$$0.131 = e^{-\left(35.0 \text{ s}^{-1}\right)t}$$

$$7.65 = e^{\left(35.0 \text{ s}^{-1}\right)t} \quad \text{or} \quad t = \frac{\ln 7.65}{35.0 \text{ s}^{-1}} = 58.1 \text{ ms}.$$ ◊

**P23.43** An air-core solenoid with 68 turns is 8.00 cm long and has a diameter of 1.20 cm. How much energy is stored in its magnetic field when it carries a current of 0.770 A?

**Solution** For a solenoid of length $\ell$, $L = \frac{\mu_0 N^2 A}{\ell}$.

Thus, since $U_B = \frac{1}{2}LI^2 = \frac{\mu_0 N^2 A I^2}{2\ell}$

$$U_B = \frac{\left(4\pi \times 10^{-7} \text{ N/A}^2\right)(68)^2 \pi\left(6.00 \times 10^{-3} \text{ m}\right)^2 (0.770 \text{ A})^2}{2(0.080\ 0 \text{ m})} = 2.44 \times 10^{-6} \text{ J}.$$ ◊

**P23.45** On a clear day at a certain location, a 100-V/m vertical electric field exists near the Earth's surface. At the same place, the Earth's magnetic field has a magnitude of $0.500 \times 10^{-4}$ T. Compute the energy densities of the two fields.

**Solution** $u_E = \dfrac{\epsilon_0 E^2}{2} = \dfrac{(8.85 \times 10^{-12}\ \text{C}^2/\text{N}\cdot\text{m}^2)(100\ \text{N/C})^2(1\ \text{J/N}\cdot\text{m})}{2} = 44.2\ \text{nJ/m}^3$ ◊

$$u_B = \frac{B^2}{2\mu_0} = \frac{(5.00 \times 10^{-5}\ \text{T})^2}{2(4\pi \times 10^{-7}\ \text{T}\cdot\text{m/A})} = 995 \times 10^{-6}\ \text{T}\cdot\text{A/m}$$

$$u_B = (995 \times 10^{-6}\ \text{T}\cdot\text{A/m})(1\ \text{N}\cdot\text{s/T}\cdot\text{C}\cdot\text{m})(1\ \text{J/N}\cdot\text{m}) = 995\ \mu\text{J/m}^3$$ ◊

Magnetic energy density is 22 500 times greater than that in the electric field.

**P23.55** The magnetic flux through a metal ring varies with time $t$ according to $\Phi_B = 3(at^3 - bt^2)\ \text{T}\cdot\text{m}^2$, with $a = 2.00\ \text{s}^{-3}$ and $b = 6.00\ \text{s}^{-2}$. The resistance of the ring is $3.00\ \Omega$. Determine the maximum current induced in the ring during the interval from $t = 0$ to $t = 2.00$ s.

**Solution** Substituting the given values, $\Phi_B = (6.00t^3 - 18.0t^2)\ \text{T}\cdot\text{m}^2$.

Therefore, the emf induced is $\mathcal{E} = -\dfrac{d\Phi_B}{dt} = -18.0t^2 + 36.0t$.

The maximum $\mathcal{E}$ occurs when $\dfrac{d\mathcal{E}}{dt} = -36.0t + 36.0 = 0$, which gives $t = 1.00$ s.

Thus, maximum current (at $t = 1.00$ s) is $I_{max} = \dfrac{\mathcal{E}}{R} = \dfrac{(-18.0 + 36.0)\ \text{V}}{3.00\ \Omega} = 6.00$ A. ◊

**P23.59** A long, straight wire carries a current that is given by $I = I_{max}\sin(\omega t + \phi)$. The wire lies in the plane of a rectangular coil of $N$ turns of wire as shown in Figure P23.59. The quantities $I_{max}$, $\omega$, and $\phi$ are all constants. Determine the emf induced in the coil by the magnetic field created by the current in the straight wire. Assume $I_{max} = 50.0$ A, $\omega = 200\pi\ \text{s}^{-1}$, $N = 100$, $h = w = 5.00$ cm, and $L = 20.0$ cm.

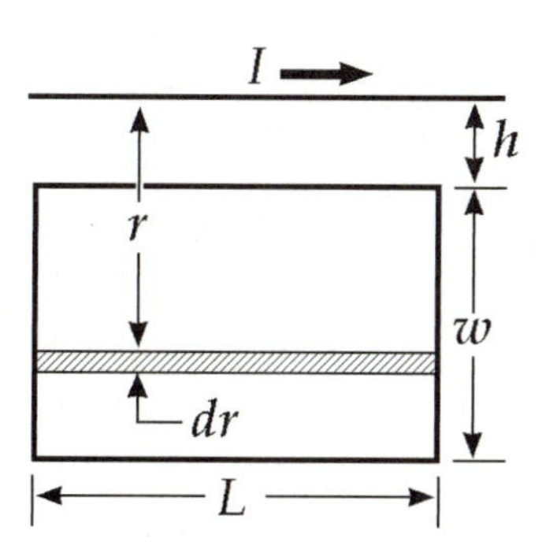


**FIG. P23.59**

*continued on next page*

**Solution** The coil is the boundary of a rectangular area. The magnetic field produced by the current in the straight wire is perpendicular to the plane of the area at all points. The field is

$$B = \frac{\mu_0 I}{2\pi r}$$ into the plane of the page.

Thus the flux is $N\Phi_B = \frac{\mu_0 NL}{2\pi} I \int_h^{h+w} \frac{dr}{r} = \frac{\mu_0 NL}{2\pi} I_{\max} \ln\left(\frac{h+w}{h}\right) \sin(\omega t + \phi)$.

Finally the induced emf is

$$\mathcal{E} = -N\frac{d\Phi_B}{dt} = -\frac{\mu_0 NL}{2\pi} I_{\max} \omega \ln\left(\frac{h+w}{h}\right) \cos(\omega t + \phi)$$

$$\mathcal{E} = -\left(4\pi \times 10^{-7}\ \text{T}\cdot\text{m/A}\right)\frac{(100)(0.200\ \text{m})}{2\pi}(50.0\ \text{A})\left(200\pi\,\text{s}^{-1}\right)\ln\left(\frac{10.0\ \text{cm}}{5.00\ \text{cm}}\right)\cos(\omega t + \phi)$$

$$\mathcal{E} = -(87.1\ \text{mV})\cos(200\pi t\ \text{rad/s} + \phi)$$ ◊

**Related Comment:** The factor $\sin(\omega t + \phi)$ in the expression for the current in the straight wire does not change appreciably when $\omega t$ changes by 0.1 rad or less. Thus, the current does not change appreciably during a time interval

$$\Delta t < \frac{0.100}{200\pi\,\text{s}^{-1}} = 1.59 \times 10^{-4}\ \text{s}.$$

We define a critical length,

$$c\Delta t = \left(3.00 \times 10^8\ \text{m/s}\right)\left(1.59 \times 10^{-4}\ \text{s}\right) = 4.77 \times 10^4\ \text{m}$$

equal to the distance to which field changes could be propagated during an interval of $1.59 \times 10^{-4}$ s. This length is so much larger than any dimension of the loop or its distance from the wire that, although we consider the straight wire to be infinitely long, we can also safely ignore the field propagation effects in the vicinity of the loop. Stated differently, the phase angle can be considered to be constant along the wire in the vicinity of the loop. If the angular frequency $\omega$ were much larger, say $200\pi \times 10^5\ \text{s}^{-1}$, the corresponding critical length would be only 48 cm. In this situation, propagation effects would be important and the above expression of $\mathcal{E}$ would require modification. As a "rule of thumb," we can consider field propagation effects for circuits of laboratory size to be negligible for frequencies, $f = \frac{\omega}{2\pi}$, that are less than about $10^6$ Hz.

# Chapter 24

# Electromagnetic Waves

## NOTES FROM SELECTED CHAPTER SECTIONS

### Section 24.2 Maxwell's Equations

Electromagnetic waves are generated by accelerating electric charges. The radiated waves consist of oscillating electric and magnetic fields, which are **at right angles to each other** and also **at right angles to the direction of wave propagation.**

The fundamental laws describing the behavior of electric and magnetic fields are Maxwell's equations. In this unified theory of electromagnetism, Maxwell showed that electromagnetic waves are a natural consequence of these fundamental laws.

The theory he developed is based upon the following pieces of information:

- A charge creates an electric field. Electric field lines originate on positive charges and terminate on negative charges. The relationship between charges and the fields they produce is described by Gauss's law.

- Magnetic field lines always form closed loops; that is, they do not begin or end anywhere.

- A varying magnetic field induces an emf and hence an electric field. This is a statement of Faraday's law.

- A moving charge (constituting a current) creates a magnetic field, as summarized in Ampère's law.

- A varying electric field creates a magnetic field. This is Maxwell's addition to Ampère's law.

## Section 24.3 Electromagnetic Waves

Following is a summary of the properties of electromagnetic waves:

- The solutions of Maxwell's third and fourth equations are wavelike, where both $\vec{\mathbf{E}}$ and $\vec{\mathbf{B}}$ satisfy the same wave equation.

- Electromagnetic waves travel through empty space with the speed of light,

$$c = \frac{1}{\sqrt{\epsilon_0 \mu_0}}$$

- The electric and magnetic field components of plane electromagnetic waves are perpendicular to each other and also perpendicular to the direction of wave propagation. The latter property can be summarized by saying that electromagnetic waves are transverse waves.

- The magnitudes of $\vec{\mathbf{E}}$ and $\vec{\mathbf{B}}$ in empty space are related by $\frac{E}{B} = c$.

- Electromagnetic waves obey the principle of superposition.

All **electromagnetic waves are produced by accelerating charges**. Types of electromagnetic waves can be characterized by "typical" ranges of frequency or wavelength.

- **Radio waves** $(\sim 10^4 \text{ m} > \lambda > \sim 0.1 \text{ m})$ are the result of electric charges accelerating through a conducting wire (antenna).

- **Microwaves** $(\sim 0.3 \text{ m} > \lambda > \sim 10^{-4} \text{ m})$ are generated by electronic devices.

- **Infrared waves** $(\sim 10^{-3} \text{ m} > \lambda > \sim 7 \times 10^{-7} \text{ m})$ are produced by high temperature objects and molecules.

- **Visible light** $(\sim 7 \times 10^{-7} \text{ m} > \lambda > \sim 4 \times 10^{-7} \text{ m})$ is produced by the rearrangement of electrons in atoms and molecules .

- **Ultraviolet (UV) light** $\left(\sim 4\times10^{-7}\text{ m} > \lambda > \sim 6\times10^{-10}\text{ m}\right)$ is an important component of radiation from the Sun.

- **X-rays** $\left(\sim 10^{-8}\text{ m} > \lambda > \sim 10^{-12}\text{ m}\right)$ are produced when high-energy electrons bombard a metal target.

- **Gamma rays** $\left(\sim 10^{-10}\text{ m} > \lambda > \sim 10^{-14}\text{ m}\right)$ are electromagnetic waves emitted by radioactive nuclei.

*Remember, the wavelength ranges stated above are approximate; regions of the electromagnetic spectrum overlap in wavelength. On the long wavelength end, radio waves can be arbitrarily long; and on the short wavelength end, gamma rays can be arbitrarily short.*

## Section 24.5 Energy Carried by Electromagnetic Waves

The magnitude of the Poynting vector represents the rate at which energy flows through a unit surface area perpendicular to the flow.

For an electromagnetic wave, the instantaneous energy density associated with the magnetic field equals the instantaneous energy density associated with the electric field. Hence, in a given volume, the energy is equally shared by the two fields.

## Section 24.6 Momentum and Radiation Pressure

Electromagnetic waves have momentum and exert pressure on surfaces on which they are incident. The pressure exerted by a normally incident wave on a **totally reflecting** surface is **double** that exerted on a surface that **completely absorbs** the incident wave.

# EQUATIONS AND CONCEPTS

Maxwell's equations are the fundamental laws governing the behavior of electric and magnetic fields. Electromagnetic waves are a natural consequence of these laws.

**Gauss's law:** The total electric flux through any closed surface equals the net charge enclosed by the surface divided by $\epsilon_0$.

$$\oint \vec{\mathbf{E}}\cdot d\vec{\mathbf{A}} = \frac{Q}{\epsilon_0} \qquad (24.4)$$

**Gauss's law for magnetism:** The net magnet flux through a closed surface is zero.

$$\oint \vec{\mathbf{B}} \cdot d\vec{\mathbf{A}} = 0 \tag{24.5}$$

**Faraday's law of induction:** The line integral of the electric field around any closed path equals the rate of change of magnetic flux through any surface area bounded by the path.

$$\oint \vec{\mathbf{E}} \cdot d\vec{\mathbf{s}} = -\frac{d\Phi_B}{dt} \tag{24.6}$$

where $\Phi_B = \vec{\mathbf{B}} \cdot \vec{\mathbf{A}}$

**Generalized form of Ampère's law:** The line integral of the magnetic field around any closed path is determined by the net current and the rate of change of electric flux through any surface bounded by the path.

$$\oint \vec{\mathbf{B}} \cdot d\vec{\mathbf{s}} = \mu_0 I + \epsilon_0 \mu_0 \frac{d\Phi_E}{dt} \tag{24.7}$$

where $\Phi_E = \vec{\mathbf{E}} \cdot \vec{\mathbf{A}}$

The **wave equations for electromagnetic waves in free space** are differential equations. Equations 24.15 and 24.16 (where $Q = 0$ and $I = 0$) represent linearly polarized waves traveling with a speed $c$.

$$\frac{\partial^2 E}{\partial x^2} = \epsilon_0 \mu_0 \frac{\partial^2 E}{\partial t^2} \tag{24.15}$$

$$\frac{\partial^2 B}{\partial x^2} = \epsilon_0 \mu_0 \frac{\partial^2 B}{\partial t^2} \tag{24.16}$$

The **speed of electromagnetic waves in vacuum** is the same as the speed of light in vacuum.

$$c = \frac{1}{\sqrt{\epsilon_0 \mu_0}} \tag{24.17}$$

The **electric and magnetic fields** propagate as sinusoidal transverse waves varying in position and time. *Their planes of vibration are perpendicular to each other and perpendicular to the direction of propagation.*

$$E = E_{\max} \cos(kx - \omega t) \tag{24.18}$$

$$B = B_{\max} \cos(kx - \omega t) \tag{24.19}$$

The **ratio of the magnitudes of the electric magnetic fields** is constant and equal to the speed of light $c$.

$$\frac{E}{B} = c \tag{24.21}$$

The **Poynting vector $\vec{\mathbf{S}}$** describes the energy flow associated with an electromagnetic wave. *The direction of $\vec{\mathbf{S}}$ is along the direction of propagation and the magnitude of $\vec{\mathbf{S}}$ is the rate at which electromagnetic energy crosses a unit surface area perpendicular to the direction of $\vec{\mathbf{S}}$.*

$$\vec{\mathbf{S}} \equiv \frac{1}{\mu_0} \vec{\mathbf{E}} \times \vec{\mathbf{B}} \tag{24.25}$$

The **wave intensity** is the time average of the magnitude of the Poynting vector. $E_{max}$ and $B_{max}$ are the maximum values of the field magnitudes.

$$I = S_{av} = \frac{E_{max}^2}{2\mu_0 c} = \frac{c B_{max}^2}{2\mu_0} \qquad (24.27)$$

The **instantaneous energy densities** of the electric and magnetic fields are equal.

$$u_B = u_E$$

$$u_E = \frac{1}{2} \epsilon_0 E^2 \qquad (24.28)$$

$$u_B = \frac{B^2}{2\mu_0} \qquad (24.29)$$

The **total instantaneous energy density** $u$ is proportional to $E^2$ and to $B^2$.

$$u = \epsilon_0 E^2 = \frac{B^2}{\mu_0}$$

The **total average energy density** is proportional to $E_{max}^2$ and to $B_{max}^2$. The average energy density is also proportional to the wave intensity.

$$u_{av} = \frac{1}{2} \epsilon_0 E_{max}{}^2 = \frac{B_{max}{}^2}{2\mu_o} \qquad (24.30)$$

$$I = S_{av} = c u_{av} \qquad (24.31)$$

The **linear momentum** $p$ **delivered to an absorbing surface** by an electromagnetic wave at normal incidence depends on the fraction of the total energy absorbed.

$$p = \frac{U}{c} \quad \text{(complete absorption)} \qquad (24.32)$$

$$p = \frac{2U}{c} \quad \text{(complete reflection)}$$

The **radiation pressure** on an absorbing surface (at normal incidence) depends on the intensity of the wave (power per unit area) and the degree of absorption.

$$P = \frac{S}{c} \quad \text{(complete absorption)} \qquad (24.33)$$

$$P = \frac{2S}{c} \quad \text{(complete reflection)}$$

## SUGGESTIONS, SKILLS, AND STRATEGIES

You should notice that:

- $\oint \vec{\mathbf{E}} \cdot d\vec{\mathbf{A}} = \frac{Q}{\epsilon_0}$ (Equation 24.4) and $\oint \vec{\mathbf{B}} \cdot d\vec{\mathbf{A}} = 0$ (Equation 24.5) are surface integrals. The normal components of electric and magnetic fields are integrated over a closed surface.

- $\oint \vec{\mathbf{E}} \cdot d\vec{\mathbf{s}} = -\frac{d\Phi_B}{dt}$ (Equation 24.6) and $\oint \vec{\mathbf{B}} \cdot d\vec{\mathbf{s}} = \mu_0 I + \epsilon_0 \mu_0 \frac{d\Phi_E}{dt}$ (Equation 24.7) are line integrals. The tangential components of electric and magnetic fields are integrated around a closed path.

## REVIEW CHECKLIST

✓ Describe the essential features of the apparatus and procedure used by Hertz in his experiments leading to the discovery and understanding of the source and nature of electromagnetic waves.

✓ For a properly described plane electromagnetic wave, calculate the values for the Poynting vector (magnitude), wave intensity, and instantaneous and average energy densities.

✓ Calculate the radiation pressure on a surface and the linear momentum delivered to a surface by an electromagnetic wave.

✓ Understand the production of electromagnetic waves and radiation of energy by an oscillating dipole. Use a diagram to show the relative directions for $\vec{\mathbf{E}}$, $\vec{\mathbf{B}}$, and $\vec{\mathbf{S}}$.

## ANSWERS TO SELECTED QUESTIONS

**Q24.1** Radio station announcers often advertise "instant news." If they mean that you can hear the news the instant they speak it, is their claim true? About how long would it take for a message to travel across this country by radio waves, assuming that the waves could be detected at this range?

**Answer** Radio waves move at the speed of light. They can travel around the curved surface of the Earth, bouncing between the ground and the ionosphere, which has an altitude that is small when compared to the radius of the Earth. The distance across the lower forty-eight states is approximately 5 000 km, requiring a time of $\frac{5 \times 10^6 \text{ m}}{3 \times 10^8 \text{ m/s}} \sim 10^{-2}$ s. To go halfway around the Earth takes only 0.07 s. In other words, a speech can be heard on the other side of the world before it is heard at the back of an auditorium.

**Q24.5** If you charge a comb by running it through your hair and then hold the comb next to a bar magnet, do the electric and magnetic fields produced constitute an electromagnetic wave?

**Answer** No. Charge on the comb creates an electric field and the bar magnet sets up a magnetic field. However, at any point these fields are constant in magnitude and direction. In an electromagnetic wave, the electric and magnetic fields must be changing in order to create each other.

**Q24.7** In the *LC* circuit shown in Figure Q24.7 the charge on the capacitor is sometimes zero, but at such instants the current in the circuit is not zero. How is that possible?

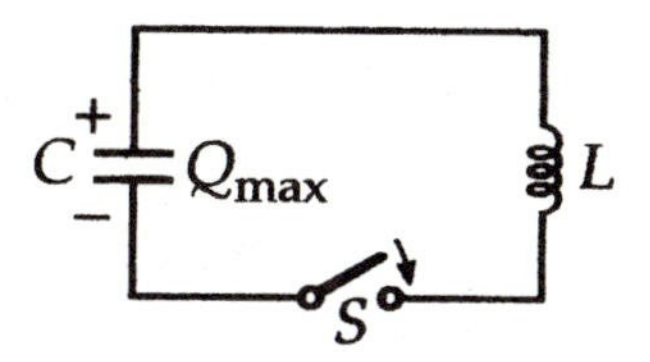


**FIG. Q24.7**

**Answer** When the capacitor is fully discharged, the current in the circuit is a maximum. The inductance of the coil is making the current continue to flow. At this time the magnetic field of the coil contains all the energy that was originally stored in the charged capacitor.

**Q24.11** What does a radio wave do to the charges in the receiving antenna to provide a signal for your car radio?

**Answer** Consider a typical metal rod antenna for a car radio. The rod detects the electric field portion of the carrier wave. Variations in the amplitude of the carrier wave cause the electrons in the rod to vibrate with amplitudes emulating those of the carrier wave. Likewise, for frequency modulation, the variations of the frequency of the carrier wave cause constant-amplitude vibrations of the electrons in the rod at frequencies that imitate those of the carrier.

**Q24.13** Suppose a creature from another planet had eyes that were sensitive to infrared radiation. Describe what the alien would see if it looked around the room you are now in. In particular, what would be bright and what would be dim?

**Answer** Lightbulbs and the toaster shine brightly in the infrared. Somewhat fainter are the back of the refrigerator and the back of the television set, while the TV screen is dark. The pipes under the sink show the same weak glow as the walls until you turn on the faucets. Then the pipe on the right gets darker while that on the left develops a rich gleam that quickly runs up along its length. The food on your plate shines; so does human skin, the same color for all races. Clothing is dark as a rule, but your seat glows like a monkey's rump when you get up from a chair, and you leave a patch of the same glow on your chair. Your face appears lit from within, like a jack-o-lantern; your nostrils and openings of your ear canals are bright; brighter still are the pupils of your eyes.

## SOLUTIONS TO SELECTED PROBLEMS

**P24.3** A proton moves through a uniform electric field $\vec{\mathbf{E}} = 50.0\hat{\mathbf{j}}$ V/m and a uniform magnetic field $\vec{\mathbf{B}} = \left(0.200\hat{\mathbf{i}} + 0.300\hat{\mathbf{j}} + 0.400\hat{\mathbf{k}}\right)$ T. Determine the acceleration of the proton when it has a velocity $\vec{\mathbf{v}} = 200\hat{\mathbf{i}}$ m/s.

**Solution** The combined force on the proton is the Lorentz force,

$$\vec{\mathbf{F}} = m\vec{\mathbf{a}} = q\vec{\mathbf{E}} + q\vec{\mathbf{v}} \times \vec{\mathbf{B}} \qquad \text{so that} \qquad \vec{\mathbf{a}} = \frac{e}{m}\left[\vec{\mathbf{E}} + \vec{\mathbf{v}} \times \vec{\mathbf{B}}\right].$$

Taking the cross product of $\vec{\mathbf{v}}$ and $\vec{\mathbf{B}}$,

$$\vec{\mathbf{v}} \times \vec{\mathbf{B}} = \begin{vmatrix} \hat{\mathbf{i}} & \hat{\mathbf{j}} & \hat{\mathbf{k}} \\ 200 & 0 & 0 \\ 0.200 & 0.300 & 0.400 \end{vmatrix} = -200(0.400)\hat{\mathbf{j}} + 200(0.300)\hat{\mathbf{k}}$$

$$\vec{\mathbf{a}} = \left(\frac{1.60 \times 10^{-19}}{1.67 \times 10^{-27}}\right)\left[50.0\hat{\mathbf{j}} - 80.0\hat{\mathbf{j}} + 60.0\hat{\mathbf{k}}\right] = 9.58 \times 10^{7}\left[-30\hat{\mathbf{j}} + 60\hat{\mathbf{k}}\right]$$

$$\vec{\mathbf{a}} = \left(2.87 \times 10^{9}\right)\left(-\hat{\mathbf{j}} + 2\hat{\mathbf{k}}\right) \text{ m/s}^2 = \left(-2.87 \times 10^{9}\hat{\mathbf{j}} + 5.75 \times 10^{9}\hat{\mathbf{k}}\right) \text{ m/s}^2$$ ◊

**P24.7** Figure 24.6 shows a plane electromagnetic sinusoidal wave propagating in the $x$ direction. Suppose that the wavelength is 50.0 m, and the electric field vibrates in the $xy$ plane with an amplitude of 22.0 V/m. Calculate

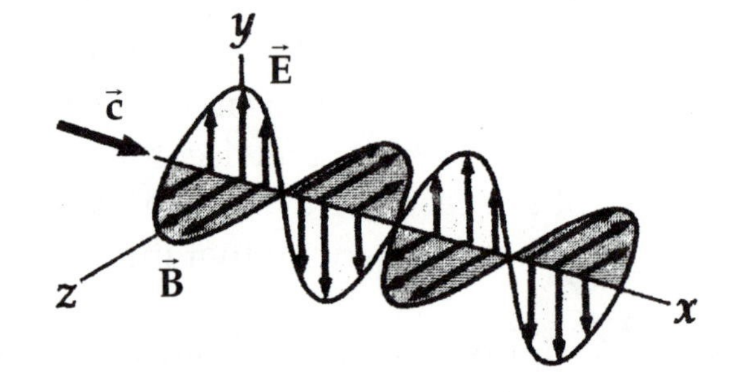


**FIG. P24.7**

(a) the frequency of the wave and

(b) the magnitude and direction $\vec{\mathbf{B}}$ when the electric field has its maximum value in the negative $y$ direction.

(c) Write an expression for $\vec{\mathbf{B}}$ with the correct unit vector, with numerical values for $B_{max}$, $k$, and $\omega$, and with its magnitude in the form $B = B_{max}\cos(kx - \omega t)$.

*continued on next page*

**Solution** (a) $c = f\lambda$: $f = \frac{c}{\lambda} = \frac{3.00\times10^8 \text{ m/s}}{50.0 \text{ m}} = 6.00\times10^6 \text{ Hz}$ ◊

(b) $c = \frac{E}{B}$: $B = \frac{E}{c} = \frac{22.0 \text{ V/m}}{3.00\times10^8 \text{ m/s}} = 7.33\times10^{-8} \text{ T} = 73.3 \text{ nT}$ ◊

$\vec{\mathbf{B}}$ is directed along **negative** $z$ **direction** when $\vec{\mathbf{E}}$ is in the negative $y$ direction, so that $\vec{\mathbf{S}} = \frac{\vec{\mathbf{E}}\times\vec{\mathbf{B}}}{\mu_0}$ will propagate in the direction of $(-\hat{\mathbf{j}})\times(-\hat{\mathbf{k}}) = +\hat{\mathbf{i}}$.

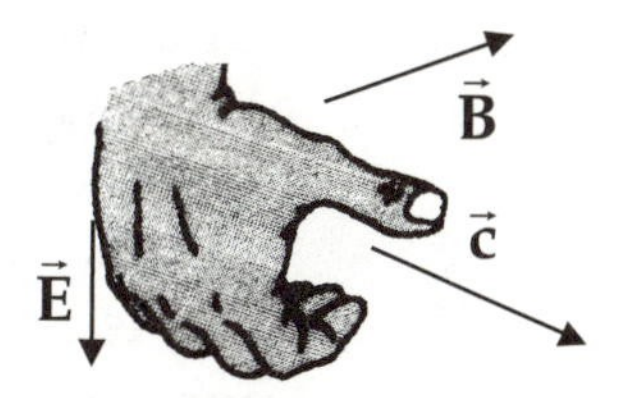


(c) $B = B_{\max}\cos(kx - \omega t)$: $k = \frac{2\pi}{\lambda} = \frac{2\pi}{50.0 \text{ m}} = 0.126 \text{ m}^{-1}$

$$\omega = 2\pi f = (2\pi \text{rad})(6.00\times10^6 \text{ Hz}) = 3.77\times10^7 \text{ rad/s}$$

Thus,

$$\vec{\mathbf{B}} = (73.3 \text{ nT})\cos\left[(0.126 \text{ rad/m})x - (3.77\times10^7 \text{ rad/s})t\right](-\hat{\mathbf{k}}).$$ ◊

Here $x$ and $t$ are variables. The function $\vec{\mathbf{B}}(x, t)$, called the *wave function*, defines the magnitude and direction of the magnetic field at every location along the $x$ axis at every instant of time.

**P24.17** A fixed inductance $L = 1.05\ \mu\text{H}$ is used in series with a variable capacitor in the tuning section of a radiotelephone on a ship. What capacitance tunes the circuit to the signal from a transmitter broadcasting at 6.30 MHz?

**Solution** **Conceptualize:** It is difficult to predict a value for the capacitance without doing the calculations, but we might expect a typical value in the $\mu$F or pF range.

**Categorize:** We want the resonance frequency of the circuit to match the broadcasting frequency, and for a simple $LC$ circuit, the resonance frequency only depends on the magnitudes of the inductance and capacitance.

**Analyze:** The resonance frequency is $f_0 = \frac{1}{2\pi\sqrt{LC}}$.

Thus, $$C = \frac{1}{(2\pi f_0)^2 L} = \frac{1}{\left[2\pi(6.30\times10^6 \text{ Hz})\right]^2(1.05\times10^{-6} \text{ H})} = 608 \text{ pF}.$$ ◊

*continued on next page*

**Finalize:** This is indeed a typical capacitance, so our calculation appears reasonable. You probably would not hear any familiar music on this broadcast frequency. The frequency range for FM radio broadcasting is 88.0-108.0 MHz, and AM radio is 535-1 605 kHz. The 6.30 MHz frequency falls in the Maritime Mobile SSB Radiotelephone range, so you might hear a ship captain instead of Top 40 tunes! This and other information about the radio frequency spectrum can be found on the National Telecommunications and Information Administration (NTIA) website, which at the time of this printing was at http://www.ntia.gov/osmhome/allochrt.html.

**P24.21** What is the average magnitude of the Poynting vector 5.00 miles from a radio transmitter broadcasting isotropically (equally in all directions) with an average power of 250 kW?

**Solution** **Conceptualize:** As the distance from the source is increased, the power per unit area will decrease, so at a distance of 5 miles from the source, the power per unit area will be a small fraction of the Poynting vector near the source.

**Categorize:** The Poynting vector is the power per unit area, where $A$ is the surface area of a sphere with a 5-mile radius.

**Analyze:** The Poynting vector is $S_{av} = \frac{\mathscr{P}}{A} = \frac{\mathscr{P}}{4\pi r^2}$.

In meters, $r = (5.00\text{ mi})(1\,609\text{ m/mi}) = 8\,045\text{ m}$

and the magnitude is $S = \frac{250 \times 10^3\text{ W}}{4\pi(8\,045)^2} = 3.07 \times 10^{-4}\text{ W/m}^2$. ◊

**Finalize:** The magnitude of the Poynting vector ten meters from the source is $199\text{ W/m}^2$, on the order of a million times larger than it is 5 miles away! It is surprising to realize how little power is actually received by a radio (at the 5-mile distance, the signal would only be about 30 nW, assuming a receiving area of about 1 $\text{cm}^2$).

**P24.23** A community plans to build a facility to convert solar radiation to electrical power. The community requires 1.00 MW of power, and the system to be installed has an efficiency of 30.0% (that is, 30.0% of the solar energy incident on the surface is converted to useful energy that can power the community). What must be the effective area of a perfectly absorbing surface used in such an installation if sunlight has a constant intensity of $1\,000\text{ W/m}^2$?

**Solution** At 30.0% efficiency, the power $\mathscr{P} = 0.300SA$:

$$A = \frac{\mathscr{P}}{0.300S} = \frac{1.00 \times 10^6\text{ W}}{0.300(1\,000\text{ W/m}^2)} = 3\,330\text{ m}^2 \approx 0.8\text{ acre}$$ ◊

**P24.25** The filament of an incandescent lamp has a 150-Ω resistance and carries a direct current of 1.00 A. The filament is 8.00 cm long and 0.900 mm in radius.

(a) Calculate the Poynting vector at the surface of the filament, associated with the static electric field producing the current and the current's static magnetic field.

(b) Find the magnitude of the static electric and magnetic fields at the surface of the filament.

**Solution** In this problem, the Poynting vector does not represent the light radiated or the energy convected away from the filament. Rather, it represents the energy flow in the static electric and magnetic fields created in the surrounding empty space, by current in the filament. The Poynting vector describes not only energy transfer by electromagnetic radiation but also energy transfer by electrical transmission. The rate at which energy is delivered to the resistor is

$$\mathcal{P} = I^2 R = (1.00\ \text{A})^2 (150\ \Omega) = 150\ \text{W}$$

and the surface area $A = 2\pi r L = 2\pi\left(0.900\times10^{-3}\ \text{m}\right)(0.080\,0\ \text{m}) = 4.52\times10^{-4}\ \text{m}^2$

(a) The Poynting vector is directed radially inward:

$$S = \frac{\mathcal{P}}{A} = \frac{150\ \text{W}}{4.52\times10^{-4}\ \text{m}^2} = 3.32\times10^5\ \text{W/m}^2 \qquad \Diamond$$

(b) $$B = \mu_0 \frac{I}{2\pi r} = \left(4\pi\times10^{-7}\ \text{T}\cdot\text{m/A}\right)\frac{1.00\ \text{A}}{2\pi\left(0.900\times10^{-3}\ \text{m}\right)} = 2.22\times10^{-4}\ \text{T} \qquad \Diamond$$

$$E = \frac{\Delta V}{\Delta x} = \frac{IR}{L} = \frac{150\ \text{V}}{0.080\,0\ \text{m}} = 1\,880\ \text{V/m} \qquad \Diamond$$

Note: We could also calculate the Poynting vector from $S = \frac{EB}{\mu_0} = 3.32\times10^5\ \text{W/m}^2$.

**P24.35** What are the wavelengths of electromagnetic waves in free space that have frequencies of

(a) $5.00 \times 10^{19}$ Hz and

(b) $4.00 \times 10^{9}$ Hz?

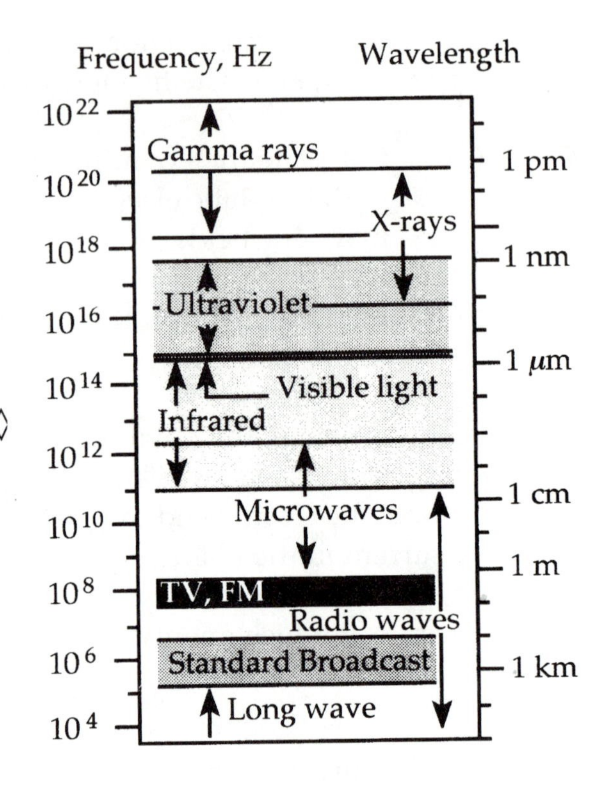


FIG. P24.35

**Solution** (a) $\lambda = \frac{c}{f} = \frac{3.00 \times 10^{8}\ \text{m/s}}{5.00 \times 10^{19}\ \text{s}^{-1}}$

$\lambda = 6.00$ pm ◊

This would be called an x-ray if it were emitted when an inner electron in an atom or an electron in a vacuum tube loses energy. It would be called a gamma ray if it were radiated by an atomic nucleus.

(b) $\lambda = \frac{c}{f} = \frac{3.00 \times 10^{8}\ \text{m/s}}{4.00 \times 10^{9}\ \text{s}^{-1}} = 7.50$ cm ◊

On the electromagnetic spectrum this finger-length wave is called a radio wave or a microwave.

**P24.41** Plane-polarized light is incident on a single polarizing disk with the direction of $\vec{\mathbf{E}}_0$ parallel to the direction of the transmission axis. Through what angle should the disk be rotated so that the intensity in the transmitted beam is reduced by a factor of

(a) 3.00?

(b) 5.00 and

(c) 10.0?

*continued on next page*

**Solution** We define the initial angle, at which all the light is transmitted to be $\theta = 0$. Turning the disk to another angle will then reduce the transmitted light by an intensity factor of $I = I_0 \cos^2 \theta$.

(a) For $I = \dfrac{I_0}{3.00}$, $\cos\theta = \dfrac{1}{\sqrt{3.00}}$ and $\theta = 54.7°$. ◊

(b) For $I = \dfrac{I_0}{5.00}$, $\cos\theta = \dfrac{1}{\sqrt{5.00}}$ and $\theta = 63.4°$. ◊

(c) For $I = \dfrac{I_0}{10.0}$, $\cos\theta = \dfrac{1}{\sqrt{10.0}}$ and $\theta = 71.6°$. ◊

**P24.51** A ruby laser delivers a 10.0-ns pulse of 1.00 MW average power. If the photons have a wavelength of 694.3 nm, how many are contained in the pulse?

**Solution** **Conceptualize:** Lasers generally produce concentrated beams that are bright (except for IR or UV lasers that produce invisible beams). Since our eyes can detect light levels as low as a few photons, there are probably at least 1 000 photons in each pulse.

**Categorize:** From the pulse width and average power, we can find the energy delivered by each pulse. The number of photons can then be found by dividing the pulse energy by the energy of each photon, which is determined from the photon wavelength.

**Analyze:** The energy in each pulse is

$$E = \mathcal{P}\Delta t = \left(1.00 \times 10^{6} \text{ W}\right)\left(1.00 \times 10^{-8} \text{ s}\right) = 1.00 \times 10^{-2} \text{ J}$$

The energy of each photon is

$$E_\gamma = hf = \frac{hc}{\lambda} = \frac{\left(6.626 \times 10^{-34} \text{ J} \cdot \text{s}\right)\left(3.00 \times 10^{8} \text{ m/s}\right)}{694.3 \times 10^{-9} \text{ m}} = 2.86 \times 10^{-19} \text{ J}$$

So $$N = \frac{E}{E_\gamma} = \frac{1.00 \times 10^{-2} \text{ J}}{2.86 \times 10^{-19} \text{ J/photon}} = 3.49 \times 10^{16} \text{ photons}$$ ◊

**Finalize:** With $10^{16}$ photons/pulse, this laser beam should produce a bright red spot when the light reflects from a surface, even though the time between pulses is generally much longer than the width of each pulse. For comparison, this laser produces more photons in a single ten-nanosecond pulse than a typical 5 mW helium-neon laser produces over a full second (about $1.6 \times 10^{16}$ photons/second). Human eyes require many more than 1 000 photons for something to appear bright.

**P24.57** **Review problem.** In the absence of cable input or a satellite dish, a television set can use a dipole-receiving antenna for VHF channels and a loop antenna for UHF channels (Fig. P24.57). The UHF antenna produces an emf from the changing magnetic flux through the loop. The TV station broadcasts a signal with a frequency $f$, and the signal has an electric-field amplitude $E_{max}$ and a magnetic-field amplitude $B_{max}$ at the location of the receiving antenna.

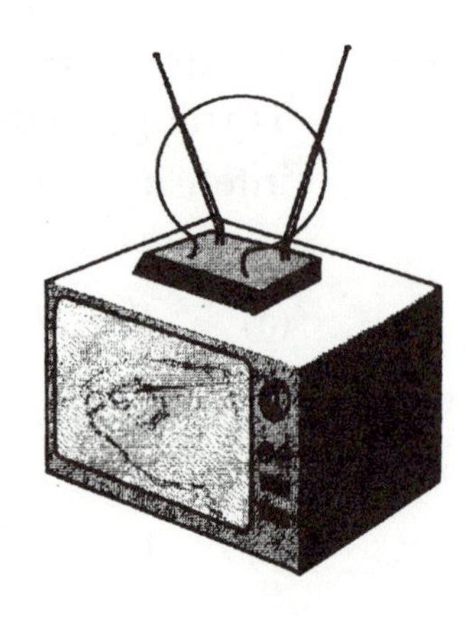

**FIG. P24.57**

(a) Using Faraday's law, derive an expression for the amplitude of the emf that appears in a single-turn circular loop antenna with a radius $r$, which is small compared with the wavelength of the wave.

(b) If the electric field in the signal points vertically, what orientation of the loop gives the best reception?

**Solution** We can approximate the magnetic field as uniform over the area of the loop while it oscillates in time as $B = B_{max}\cos\omega t$. The induced voltage is

$$\mathcal{E} = -\frac{d\Phi_B}{dt} = -\frac{d}{dt}(BA\cos\theta) = -A\frac{d}{dt}(B_{max}\cos\omega t\cos\theta)$$

$$\mathcal{E} = AB_{max}\omega(\sin\omega t\cos\theta)$$

$$\mathcal{E}(t) = 2\pi fB_{max}A\sin(2\pi ft)\cos\theta = 2\pi^2r^2fB_{max}\cos\theta\sin(2\pi ft)$$

(a) The amplitude of this emf is $\mathcal{E}_{max} = 2\pi^2r^2fB_{max}\cos\theta$, where $\theta$ is the angle between the magnetic field and the normal to the loop. ◊

(b) If $\vec{\mathbf{E}}$ is vertical, then $\vec{\mathbf{B}}$ is horizontal, so the plane of the loop should be vertical, and the plane should contain the line of sight to the transmitter. This will make $\theta = 0°$, so $\cos\theta$ takes on its maximum value. ◊

The induced emf oscillates sinusoidally at the frequency of the radio wave. Its amplitude is proportional to the amplitude of the wave $B_{max}$ and to the area of the loop. The amplitude is also proportional to the frequency of the wave. But a higher assigned frequency may not be an advantage for a broadcasting TV station. A higher-frequency wave may be more strongly absorbed by electrically conducting obstacles between the station and the receiver.

**P24.59** A dish antenna having a diameter of 20.0 m receives (at normal incidence) a radio signal from a distant source, as shown in Figure P24.59. The radio signal is a continuous sinusoidal wave with amplitude $E_{max} = 0.200\ \mu V/m$. Assume the antenna absorbs all the radiation that falls on the dish.

**FIG. P24.59**

(a) What is the amplitude of the magnetic field in this wave?

(b) What is the intensity of the radiation received by this antenna?

(c) What is the power received by the antenna?

(d) What force is exerted by the radio waves on the antenna?

**Solution** (a) $B_{max} = \frac{E_{max}}{c} = 6.67 \times 10^{-16}\ T$ ◊

(b) $S_{av} = \frac{E_{max}^2}{2\mu_0 c} = 5.31 \times 10^{-17}\ W/m^2$ ◊

(c) $\mathscr{P}_{av} = S_{av}A = 1.67 \times 10^{-14}\ W$ ◊

(Do not confuse this power with the expression for pressure $P = \frac{S}{c}$, which we use to find the force.)

(d) $F = PA = \left(\frac{S_{av}}{c}\right)A = 5.56 \times 10^{-23}$ N (~3 000 Hydrogen atoms' weight!) ◊

**P24.61** In 1965, Arno Penzias and Robert Wilson discovered the cosmic microwave radiation left over from the Big Bang expansion of the Universe. Suppose the energy density of this background radiation is $4.00 \times 10^{-14}\ J/m^3$. Determine the corresponding electric field amplitude.

*continued on next page*

**Solution** The energy density can be written as $u = \frac{1}{2} \epsilon_0 E_{\max}^2$

so $$E_{\max} = \sqrt{\frac{2u}{\epsilon_0}} = \sqrt{\frac{2\left(4.00 \times 10^{-14} \text{ N/m}^2\right)}{8.85 \times 10^{-12} \text{ C}^2/\text{N} \cdot \text{m}^2}} = 95.1 \text{ mV/m}.$$ ◊

Chapter 25

# Reflection and Refraction of Light

## NOTES FROM SELECTED CHAPTER SECTIONS

### Section 25.2 The Ray Model in Geometric Optics

In the simplification model called the ray model or ray approximation, it is assumed that a wave travels through a medium along a straight line in the direction of its rays. A ray for a given wave is a straight line perpendicular to the wave front; geometric models are based on these straight lines. This approximation also neglects diffraction effects.

### Section 25.3 The Wave Under Reflection

**Specular reflection** occurs when the reflecting surface is smooth; reflection from a rough surface is called **diffuse reflection**. In the case of reflection from a smooth surface, the angle of incidence equals the angle of reflection; the angles are measured between the normal to the surface and the respective rays. *The path of a light ray is reversible, an important property in geometric optics.*

### Section 25.4 The Wave Under Refraction

Refraction refers to the change in direction of a light ray upon striking obliquely the interface between two transparent media. The angle of refraction (between the refracted ray and the normal) depends on the properties of the two media and the angle of incidence as expressed by Snell's law, Eq. 25.7. The incident ray, reflected ray, and refracted (transmitted ) ray lie in the same plane. *As a ray travels from one medium to another, the speed of the wave changes but the frequency does not change.*

## Section 25.5 Dispersion and Prisms

For a given material, the index of refraction is a function of the wavelength of the light passing through the material. (The wavelength depends in turn on the speed.) This effect is called dispersion. In particular, when light passes through a prism, a given ray is refracted at two surfaces and emerges bent away from its original direction by an angle of deviation, $\delta$. Due to dispersion, $\delta$ is different for different wavelengths.

## Section 25.6 Huygens's Principle

Every point on a given wave front can be considered as a point source for a secondary wavelet. At some later time, the new position of the wave front is determined by the surface tangent to the set of secondary wavelets.

## Section 25.7 Total Internal Reflection

Total internal reflection, illustrated in Figure 25.1, is possible only when light rays traveling in one medium are incident on a boundary between the first medium and a second medium of lesser index of refraction. The angle $\theta_c$ shown in the figure is called the critical angle.

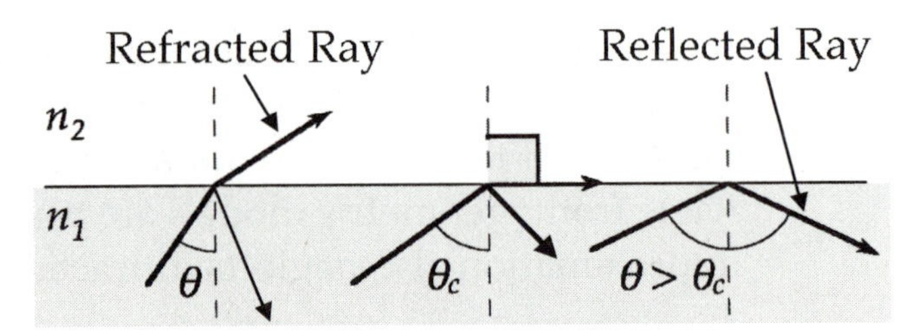


Total internal reflection of light occurs at angles of incidence $\theta_1 \geq \theta_c$ where $n_1 > n_2$

**FIG. 25.1**

## EQUATIONS AND CONCEPTS

Consider the situation in the figure. A light ray is incident obliquely on a smooth, planar surface that forms the boundary between two transparent media of different indices of refraction. A portion of the ray will be reflected back into the original medium, while the remaining fraction will be transmitted into the second medium. *The incident, reflected, and refracted rays and the normal to the interface are all in the same plane.*

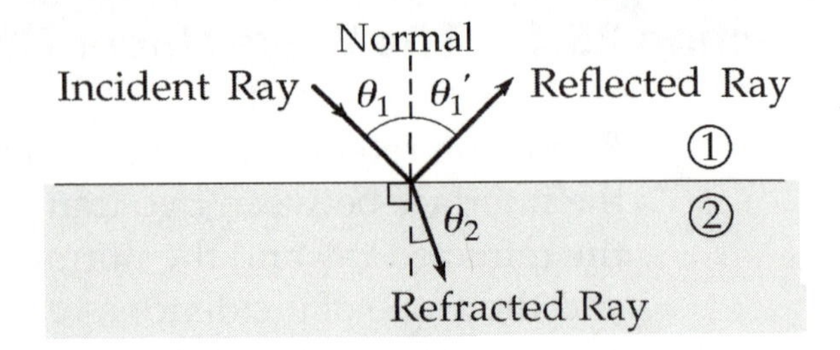

The **law of reflection** states that the angle of incidence (the angle measured between the incident ray and the normal to the surface) equals the angle of reflection (the angle measured between the reflected ray and the normal to the surface).

$$\theta_1' = \theta_1 \tag{25.1}$$

**Snell's law of refraction** can be expressed in terms of the speeds of light in the media on either side of the refracting surface (Equation 25.2); or in terms of the indices of refraction of the two media (Equation 25.7).

$$\frac{\sin\theta_2}{\sin\theta_1} = \frac{v_2}{v_1} = \text{constant} \tag{25.2}$$

The **most practical form of Snell's law** is given in Equation 25.7. This equation involves a parameter called the index of refraction, $n$, the value of which is characteristic of a particular medium.

$$n_1 \sin\theta_1 = n_2 \sin\theta_2 \tag{25.7}$$

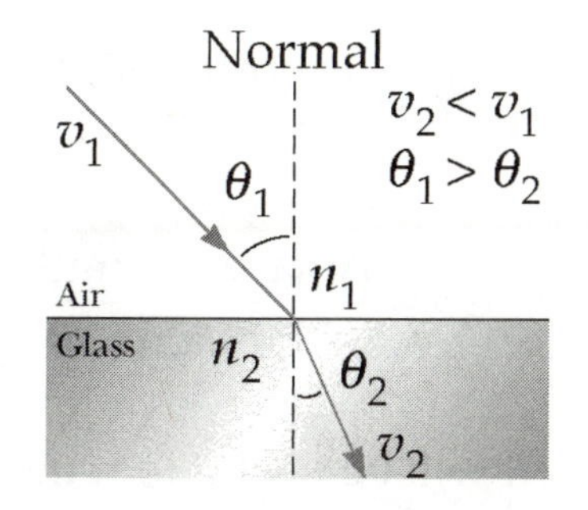


The **index of refraction** of a transparent medium is defined as the ratio of the speed of light in vacuum to the speed of light in the medium.

$$n \equiv \frac{\text{Speed of light in vacuum}}{\text{Speed of light in a medium}} = \frac{c}{v} \tag{25.3}$$

The index of refraction of a given medium can be expressed as the ratio of the wavelength of light in vacuum to the wavelength in that medium.

$$n = \frac{\lambda_0}{\lambda_n} \tag{25.6}$$

*The frequency of a wave is characteristic of the source. Therefore, as light travels from one medium into another of different index of refraction, the frequency remains constant but the wavelength changes.*

The **critical angle for total internal reflection** is the minimum angle of incidence for which total internal reflection can occur. *Total internal reflection is possible only when a light ray is directed from a medium of higher index of refraction into a medium of lower index of refraction.*

$$\sin\theta_c = \frac{n_2}{n_1} \text{ (for } n_1 > n_2\text{)} \qquad (25.8)$$

## SUGGESTIONS, SKILLS, AND STRATEGIES

It is helpful in this chapter to review the laws of trigonometry. Most important, of course, is the definition of the trigonometric functions:

$$\tan\theta = \frac{\text{Opposite length}}{\text{Adjacent length}}$$

$$\sin\theta = \frac{\text{Opposite length}}{\text{Hypotenuse}} \qquad \sin(-\theta) = -\sin\theta$$

$$\cos\theta = \frac{\text{Adjacent length}}{\text{Hypotenuse}} \qquad \cos(-\theta) = \cos\theta$$

Hypotenuse
Opposite
$\theta$
90°
Adjacent

(Valid for right triangles only)

It will be useful to remember that the interior angles of a triangle add to 180°. It may also be useful to convert trigonometric functions from one form to another, using the trigonometric versions of the Pythagorean theorem:

$$\sin^2\theta = 1 - \cos^2\theta \qquad \tan^2\theta = \frac{\sin^2\theta}{1-\sin^2\theta}$$

Finally, you should review the angle-sum relations:

$$\sin(\theta+\phi) = \sin\theta\cos\phi + \cos\theta\sin\phi \qquad \cos(\theta+\phi) = \cos\theta\cos\phi - \sin\theta\sin\phi$$

## REVIEW CHECKLIST

✓ Understand Huygens's principle and the use of this technique to construct the subsequent position and shape of a given wave front.

✓ Determine the directions of the reflected and refracted rays when a light ray is incident obliquely on the interface between two optical media.

✓ Understand the conditions under which total internal reflection can occur in a medium and determine the critical angle for a given pair of adjacent media.

## ANSWERS TO SELECTED QUESTIONS

**Q25.5** As light travels from one medium to another, does the wavelength of the light change? Does the frequency change? Does the speed change? Explain.

**Answer** You can think about this based upon the fact that the light wave must be continuous. Therefore, the frequency of the light must remain constant, since any one crest coming up to the interface must create one crest in the new medium. However, the speed of light does vary in different media, as described by Snell's law. In addition, since the speed of the light wave is equal to the light wave's frequency times its wavelength, and the frequency does not change, we can be sure that the wavelength must change. The speed decreases upon entering a new medium from vacuum; therefore the wavelength decreases as well.

**Q25.7** Explain why a diamond sparkles more than a glass crystal of the same shape and size.

**Answer** Diamond has a larger index of refraction than glass, and consequently has a smaller critical angle for internal reflection. A brilliant-cut diamond is shaped to admit light from above, reflect it totally at the converging facets on the underside of the jewel, and let the light escape only at the top. Glass will have less light internally reflected.

**Q25.9** Explain why a diamond loses most of its sparkle when submerged in carbon disulfide and why an imitation diamond of cubic zirconia loses all its sparkle in corn syrup.

**Answer** A ball covered with mirrors sparkles by reflecting light from its surface. On the other hand, a faceted diamond lets in light at the top, reflects it by total internal reflection in the bottom half, and sends the light out through the top again. Its high index of refraction means that in air, the critical angle for total internal reflection, $\theta_c = \sin^{-1}\left(\dfrac{n_{\text{air}}}{n_{\text{diamond}}}\right)$, is small. Light rays can enter through a large range of directions, and exit through a small range with higher intensity. When a diamond is immersed in carbon disulfide, the critical angle is increased to $\theta_c = \sin^{-1}\left(\dfrac{n_{\text{CS}_2}}{n_{\text{diamond}}}\right)$. As a result, more light escapes through the bottom of the gem and the reflected light is less intense.

An imitation diamond of cubic zirconia loses all of its sparkle in corn syrup because its index of refraction is the same as that of corn syrup. This stone will therefore reflect almost no light, either internally or externally. If the absorption spectrum (color) of the stone is the same as that of corn syrup, we can expect the stone to be invisible.

**Q25.11** When two colors of light (X and Y) are sent through a glass prism, X is bent more than Y. Which color travels more slowly in the prism?

**Answer** If the light slows down upon entering the prism, a light ray that is bent more suffers a greater loss of speed as it enters the new medium. Therefore, the X light rays travel more slowly.

## SOLUTIONS TO SELECTED PROBLEMS

**P25.7** An underwater scuba diver sees the Sun at an apparent angle of 45.0° above the horizon. What is the actual elevation angle of the Sun above the horizon?

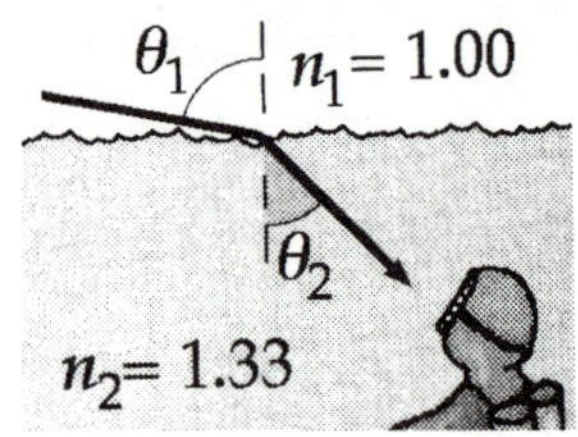


FIG. P25.7

**Solution** **Conceptualize:** The sunlight refracts as it enters the water from the air. Because the water has a higher index of refraction, the light slows down and bends toward the vertical line that is normal to the interface. Therefore, the elevation angle of the Sun above the water will be less than 45° as shown in the diagram to the right, even though it appears to the diver that the sun is 45° above the horizon.

**Categorize:** We can use Snell's law of refraction to find the precise angle of incidence.

**Analyze:** Snell's law is $n_1 \sin\theta_1 = n_2 \sin\theta_2$

which gives $\sin\theta_1 = 1.333 \sin 45.0°$

$$\sin\theta_1 = (1.333)(0.707) = 0.943$$

The sunlight is at $\theta_1 = 70.5°$ to the vertical, so the Sun is 19.5° above the horizon. ◊

**Finalize:** The calculated result agrees with our prediction. When applying Snell's law, it is easy to mix up the index values and to confuse angles-with-the-normal and angles-with-the-surface. Making a sketch and a prediction as we did here helps avoid careless mistakes.

**P25.13** A ray of light strikes a flat block of glass ($n = 1.50$) of thickness 2.00 cm at an angle of 30.0° with the normal. Trace the light beam through the glass, and find the angles of incidence and refraction at each surface.

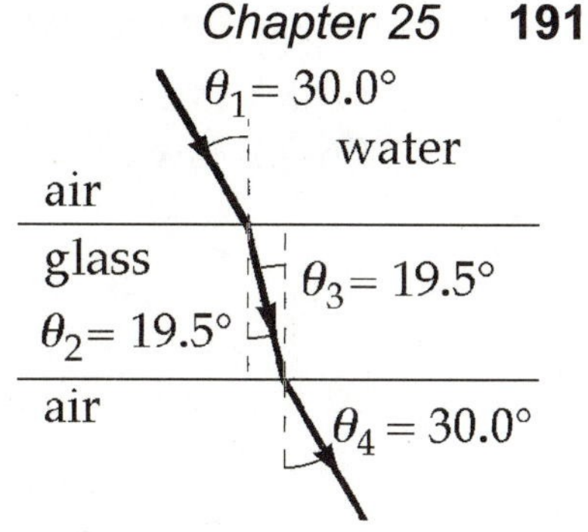


**FIG. P25.13**

**Solution** At entry, $n_1 \sin\theta_1 = n_2 \sin\theta_2$:

$$1.00 \sin 30° = 1.50 \sin\theta_2$$

and $$\theta_2 = \sin^{-1}\left(\frac{0.500}{1.50}\right) = 19.5° \quad \Diamond$$

To do geometry optics, you must remember some geometry. The surfaces of entry and exit are parallel so their normals are parallel. Then angle $\theta_2$ of refraction at entry and the angle $\theta_3$ of incidence at exit are alternate interior angles formed by the ray as transversal cutting parallel lines.

So $$\theta_3 = \theta_2 = 19.5° \quad \Diamond$$

At the exit, $n_2 \sin\theta_3 = n_1 \sin\theta_4$: $1.50 \sin 19.5° = 1.00 \sin\theta_4$

and $$\theta_4 = 30.0° \quad \Diamond$$

The exiting ray in air is parallel to the original ray in air. Thus, a car windshield of uniform thickness will not distort, but shows the driver the actual direction to every object outside.

**P25.25** A triangular glass prism with apex angle $\Phi = 60.0°$ has an index of refraction $n = 1.50$ (Fig. P25.25). What is the smallest angle of incidence $\theta_1$ for which a light ray can emerge from the other side?

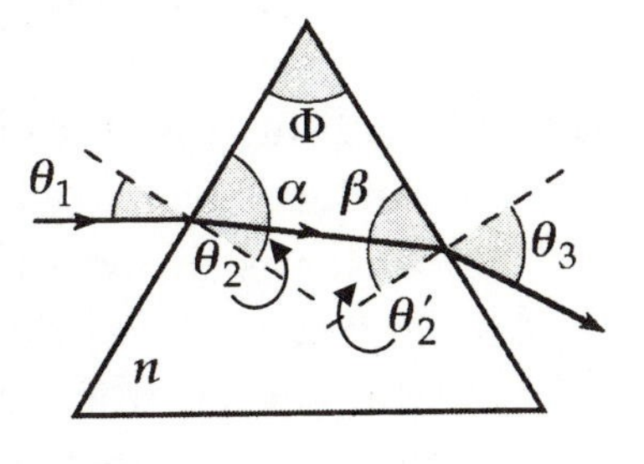


**FIG. P25.25**

**Solution** Call the angles of incidence and refraction, at the surfaces of entry and exit, $\theta_1, \theta_2, \theta_2'$, and $\theta_3$, in order as shown. The apex angle $\Phi$ is the angle between the surfaces of entry and exit. The ray in the glass forms a triangle with these surfaces, in which the interior angles must add to 180°. Thus, with

$$\Phi = 60.0°,$$

$$(90° - \theta_2) + 60° + (90° - \theta_2') = 180°$$

so

$$\theta_2 + \theta_2' = 60.0° \quad (1)$$

*continued on next page*

which is a general rule for light going through prisms. At the first refraction, Snell's law gives

$$\sin\theta_1 = 1.50\sin\theta_2. \tag{2}$$

At the second boundary, we want to almost reach the condition for total internal reflection:

$$1.50\sin\theta_2' = 1.00\sin 90° = 1.00$$

or
$$\theta_2' = \sin^{-1}\left(\frac{1.00}{1.50}\right) = 41.8°.$$

Now by Equation (1) above,

$$\theta_2 = 60.0° - 41.8° = 18.2°$$

while by Equation (2), we find that

$$\theta_1 = \sin^{-1}(1.50\sin 18.2°).$$

So
$$\theta_1 = 27.9°.$$ ◊

Figure 25.23 in the text shows the effect nicely.

**P25.26** A triangular glass prism with apex angle $\Phi$ has index of refraction $n$. (See Fig. P25.26). What is the smallest angle of incidence $\theta_1$ for which a light ray can emerge from the other side?

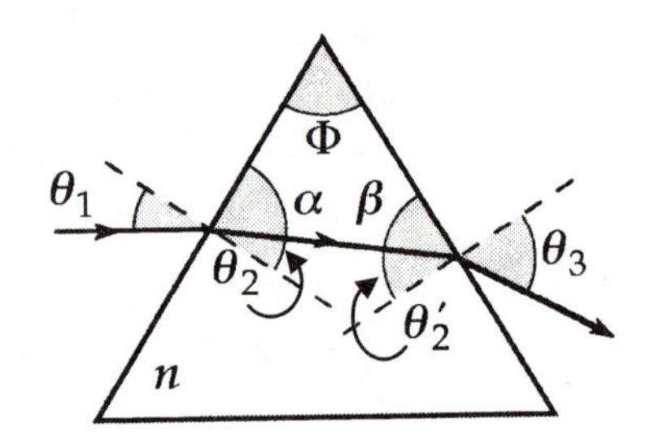

**FIG. P25.26**

**Solution** Call the angles of incidence and refraction, at the surfaces of entry and exit, $\theta_1$, $\theta_2$, $\theta_2'$, and $\theta_3$, in order as shown. The apex angle $\Phi$ is the angle between the surfaces of entry and exit. The ray in the glass forms a triangle with these surfaces, in which the interior angles must add to 180°. Thus,

$$(90° - \theta_2) + \Phi + (90° - \theta_2') = 180°$$

so
$$\theta_2 + \theta_2' = \Phi \tag{1}$$

which is a general rule for light going through prisms. At the first interface between the air and prism,

$$\sin\theta_1 = n\sin\theta_2 \tag{2}$$

*continued on next page*

At the second interface, we want to almost reach the condition for total internal reflection:

$$n \sin\theta_2' = \sin 90° = 1$$

or $$\theta_2' = \sin^{-1}\left(\frac{1}{n}\right)$$

Now by Equation (1) above,

$$\theta_2 = \Phi - \theta_2' = \Phi - \sin^{-1}\left(\frac{1}{n}\right)$$

while by Equation (2), we find that

$$\theta_1 = \sin^{-1}\left[n \sin\left(\Phi - \sin^{-1}\left(\frac{1}{n}\right)\right)\right]$$

Looking at the drawing of a right triangle, remember the identities that

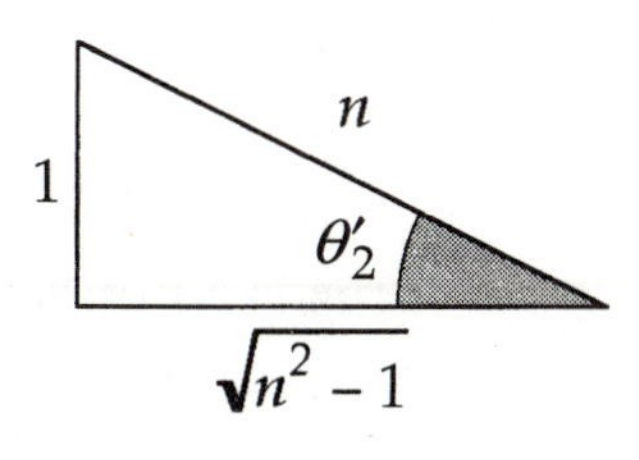


$$\sin(\alpha - \beta) = \sin\alpha\cos\beta - \cos\alpha\sin\beta$$

$$\cos\left[\sin^{-1}\left(\frac{1}{n}\right)\right] = \frac{\sqrt{n^2-1}}{n}$$

$\theta_1$ therefore simplifies to

$$\theta_1 = \sin^{-1}\left[\left(\sqrt{n^2-1}\right)\sin\Phi - \cos\Phi\right].$$ ◊

Note: The numerical solution of Problem 25.25 is shorter than the symbolic solution of Problem 25.26. The answer to Problem 25.26 shows that as $n$ increases and also as $\Phi$ increases, $\theta_1$ increases, to make it easier to produce total internal reflection. If the quantity $\sqrt{n^2-1}\sin\Phi - \cos\Phi$ is greater than 1, light can never emerge from the second surface, for any value of $\theta_1$. If the quantity is negative, $\theta_1$, is negative. Then the critical incident ray must be on the other side of the normal to the first surface from that shown, but the equation is still true.

**P25.27** The index of refraction for violet light in silica flint glass is 1.66, and that for red light is 1.62. What is the angular dispersion of visible light passing through a prism of apex angle 60.0° if the angle of incidence is 50.0°? (See Fig. P25.27.)

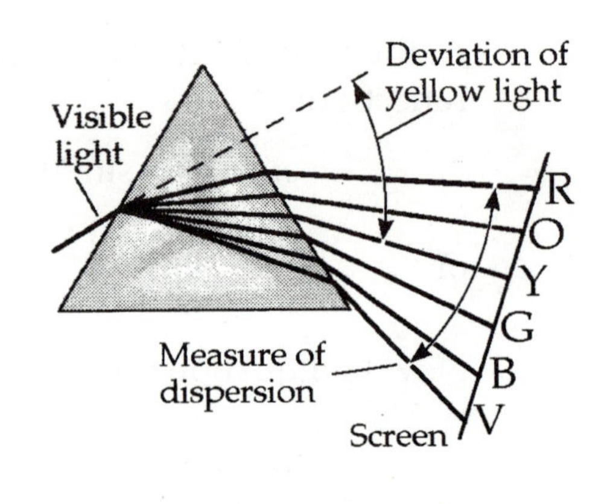


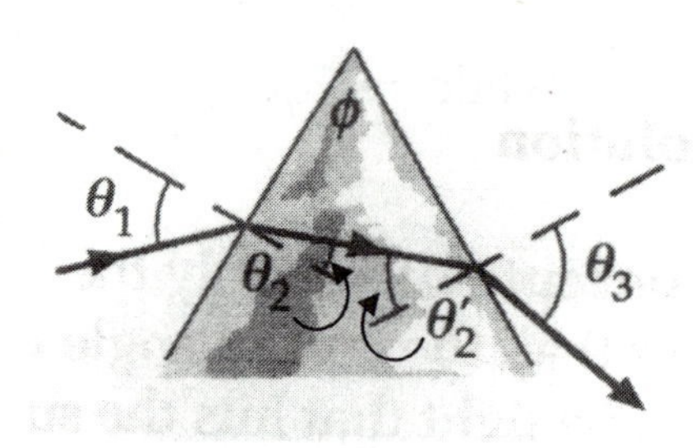


FIG. P25.27

**Solution** Call the angles of incidence and refraction, at the surfaces of entry and exit, $\theta_1$, $\theta_2$, $\theta_2'$, and $\theta_3$, in the order as shown. The apex angle ($\phi = 60.0°$) is the angle between the surfaces of entry and exit. The ray in the glass forms a triangle with these surfaces, in which the interior angles must add to 180°.

Thus, $$(90°-\theta_2)+\phi+(90°-\theta_2') = 180°$$

and $$\theta_2' = \phi - \theta_2.$$

This is a general rule for light going through prisms.

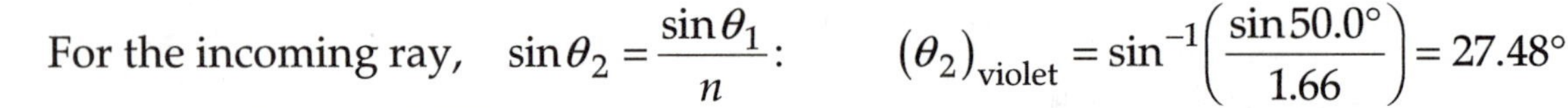

For the incoming ray, $\sin\theta_2 = \dfrac{\sin\theta_1}{n}$: $$(\theta_2)_{\text{violet}} = \sin^{-1}\left(\frac{\sin 50.0°}{1.66}\right) = 27.48°$$

$$(\theta_2)_{\text{red}} = \sin^{-1}\left(\frac{\sin 50.0°}{1.62}\right) = 28.22°.$$

For the outgoing ray, $\theta_2' = 60° - \theta_2$ and $\sin\theta_3 = n\sin\theta_2'$

$$(\theta_3)_{\text{violet}} = \sin^{-1}[1.66\sin 32.52°] = 63.17°$$

$$(\theta_3)_{\text{red}} = \sin^{-1}[1.62\sin 31.78°] = 58.56°.$$

The dispersion is the difference between these two angles:

$$\Delta\theta_3 = 63.17° - 58.56° = 4.61°.$$ ◊

**P25.31** Consider a common mirage formed by super-heated air just above a roadway. A truck driver whose eyes are 2.00 m above the road, where $n = 1.000\,3$, looks forward. She perceives the illusion of a patch of water ahead on the road, where her line of sight makes an angle of 1.20° below the horizontal. Find the index of refraction of the air just above the road surface. (*Suggestion:* Treat this problem as one about total internal reflection.)

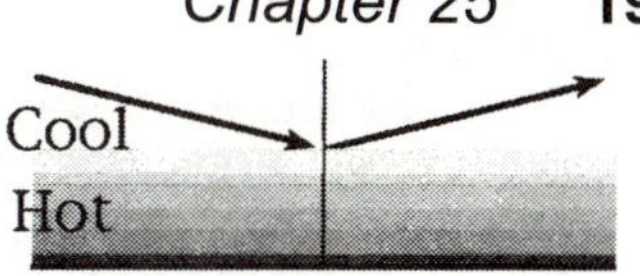


**FIG. P25.31**

**Solution** Think of the air as in two discrete layers, the first medium being cooler air with $n_1 = 1.000\,3$ and the second medium being hot air with a lower index, which reflects light from the sky by total internal reflection.

Use $n_1 \sin\theta_1 \geq n_2 \sin 90°$: $1.000\,3 \sin 88.8° \geq n_2$ so $n_2 \leq 1.000\,08$. ◊

**P25.35** A laser beam strikes one end of a slab of material, as shown in Figure P25.35. The index of refraction of the slab is 1.48. Determine the number of internal reflections of the beam before it emerges from the opposite end of the slab.

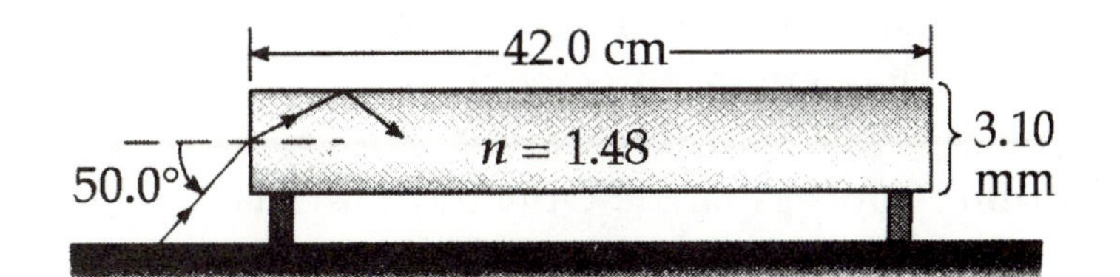


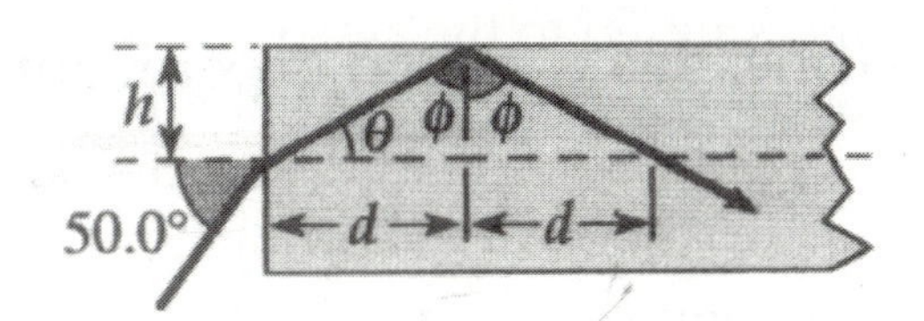


**FIG. P25.35**

**Solution** As the beam enters the end of the slab, the angle of refraction is found from Snell's law:

$$\sin\theta = \left(\frac{n_{\text{air}}}{n_{\text{slab}}}\right)\sin 50.0° = \left(\frac{1.00}{1.48}\right)\sin 50.0° = 0.518 \quad \text{and} \quad \theta = 31.2°.$$

Note that the two normal lines are perpendicular to each other. Using the right triangle having these lines as two of its sides, observe that the angle of incidence at the top of the slab is $\phi = 90.0° - \theta = 58.8°$. The critical angle as the light tries to go from the slab back into air is

$$\theta_c = \sin^{-1}\left(\frac{n_{\text{air}}}{n_{\text{slab}}}\right) = \sin^{-1}\left(\frac{1.00}{1.48}\right) = 42.5°.$$

*continued on next page*

Since $\phi > \theta_c$, total internal reflection will indeed occur at the top and bottom of the slab. The distance the beam travels down the length of the slab for each reflection is $2d$, where $d$ is the base of the right triangle shown in the sketch. Given that the altitude of the triangle, $h$, is one half the thickness of the slab,

$$d = \frac{h}{\tan\theta} = \frac{(3.10 \text{ mm})/2}{\tan\theta} \quad \text{and} \quad 2d = \frac{3.10 \times 10^{-1} \text{ cm}}{\tan 31.2°} = 5.12 \times 10^{-1} \text{ cm}.$$

The number of internal reflections made before reaching the opposite end of the slab is then

$$N = \frac{\text{length of slab}}{2d} = \frac{42.0 \text{ cm}}{5.12 \times 10^{-1} \text{ cm}} = 82 \text{ reflections}. \quad ◊$$

**P25.41** A small light fixture is on the bottom of a swimming pool, 1.00 m below the surface. The light emerging from the water forms a circle on the still water surface. What is the diameter of this circle?

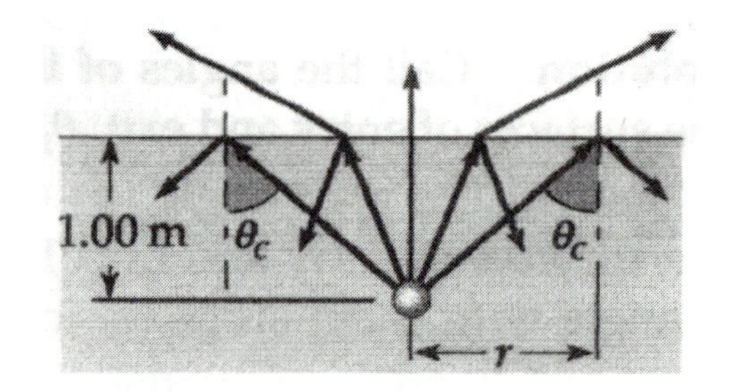


FIG. P25.41

**Solution** **Conceptualize:** Only the light that is directed upwards and hits the water's surface at less than the critical angle will be transmitted to the air so that someone outside can see it. The light that hits the surface farther from the center at an angle greater than $\theta_c$ will be totally reflected within the water, unable to be seen from the outside. From the diagram to the right, the diameter of this circle of light appears to be about 2 m.

**Categorize:** We can apply Snell's law to find the critical angle, and the diameter can then be found from the geometry.

**Analyze:** The critical angle is found when the refracted ray just grazes the surface ($\theta_2 = 90°$). The index of refraction of water is $n_1 = 1.333$, and $n_2 = 1.00$ for air, so

$n_1 \sin\theta_c = n_2 \sin 90°$ gives $\quad \theta_c = \sin^{-1}\left(\frac{1}{1.333}\right) = \sin^{-1}(0.750) = 48.6°$

The radius then satisfies $\quad \tan\theta_c = \frac{r}{1.00 \text{ m}}.$

So the diameter is $\quad d = 2r = 2(1.00 \text{ m})(\tan 48.6°) = 2.27 \text{ m}. \quad ◊$

*continued on next page*

**Finalize:** Only the light rays within a 97.2° cone above the lamp escape the water and can be seen by an outside observer (Note: this angle does not depend on the depth of the light source). The path of a light ray is always reversible, so if a person were located beneath the water, they could see the whole hemisphere above the water surface within this cone; this is a good experiment to try the next time you go swimming!

**P25.45** A hiker stands on an isolated mountain peak near sunset and observes a rainbow caused by water droplets in the air 8.00 km away. The valley is 2.00 km below the mountain peak and entirely flat. What fraction of the complete circular arc of the rainbow is visible to the hiker? (See Fig. 25.17.)

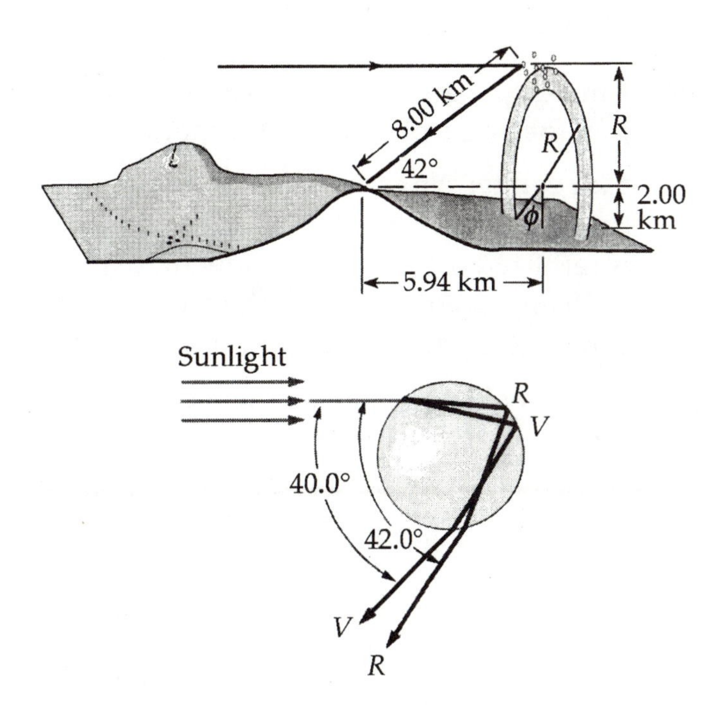


**FIG. P25.45**

**Solution** Horizontal light rays from the setting Sun pass above the hiker. The light rays are twice refracted and once reflected, as in Figure 25.17. The most intense light reaching the hiker, that which represents the visible rainbow, is located between angles of 40.0° and 42.0° from the hiker's shadow. The hiker sees a greater percentage of the violet inner edge, so we consider the red outer edge. The radius $R$ of the circle of droplets is

$$R = (8.00 \text{ km})\sin 42.0° = 5.35 \text{ km}.$$

Then the angle $\phi$ between the vertical and the radius where the bow touches the ground, is given by

$$\cos\phi = \frac{2.00 \text{ km}}{R} = \frac{2.00 \text{ km}}{5.35 \text{ km}} = 0.374 \qquad \text{or} \qquad \phi = 68.1°.$$

The angle filled by the visible bow is $360° - 2(68.1°) = 224°$, so the visible bow is

$$\frac{224°}{360°} = 62.2\% \text{ of a circle.} \qquad \Diamond$$

*continued on next page*

This striking view motivated Charles Wilson's 1906 invention of the cloud chamber, a standard tool of nuclear physics. Look for a full circle of color around your shadow when you fly in an airplane. The effect is mentioned in the Bible, Ezekiel 1:29.

**P25.51** The light beam in Figure P25.51 strikes surface 2 at the critical angle. Determine the angle of incidence $\theta_1$.

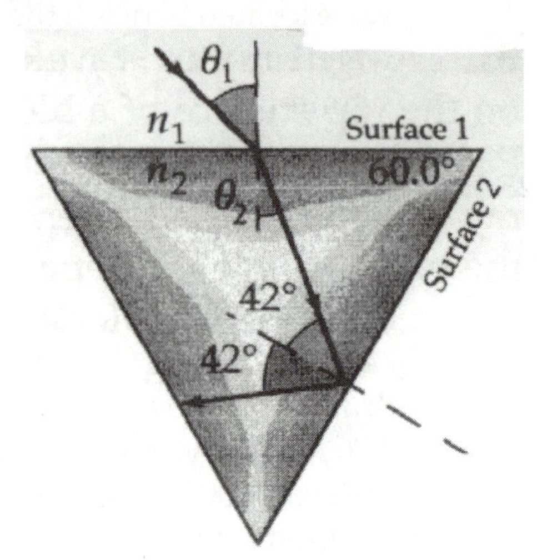


FIG. P25.51

**Solution** **Conceptualize:** From the diagram it appears that the angle of incidence is about 40°.

**Categorize:** We can find $\theta_1$ by applying Snell's law at the first interface where the light is refracted. At surface 2, knowing that the 42.0° angle of reflection is the critical angle, we can work backwards to find $\theta_1$.

**Analyze:** Define $n_1$ to be the index of refraction of the surrounding medium and $n_2$ to be that for the prism material. We can use the critical angle of 42.0° to find the ratio $\frac{n_2}{n_1}$:

$$n_2 \sin 42.0° = n_1 \sin 90.0°$$

So, $$\frac{n_2}{n_1} = \frac{1}{\sin 42.0°} = 1.49$$

Call the angle of refraction $\theta_2$ at the surface 1. The ray inside the prism forms a triangle with surfaces 1 and 2, so the sum of the interior angles of this triangle must be 180°.

Thus, $$(90.0° - \theta_2) + 60.0° + (90.0° - 42.0°) = 180°$$

Therefore, $$\theta_2 = 18.0°$$

Applying Snell's law at surface 1,

$$n_1 \sin \theta_1 = n_2 \sin 18.0°$$

$$\sin \theta_1 = \left(\frac{n_2}{n_1}\right) \sin \theta_2 = 1.49 \sin 18.0° \qquad \theta_1 = 27.5°$$ ◊

**Finalize:** The result is a bit less than the 40.0° we expected, but this is probably because the figure is not drawn to scale. This problem was a bit tricky because it required four key concepts (refraction, reflection, critical angle, and geometry) in order to find the solution. One practical extension of this problem is to consider what would happen to the exiting light if the angle of incidence were varied slightly. Would all the light still be reflected off surface 2, or would some light be refracted and pass through this second surface? See Figure 25.23 in the textbook.

**P25.53** A light ray of wavelength 589 nm is incident at an angle $\theta$ on the top surface of a block of polystyrene as shown in Figure P25.53.

(a) Find the maximum value of $\theta$ for which the refracted ray undergoes total internal reflection at the left vertical face of the block.

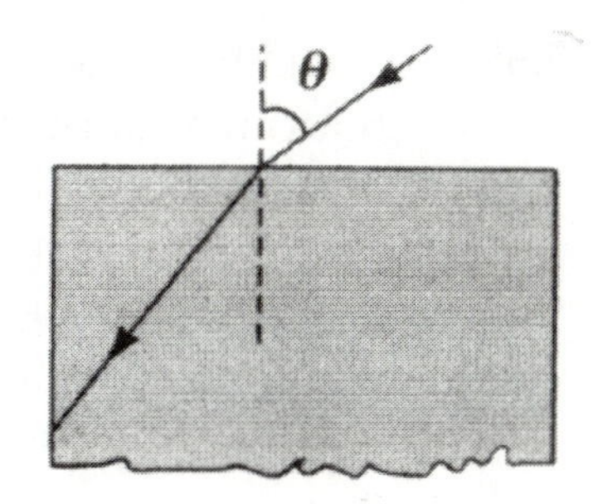


Repeat the calculation for the case in which the polystyrene block is immersed in

(b) water and

(c) carbon disulfide.

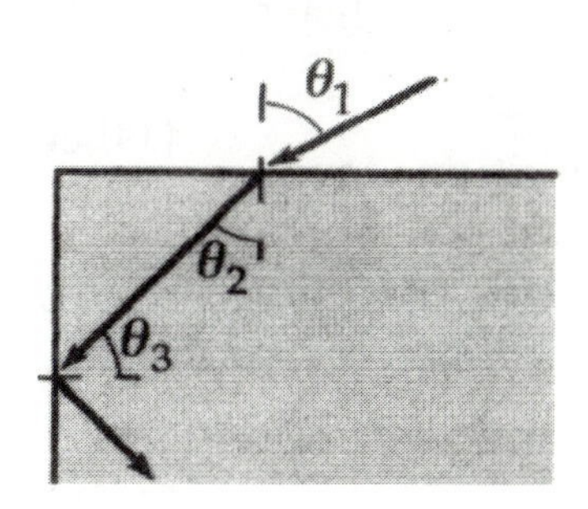


**FIG. P25.53**

**Solution** (a) The index of refraction (for 589 nm light) of each material is listed below:

| | |
|---|---|
| Air: | 1.00 |
| Water: | 1.33 |
| Polystyrene: | 1.49 |
| Carbon disulfide: | 1.63 |

For polystyrene **surrounded by air,**

the critical angle for total internal reflection is $\theta_3 = \sin^{-1}\left(\frac{1}{1.49}\right) = 42.2°$

and then from geometry, $\theta_2 = 90.0° - \theta_3 = 47.8°$

From Snell's law, $\sin\theta_1 = 1.49 \sin 47.8°$

This has no solution; thus, the real maximum value for $\theta_1$ is 90.0°.

Total internal reflection always occurs. ◊

(b) For polystyrene **surrounded by water,** we have $\theta_3 = \sin^{-1}\left(\frac{1.33}{1.49}\right) = 63.2°$

and $\theta_2 = 26.8°$

From Snell's law, $n_1 \sin\theta_1 = n_2 \sin\theta_2$: $1.33 \sin\theta_1 = 1.49 \sin 26.8°$

and $\theta_1 = 30.3°$ 

(c) **This is not possible** since the beam is initially traveling in a medium of lower index of refraction. ◊

**P25.55** A shallow glass dish is 4.00 cm wide at the bottom as shown in Figure P25.55. When an observer's eye is placed as shown, the observer sees the edge of the bottom of the empty dish. When this dish is filled with water, the observer sees the center of the bottom of the dish. Find the height of the dish.

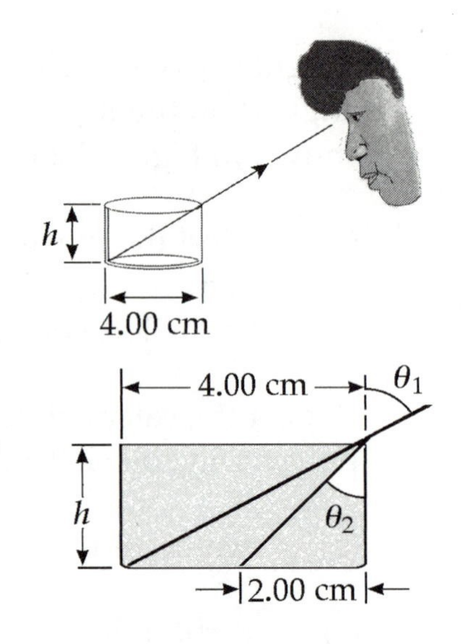


**FIG. P25.55**

**Solution** $\tan\theta_1 = \dfrac{4.00 \text{ cm}}{h}$ and $\tan\theta_2 = \dfrac{2.00 \text{ cm}}{h}$

$$\tan^2\theta_1 = (2.00\tan\theta_2)^2 = 4.00\tan^2\theta_2$$

$$\frac{\sin^2\theta_1}{1-\sin^2\theta_1} = 4.00\left(\frac{\sin^2\theta_2}{1-\sin^2\theta_2}\right) \quad (1)$$

Snell's law in this case is: $n_1\sin\theta_1 = n_2\sin\theta_2$

$$\sin\theta_1 = 1.333\sin\theta_2$$

Squaring both sides, $\sin^2\theta_1 = 1.777\sin^2\theta_2$ (2)

Substituting [2] into [1] yields $\dfrac{1.777\sin^2\theta_2}{1-1.777\sin^2\theta_2} = 4.00\left(\dfrac{\sin^2\theta_2}{1-\sin^2\theta_2}\right)$

Defining $x = \sin^2\theta_2$

and solving for $x$, $\dfrac{0.444}{1-1.777x} = \dfrac{1}{1-x}$

$$0.444 - 0.444x = 1 - 1.777x$$

$$x = 0.417$$

From $x$ we can solve for $\theta_2$: $\theta_2 = \sin^{-1}\sqrt{0.417} = 40.2°$

and $h = \dfrac{2.00 \text{ cm}}{\tan\theta_2} = \dfrac{2.00 \text{ cm}}{\tan 40.2°} = 2.36 \text{ cm}$ ◊

# Chapter 26

# Image Formation by Mirrors and Lenses

## NOTES FROM SELECTED CHAPTER SECTIONS

### Section 26.1 Images Formed by Flat Mirrors

An image formed by a flat mirror has the following properties:

- The image is as far behind the mirror as the object is in front.
- The image is actual-size, virtual, and upright. (By upright, we mean that, if the object arrow points upward, so does the image arrow.)
- The image has an apparent right-left reversal.

### Section 26.2 Images Formed by Spherical Mirrors

A spherical mirror is a reflecting surface, that has the shape of a segment of a sphere. A concave mirror is one that reflects light from the inner concave surface; a convex mirror reflects light from the outer convex surface. When using the mirror equation, Eq. 26.6, carefully adhere to the sign conventions stated in the Suggestions, Skills, and Strategies section.

**Real images** are formed at a point when reflected light actually passes through the image point. **Virtual images** are formed at a point when light rays appear to diverge from the image point.

The point of intersection of any two of the following rays in a ray diagram for mirrors locates the image:

- The first ray is drawn from the top of the object parallel to the principal axis and is reflected back through the focal point, $F$.

- The second ray is drawn from the top of the object through the focal point and is reflected parallel to the axis.

- The third ray is drawn from the top of the object through the center of curvature, C, and is reflected back on itself.

- The fourth ray is drawn from the top of the object to the center of the mirror's surface and reflects on the other side of the principal axis, with an angle of reflection equal to its angle of incidence.

## Section 26.4 Thin Lenses

Images formed by thin lenses can be located and described by pictorial representations called ray diagrams.

To locate the position of an image formed by a converging lens, draw the following three rays from the top of the object:

- Ray 1 is drawn parallel to the principal axis. After being refracted by the lens, this ray passes through the focal point on the back side of the lens.

- Ray 2 is drawn through the center of the lens. This ray continues in a straight line.

- Ray 3 is drawn through the focal point on the front side of the lens (or as if coming from the focal point if $p < f$) and emerges from the lens parallel to the principal axis.

In the case of a diverging lens, the following three rays are drawn from the top of the object:

- Ray 1 is drawn parallel to the principal axis, and after refraction emerges directed away from the focal point on the front side of the lens.

- Ray 2 is drawn through the center of the lens and continues in a straight line.

- Ray 3 is drawn in the direction toward the focal point on the back side of the lens and emerges from the lens parallel to the principle axis.

Note that any two rays can be used to locate the object; the third ray is a useful check on the accuracy of your diagram.

## EQUATIONS AND CONCEPTS

The **sign conventions** appropriate for the form of the equations stated here are summarized in SUGGESTIONS, SKILLS AND STRATEGIES.

The **mirror equation** is used to locate the position of an image formed by reflection of paraxial rays. The focal point of a spherical mirror is located midway between the center of curvature and the center of the mirror.

$$\frac{1}{p}+\frac{1}{q}=\frac{1}{f} \qquad (26.6)$$

$$f=\frac{R}{2} \qquad (26.5)$$

The **lateral magnification of a spherical mirror** can be stated either as a ratio of image size to object size or in terms of the ratio of image distance to object distance.

$$M=\frac{h'}{h}=-\frac{q}{p} \qquad (26.2)$$

A magnified image of an object can be formed by a **single spherical refracting surface** of radius $R$, which separates two media whose indices of refraction are $n_1$ and $n_2$.

$$\frac{n_1}{p}+\frac{n_2}{q}=\frac{n_2-n_1}{R} \qquad (26.8)$$

$$M=-\frac{n_1 q}{n_2 p} \qquad (26.9)$$

A special case is that of the virtual image formed by a **planar refracting surface** ($R=\infty$).

$$\frac{n_1}{p}=-\frac{n_2}{q}$$

$$q=-\frac{n_2}{n_1}p \qquad (26.10)$$

The **focal length of a thin lens** is determined by the characteristic properties of the lens (index of refraction $n$ and radii of curvature $R_1$ and $R_2$).

$$\frac{1}{f}=(n-1)\left(\frac{1}{R_1}-\frac{1}{R_2}\right) \qquad (26.13)$$

The **lateral magnification**, $M$, will be negative when the image is inverted.

$$M=\frac{h'}{h}=-\frac{q}{p} \qquad (26.11)$$

The **thin lens equation** is identical to the mirror equation and can be used with both converging (positive $f$) and diverging (negative $f$) lenses.

$$\frac{1}{p}+\frac{1}{q}=\frac{1}{f} \qquad (26.12)$$

# SUGGESTIONS, SKILLS, AND STRATEGIES

A major portion of this chapter is devoted to the development and presentation of equations, which can be used to determine the location and nature of images formed by various optical components acting either singly or in combination. It is essential that these equations be used with the correct algebraic sign associated with each quantity involved. You must understand clearly the sign conventions for mirrors, refracting surfaces, and lenses. The following discussion represents a review of these sign conventions.

## Sign Conventions for Mirrors

Equations: $\frac{1}{p}+\frac{1}{q}=\frac{1}{f}=\frac{2}{R}$ $\qquad M=\frac{h'}{h}=-\frac{q}{p}$

The front side of the mirror is the region on which light rays are incident and reflected.

$p$ is + if the object is in front of the mirror (real object).
$p$ is – if the object is in back of the mirror (virtual object).

$q$ is + if the image is in front of the mirror (real image).
$q$ is – if the image is in back of the mirror (virtual image).

Both $f$ and $R$ are + if the center of curvature is in front of the mirror (concave mirror).
Both $f$ and $R$ are – if the center of curvature is in back of the mirror (convex mirror).

If $M$ is positive, the image is upright.
If $M$ is negative, the image is inverted.

You should check the sign conventions as stated against the situations described in Figure 26.1. You may also wish to check the image location in each diagram by drawing in ray number two, as described in Section 26.2.

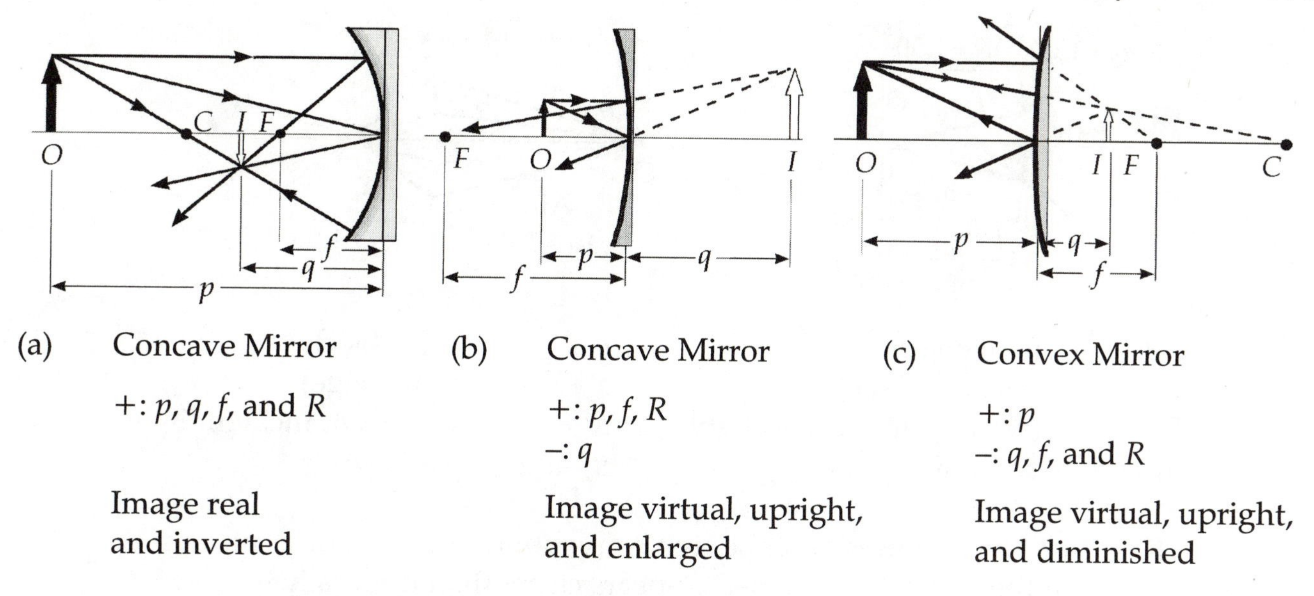


**FIG. 26.1** Figures describing sign conventions for mirrors.

## Sign Conventions for Refracting Surfaces

Equations: $$\frac{n_1}{p}+\frac{n_2}{q}=\frac{n_2-n_1}{R} \qquad M=\frac{h'}{h}=-\frac{n_1 q}{n_2 p}$$

In the following table, the front side of the surface is the side from which the light is incident.

| | |
|---|---|
| $p$ | is + if the object is in front of the surface (real object). |
| $p$ | is – if the object is in back of the surface (virtual object). |
| $q$ | is + if the image is in back of the surface (real image). |
| $q$ | is – if the image is in front of the surface (virtual image). |
| $R$ | is + if the center of curvature is in back of the surface. |
| $R$ | is – if the center of curvature is in front of the surface. |
| $n_1$ | refers to the index of refraction of the medium on the side of the interface from which the light comes. |
| $n_2$ | is the index of refraction of the medium into which the light is transmitted after refraction at the interface. |

Review the above sign conventions for the situations shown in Figure 26.2.

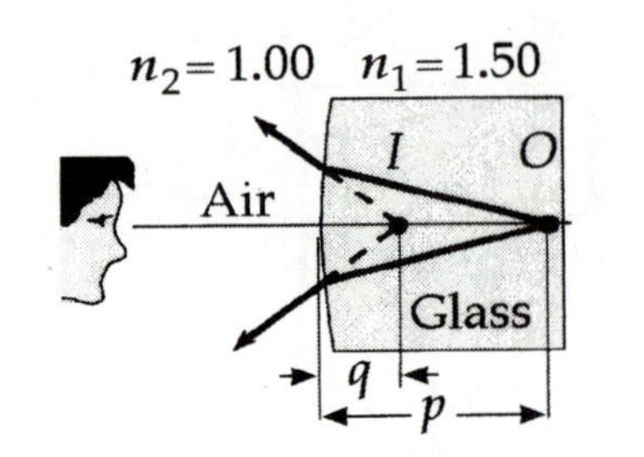


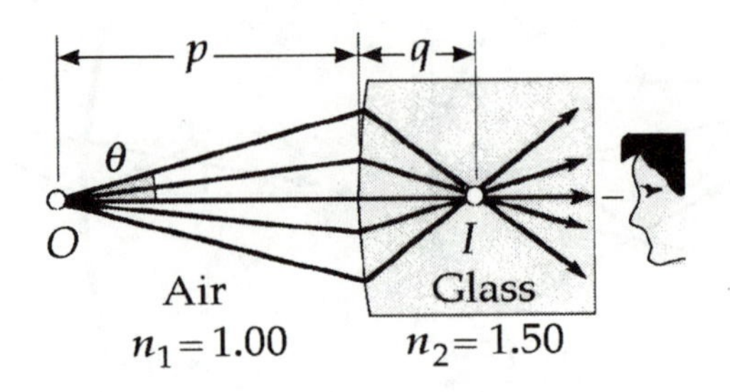


| | | | |
|---|---|---|---|
| $p$ + | (real object) | $p$ + | (real object) |
| $q$ – | (virtual image) | $q$ + | (real image) |
| $R$ – | (concave to incident light) | $R$ + | (convex to incident light) |
| $n_1$ and $n_2$ as shown | | $n_1$ and $n_2$ as shown | |

**FIG. 26.2** These figures describe sign conventions for refracting surfaces. In the first case, the object appears closer than it actually is. In the second case, an object beyond the glass appears to be in the glass.

## Sign Conventions for Thin Lenses

Equations: $$\frac{1}{p}+\frac{1}{q}=\frac{1}{f}=(n-1)\left(\frac{1}{R_1}-\frac{1}{R_2}\right) \qquad M=\frac{h'}{h}=-\frac{q}{p}$$

In the following table, the front of the lens is the side from which the light is incident.

| | |
|---|---|
| $p$ | is + if the object is in front of the lens. |
| $p$ | is – if the object is in back of the lens. |
| $q$ | is + if the image is in back of the lens. |
| $q$ | is – if the image is in front of the lens. |
| $f$ | is + if the lens is thickest at the center. |
| $f$ | is – if the lens is thickest at the edges. |

$R_1$ and $R_2$ are + if the center of curvature is in back of the lens.
$R_1$ and $R_2$ are – if the center of curvature is in front of the lens.

The sign conventions for thin lenses are illustrated by the examples shown in Figure 26.3.

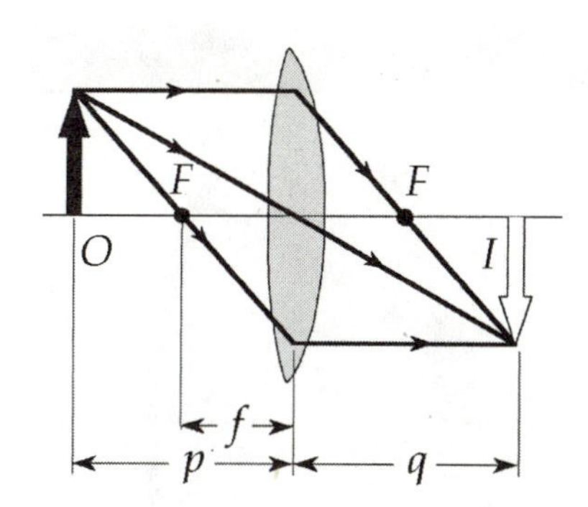


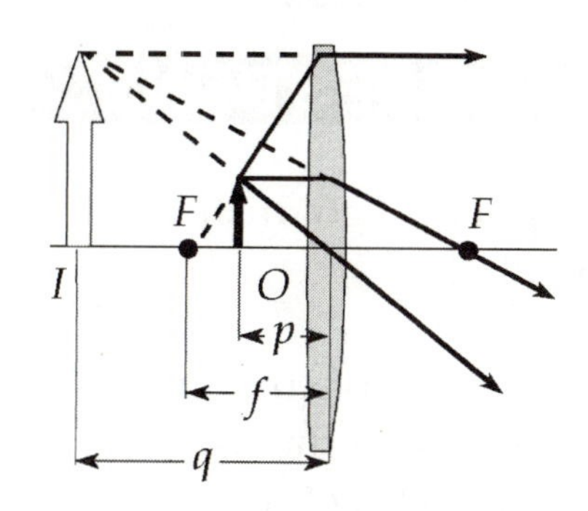


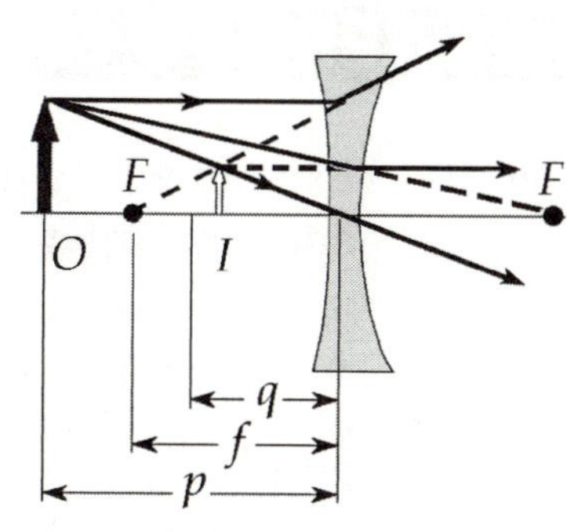


(a) Double-Convex Lens (Converging)

+: $p, q, f, R_1$
–: $R_2$

Image real and inverted

(b) Double-Convex Lens (Converging)

+: $p, R_1, f,$
–: $q, R_2$

Image virtual, upright, and enlarged

(c) Double-Concave Lens (Diverging)

+: $p, R_2$
–: $q, R_1, f$

Image virtual, upright, and diminished

**FIG. 26.3** Figures describing the sign conventions for various thin lenses.

## REVIEW CHECKLIST

✓ Identify the following properties that characterize an image formed by a lens or mirror system with respect to an object: position, magnification, orientation (i.e. inverted or upright) and whether real or virtual.

✓ Understand the relationship of the algebraic signs associated with calculated quantities to the nature of the image and object: real or virtual, upright or inverted.

✓ Calculate the location of the image of a specified object as formed by a plane mirror, spherical mirror, plane refracting surface, spherical refracting surface, thin lens, or a combination of two or more of these devices. Determine the magnification and character of the image in each case.

✓ Construct ray diagrams to determine the location and nature of the image of a given object when the geometrical characteristics of the optical device (lens or mirror) are known.

## ANSWERS TO SELECTED QUESTIONS

**Q26.10** Explain why a fish in a spherical goldfish bowl appears larger than it really is.

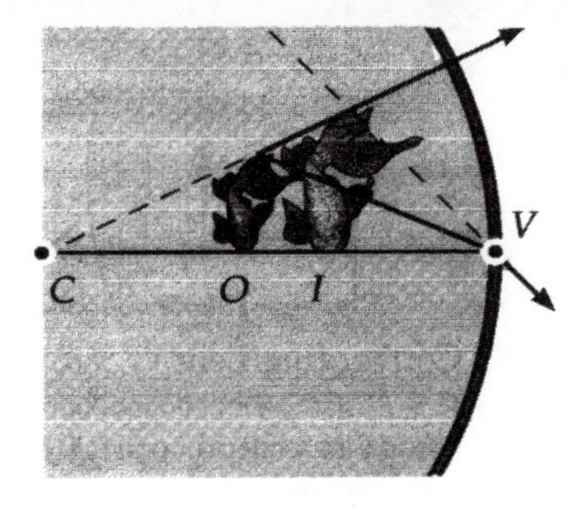


FIG. Q26.10

**Answer** As in the diagram, let the center of curvature $C$ of the fishbowl and the bottom of the fish define the optical axis, intersecting the fishbowl at vertex $V$. A ray from the top of the fish that reaches the bowl along a radial line through $C$ has angle of incidence zero and angle of refraction zero. A ray from the top of the fish to $V$ is refracted to bend away from the normal. Its extension back inside the fishbowl determines the location of the image and the characteristics of the image. It is upright, virtual, and enlarged.

**Q26.12** Lenses used in eyeglasses, whether converging or diverging, are always designed so that the middle of the lens curves away from the eye, like the center lenses of Figure 26.20 a and b. Why?

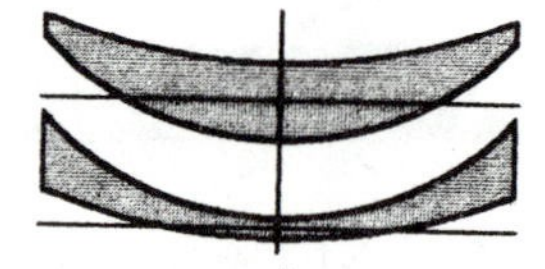

FIG. Q26.12

**Answer** With the meniscus design, when you direct your gaze near the outer circumference of the lens you receive a ray that has passed through glass with more nearly parallel surfaces of entry and exit. Thus, the lens minimally distorts the direction to the object you are looking at. If you wear glasses, you can demonstrate this by turning them around and looking through them the wrong way, maximizing the distortion.

## SOLUTIONS TO SELECTED PROBLEMS

**P26.3** Determine the minimum height of a vertical flat mirror in which a person 5′10″ in height can see his or her full image. (A ray diagram would be helpful.)

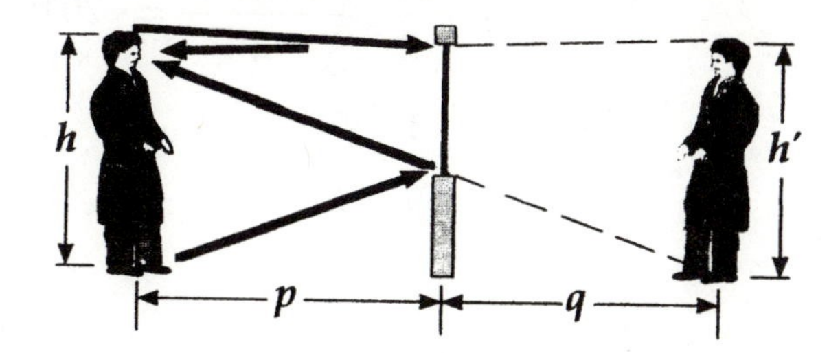


**FIG. P26.3**

**Solution** The flatness of the mirror is described by $R=\infty$, $f=\infty$, and $\frac{1}{f}=0$. By our general mirror equation,

$$\frac{1}{p}+\frac{1}{q}=\frac{1}{f} \qquad \text{or} \qquad q=-p.$$

Thus, the image is as far behind the mirror as the person is in front. The magnification is then

$$M=\frac{-q}{p}=1=\frac{h'}{h} \qquad \text{so} \qquad h'=h=70.0 \text{ inches}.$$

The required height of the mirror is defined by the triangle from the person's eye to the top and bottom of the image as shown. From the geometry of the triangle, we see that the mirror height must be:

$$h'\left(\frac{p}{p-q}\right)=h'\left(\frac{p}{2p}\right)=\frac{h'}{2}.$$

Thus, the mirror must be at least 35.0 inches high. ◊

**P26.9** A spherical convex mirror (Fig. P26.9) has a radius of curvature with a magnitude of 40.0 cm. Determine the position of the virtual image and the magnification for object distances of

(a) 30.0 cm and

(b) 60.0 cm.

(c) Are the images upright or inverted?

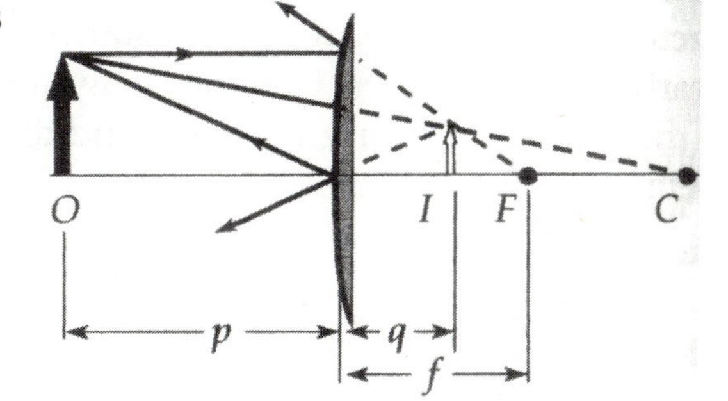


**FIG. P26.9(a)**

*continued on next page*

**Solution** The convex mirror is described by

$$f = \frac{R}{2} = \frac{-40.0 \text{ cm}}{2} = -20.0 \text{ cm}$$

and $\frac{1}{p} + \frac{1}{q} = \frac{1}{f}$.

(a) $\frac{1}{30.0 \text{ cm}} + \frac{1}{q} = \frac{1}{-20.0 \text{ cm}}$: $q = -12.0 \text{ cm}$ ◊

$$M = -\frac{q}{p} = -\left(\frac{-12.0 \text{ cm}}{30.0 \text{ cm}}\right) = +0.400$$ ◊

(b) $\frac{1}{60.0 \text{ cm}} + \frac{1}{q} = \frac{1}{-20.0 \text{ cm}}$: $q = -15.0 \text{ cm}$ ◊

$$M = \frac{-q}{p} = -\left(\frac{-15.0 \text{ cm}}{60.0 \text{ cm}}\right) = +0.250$$ ◊

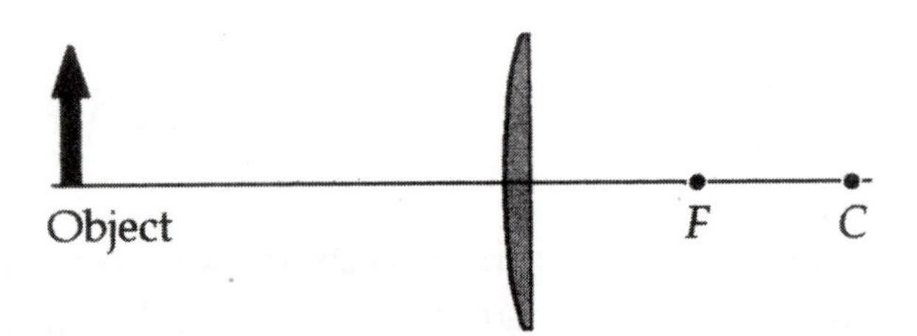


**FIG. P26.9(b)**

(c) As shown by FigureP26.9(a) on the previous page, the image for part (a) is behind the mirror, upright, virtual, and diminished. ◊

The principal ray diagram is an essential complement to the numerical description of the image. Add rays on Figure P26.9(b) for a 60-cm object distance.

Use your diagram to confirm that the image is behind the mirror, upright, virtual and diminished. ◊

**P26.11** A concave mirror has a radius of curvature of 60.0 cm. Calculate the image position and magnification of an object placed in front of the mirror at distance of

(a) 90.0 cm and

(b) 20.0 cm.

(c) Draw ray diagrams to obtain the image characteristics in each case.

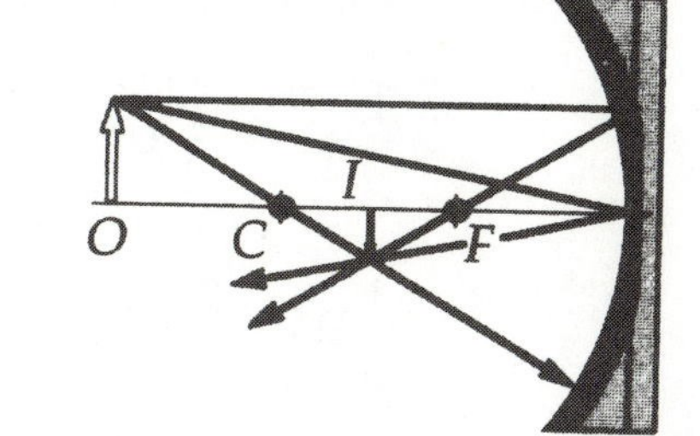


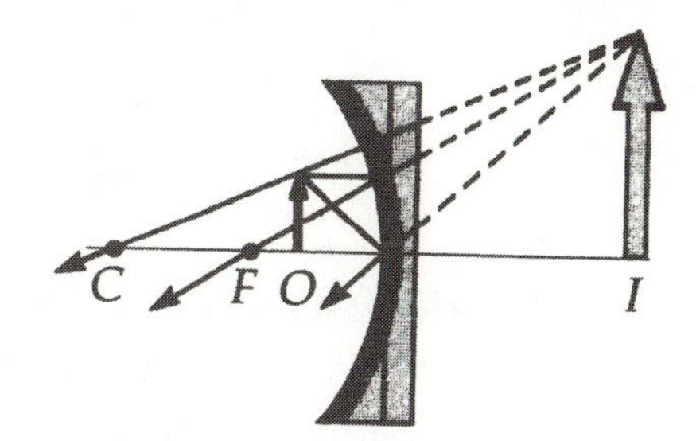


**FIG. P26.11**

**Solution** **Conceptualize:** It is always a good idea to first draw a ray diagram for any optics problem. This gives a qualitative sense of how the image appears relative to the object. From the ray diagrams above, we see that when the object is 90 cm from the mirror, the image will be real, inverted, diminished, and located about 45 cm in front of the mirror, midway between the center of curvature and the focal point. When the object is 20 cm from the mirror, the image will be virtual, upright, magnified, and located about 50 cm behind the mirror.

**Categorize:** The mirror equation can be used to find precise values.

**Analyze:**

(a) The mirror equation is applied using the sign conventions listed in **Suggestions, Skills, and Strategies**:

$$\frac{1}{p}+\frac{1}{q}=\frac{2}{R} \text{ or } \frac{1}{90.0 \text{ cm}}+\frac{1}{q}=\frac{2}{60.0 \text{ cm}}$$

$$\frac{1}{q}=\frac{2}{60.0 \text{ cm}}-\frac{1}{90.0 \text{ cm}}=0.022\,2 \text{ cm}^{-1}$$

$q = 45.0$ cm (real, in front of the mirror) ◊

$$M=\frac{-q}{p}=-\frac{45.0 \text{ cm}}{90.0 \text{ cm}}=-0.500 \text{ (inverted)}$$ ◊

*continued on next page*

(b) We again use the mirror equation:

$$\frac{1}{p}+\frac{1}{q}=\frac{2}{R} \text{ or } \frac{1}{20.0 \text{ cm}}+\frac{1}{q}=\frac{2}{60.0 \text{ cm}}$$

$$\frac{1}{q}=\frac{2}{60.0 \text{ cm}}-\frac{1}{20.0 \text{ cm}}=-0.0167 \text{ cm}^{-1}$$

$q=-60.0$ cm (virtual, behind the mirror) ◊

$$M=-\frac{q}{p}=\frac{-60.0 \text{ cm}}{20.0 \text{ cm}}=3.00 \text{ (upright)}$$ ◊

**Finalize:** The calculated image characteristics agree well with our predictions. It is easy to miss a minus sign or to make a computational mistake when using the mirror-lens equation, so the qualitative values obtained from the ray diagrams are useful for a check on the reasonableness of the calculated values.

**P26.17** A spherical mirror is to be used to form, on a screen located 5.00 m from the object, an image five times the size of the object.

(a) Describe the type of mirror required.

(b) Where should the mirror be positioned relative to the object?

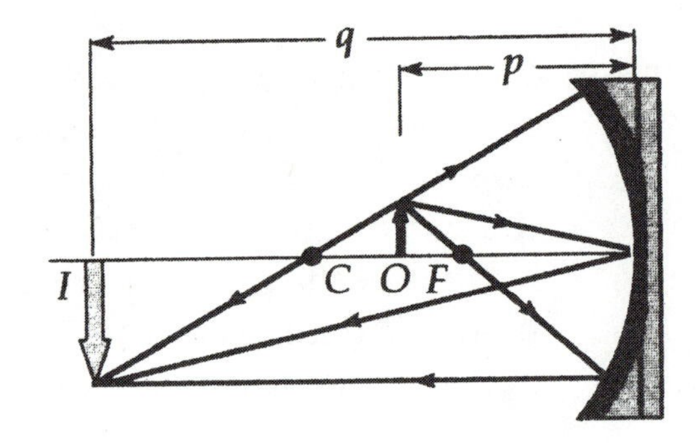


**FIG. P26.17**

**Solution** We are given that $q=(p+5.00 \text{ m})$, $M=-\frac{q}{p}$ and $|M|=5.00$

Since the image must be real, $q$ must be positive, and $M=-5.00$ or $q=5.00p$

(b) Solving for the distance of the mirror from the object, $p+5.00 \text{ m}=5.00p$

$p=1.25$ m ◊

(a) Applying the mirror-lens equation, $\frac{1}{f}=\frac{1}{p}+\frac{1}{q}$: $\frac{1}{f}=\frac{1}{1.25 \text{ m}}+\frac{1}{6.25 \text{ m}}$

so that the focal length of the mirror is $f=1.04$ m

*continued on next page*

Noting that the image is real, inverted and enlarged, we can say that the mirror must be concave, and must have a radius of curvature of $R = 2f = 2.08$ m. ◊

**P26.23** A glass sphere ($n = 1.50$) with a radius of 15.0 cm has a tiny air bubble 5.00 cm above its center. The sphere is viewed looking down along the extended radius containing the bubble. What is the apparent depth of the bubble below the surface of the sphere?

C

**FIG. P26.23**

**Solution** For refraction at a curved surface $\frac{n_1}{p} + \frac{n_2}{q} = \frac{n_2 - n_1}{R}$.

We solve $q$ to find $q = \frac{n_2 R p}{p(n_2 - n_1) - n_1 R}$.

In this case, $n_1 = 1.50$, $n_2 = 1.00$, $p = 10.0$ cm, $R = -15.0$ cm.

So the apparent depth is

$$q = \frac{(1.00)(-15.0 \text{ cm})(10.0 \text{ cm})}{(10.0 \text{ cm})(1.00 - 1.50) - (1.50)(-15.0 \text{ cm})} = -8.57 \text{ cm}.$$ ◊

To sketch a ray diagram we use the wave under refraction model, as shown. The image is virtual, upright, and enlarged.

**P26.27** The left face of a bioconvex lens has a radius of curvature of magnitude 12.0 cm, and the right face has a radius of curvature of magnitude 18.0 cm. The index of refraction of the glass is 1.44.

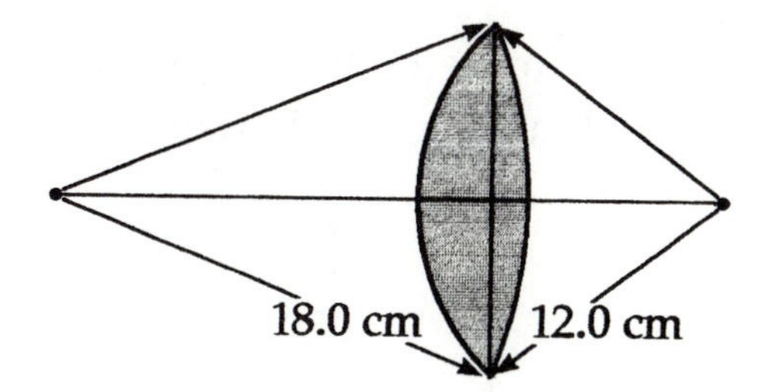


**FIG. P26.27**

(a) Calculate the focal length of the lens.

(b) Calculate the focal length the lens has after it is turned around to interchange the radii of curvature of the two faces.

*continued on next page*

**Solution** **Conceptualize:** Since this is a bioconvex lens, the center is thicker than the edges, and the lens will tend to converge incident light rays. Therefore, it has a positive focal length. Exchanging the radii of curvature amounts to turning the lens around so the light enters the opposite side first. However, this does not change the fact that the center of the lens is still thicker than the edges, so we should not expect the focal length of the lens to be different (assuming the thin-lens approximation is valid).

**Categorize:** The lens makers' equation can be used to find the focal length of this lens.

**Analyze:** The centers of curvature of the lens surfaces are on opposite sides, so the second surface has a negative radius:

(a) $$\frac{1}{f}=(n-1)\left(\frac{1}{R_1}-\frac{1}{R_2}\right)=(1.44-1.00)\left(\frac{1}{12.0\text{ cm}}-\frac{1}{-18.0\text{ cm}}\right) \quad \text{so } f=16.4\text{ cm} \quad \lozenge$$

(b) $$\frac{1}{f}=0.440\left(\frac{1}{18.0\text{ cm}}-\frac{1}{-12.0\text{ cm}}\right) \quad \text{so } f=16.4\text{ cm} \quad \lozenge$$

**Finalize:** As expected, reversing the orientation of the lens does not change what it does to the light, as long as the lens is relatively thin (variations may be noticed with a thick lens). The fact that light rays can be traced forward or backward through an optical system is sometimes referred to the **principle of reversibility**. We can see that the focal length of this bioconvex lens is about the same magnitude as the average radius of curvature. A few approximations, useful as checks, are that a symmetric bioconvex lens with radii of magnitude $R$ will have focal length $f\approx R$; a plano-convex lens with radius $R$ will have $f=\frac{R}{2}$; and a symmetric bioconcave lens has $f\approx -R$. These approximations apply when the lens has $n\approx 1.5$, which is typical of many types of clear glass and plastic.

**P26.31** The nickel's image in Figure P26.31 has twice the diameter of the nickel and is 2.84 cm from the lens. Determine the focal length of the lens.

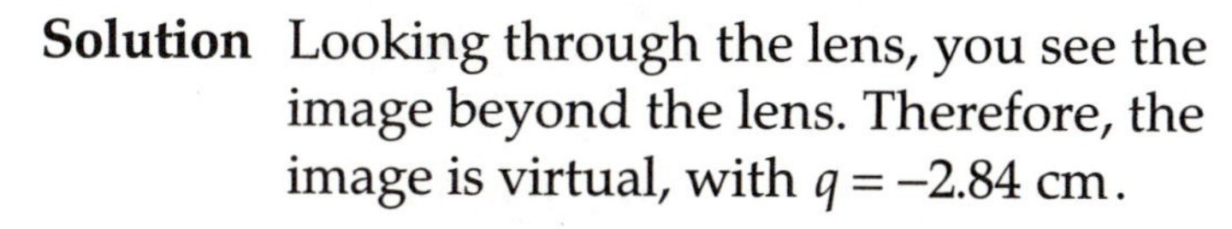

**Solution** Looking through the lens, you see the image beyond the lens. Therefore, the image is virtual, with $q = -2.84$ cm.

Now, $$M = \frac{h'}{h} = 2 = -\frac{q}{p}$$

so $$p = -\frac{q}{2} = 1.42 \text{ cm}$$

Thus,

$$f = \left(\frac{1}{p} + \frac{1}{q}\right)^{-1} = \left[\frac{1}{1.42 \text{ cm}} + \frac{1}{(-2.84 \text{ cm})}\right]^{-1}$$

$f = 2.84$ cm ◊

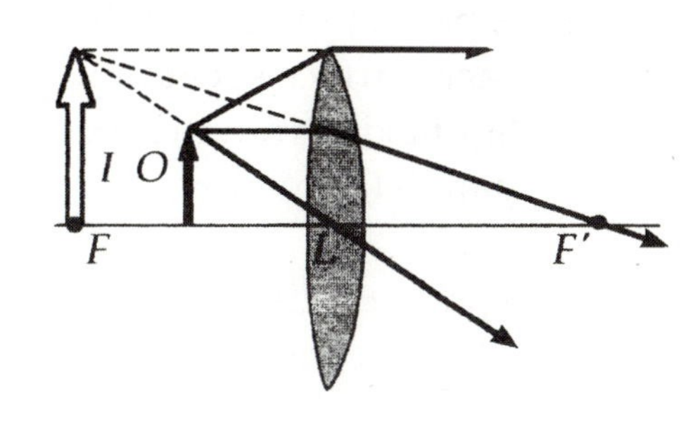


**FIG. P26.31**

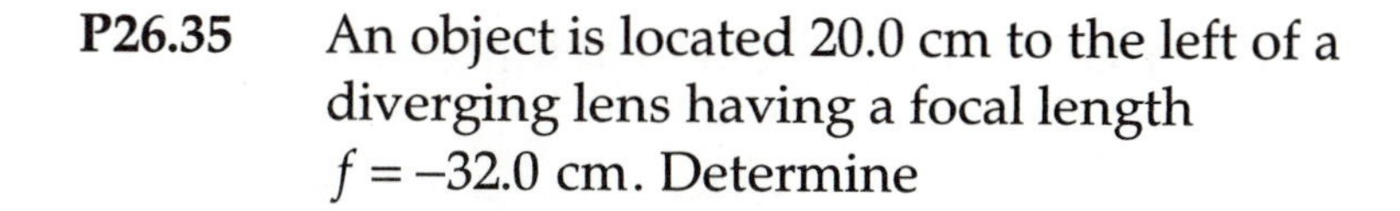

**P26.35** An object is located 20.0 cm to the left of a diverging lens having a focal length $f = -32.0$ cm. Determine

(a) the location of the image and

(b) the magnification of the image.

(c) Construct a ray diagram for this arrangement.

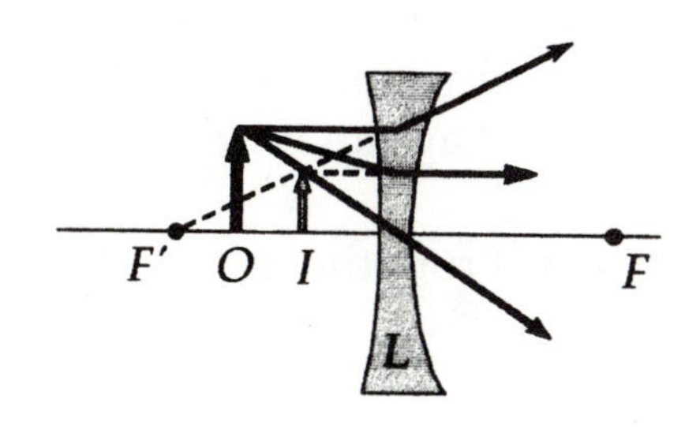


**FIG. P26.35**

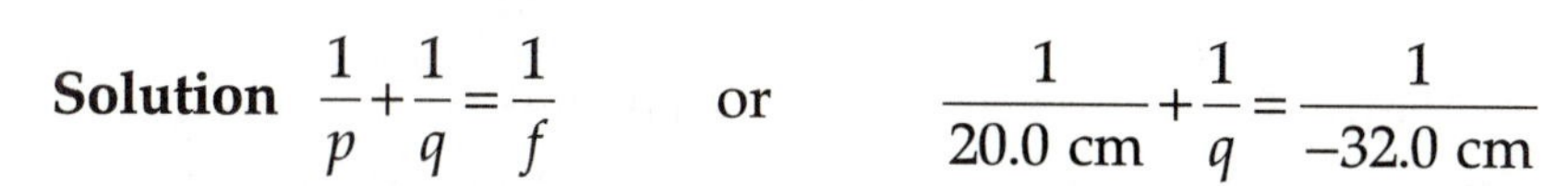

**Solution** $$\frac{1}{p} + \frac{1}{q} = \frac{1}{f} \quad \text{or} \quad \frac{1}{20.0 \text{ cm}} + \frac{1}{q} = \frac{1}{-32.0 \text{ cm}}$$

(a) So $q = -\left(\dfrac{1}{20.0 \text{ cm}} + \dfrac{1}{32.0 \text{ cm}}\right)^{-1} = -12.3$ cm ◊

(b) $M = -\dfrac{q}{p} = -\dfrac{(-12.3 \text{ cm})}{20.0 \text{ cm}} = 0.615$ ◊

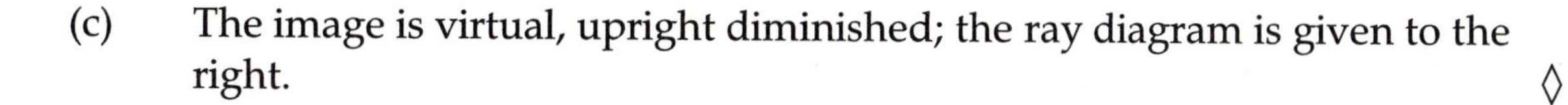

(c) The image is virtual, upright diminished; the ray diagram is given to the right. ◊

**P26.49** A parallel beam of light enters a glass hemisphere perpendicular to the flat face, as shown in Figure P26.49. The magnitude of the radius is 6.00 cm, and the index of refraction is 1.560. Determine the point at which the beam is focused. (Assume paraxial rays.)

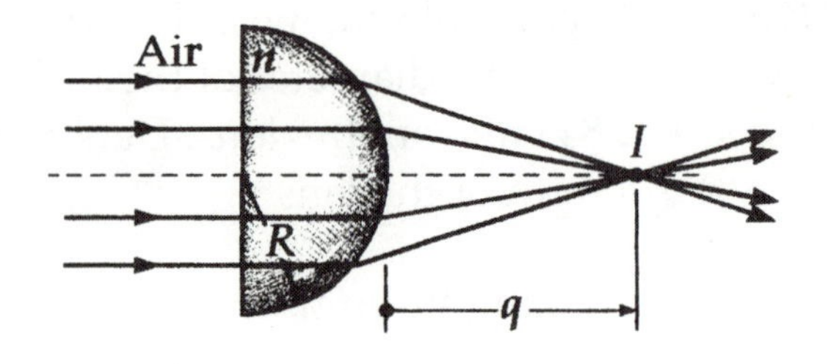


**FIG. P26.49**

**Solution** A hemisphere is too thick to be described as a thin lens. We consider refraction at each surface separately. The light is undeviated on entry into the flat face. The light exits from the second surface, for which $R = -6.00$ cm. The incident rays are parallel, so $p = \infty$.

Then $\frac{n_1}{p} + \frac{n_2}{q} = \frac{n_2 - n_1}{R}$ becomes $0 + \frac{1}{q} = \frac{1 - 1.56}{-6.00 \text{ cm}}$

and $q = 10.7$ cm. ◊

**P26.51** An object is placed 12.0 cm to the left of a diverging lens of focal length –6.00 cm. A converging lens of focal length 12.0 cm is placed a distance $d$ to the right of the diverging lens. Find the distance $d$ so that the final image is at infinity. Draw a ray diagram for this case.

**Solution** From the mirror-lens equation for the first lens,

$$q_1 = \frac{f_1 p_1}{p_1 - f_1} = \frac{(-6.00 \text{ cm})(12.0 \text{ cm})}{12.0 \text{ cm} - (-6.00 \text{ cm})} = -4.00 \text{ cm}.$$

When we require that $q_2 \to \infty$.

The mirror-lens equation for the second lens becomes $p_2 = f_2 = 12.0$ cm.

Since the object for the converging lens must be 12.0 cm to its left, and since this is the image for the diverging lens that is 4.00 cm to **its** left, the two lenses must be separated by 8.00 cm.

Mathematically, $p_2 = d - (-4.00 \text{ cm})$

$d + 4.00 \text{ cm} = f_2 = 12.0 \text{ cm}$ and $d = 8.00$ cm. ◊

*continued on next page*

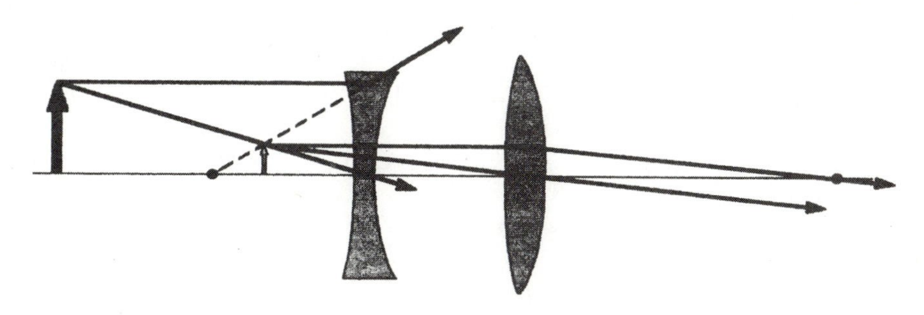

◊

FIG. P26.51

**P26.53** The disk of the Sun subtends an angle of 0.533° at the Earth. What are the position and diameter of the solar image formed by a concave spherical mirror with a radius of curvature of 3.00 m?

**Solution** For the mirror, $f = \frac{R}{2} = +1.50\text{ m}$. In addition, because the distance to the Sun is so much larger than any other figures, we can take $p = \infty$.

The mirror equation, $\frac{1}{p} + \frac{1}{q} = \frac{1}{f}$ then gives $q = f = 1.50\text{ m}$. ◊

Now, in $M = -\frac{q}{p} = \frac{h'}{h}$, the magnification is nearly zero, but we can be more precise: the definition of radian measure means that $\frac{h}{p}$ is the angular diameter of the object. Thus the image diameter is

$$h' = -\frac{hq}{p} = (-0.533°)\left(\frac{\pi}{180}\text{ rad/deg}\right)(1.50\text{ m}) = -0.014\,0\text{ m} = -1.40\text{ cm}.$$ ◊

**P26.55** In a darkened room, a burning candle is placed 1.50 m from a white wall. A lens is placed between candle and wall at a location that causes a larger, inverted image to form on the wall. When the lens is moved 90.0 cm toward the wall, another image of the candle is formed. Find

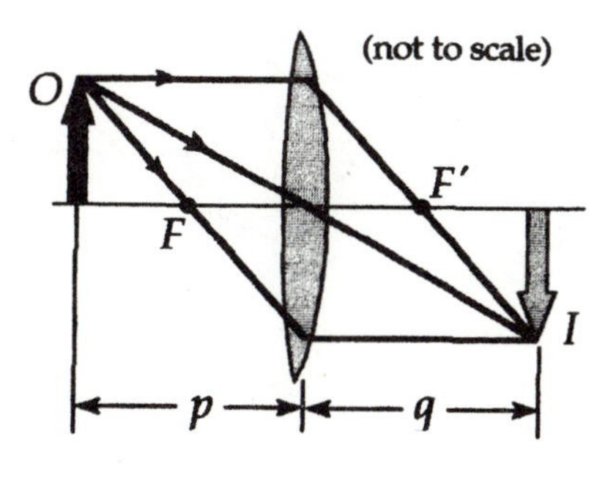


FIG. P26.55

(a) the two object distances that produce the specified images and

(b) the focal length of the lens.

(c) Characterize the second image.

*continued on next page*

**Solution** Originally, $p_1 + q_1 = 1.50\text{ m}$

In the final situation, $p_2 = p_1 + 0.900\text{ m}$

and $q_2 = q_1 - 0.900\text{ m} = (1.50\text{ m} - p_1) - 0.900\text{ m} = 0.600\text{ m} - p_1$

Our lens equation is
$$\frac{1}{p_1} + \frac{1}{q_1} = \frac{1}{f} = \frac{1}{p_2} + \frac{1}{q_2}$$

Substituting, we have
$$\frac{1}{p_1} + \frac{1}{1.50\text{ m} - p_1} = \frac{1}{p_1 + 0.900} + \frac{1}{0.600 - p_1}$$

Adding the fractions,
$$\frac{1.50\text{ m} - p_1 + p_1}{p_1(1.50\text{ m} - p_1)} = \frac{0.600 - p_1 + p_1 + 0.900}{(p_1 + 0.900)(0.600 - p_1)}$$

Simplified, this becomes $p_1(1.50\text{ m} - p_1) = (p_1 + 0.900)(0.600 - p_1)$

(a) Thus, $p_1 = \dfrac{0.540}{1.80}\text{ m} = 0.300\text{ m}$ ◊

$p_2 = p_1 + 0.90\text{ m} = 1.20\text{ m}$ ◊

(b) $$\frac{1}{f} = \frac{1}{0.300\text{ m}} + \frac{1}{1.50\text{ m} - 0.300\text{ m}}$$

and $f = 0.240\text{ m}$ ◊

(c) The second image is real, inverted, and diminished, with

$$M = \frac{-q_2}{p_2} = -0.250.$$ ◊

# Chapter 27

# Wave Optics

## NOTES FROM SELECTED CHAPTER SECTIONS

### Section 27.1 Conditions for Interference

In order to observe **sustained interference** in light waves, the following conditions must be met:

- The sources must be **coherent**; they must maintain a **constant phase** with respect to each other.
- The sources must be **monochromatic—**of a **single wavelength.**

### Section 27.2 Young's Double-Slit Experiment

A schematic diagram illustrating the geometry used in Young's double-slit experiment is shown in Figure 27.1 below. The two slits $S_1$ and $S_2$ serve as coherent monochromatic sources. When $L$ (the distance from source plane to viewing screen) is much greater than $d$ (the distance between sources), the **path difference** $\delta = r_2 - r_1 = d\sin\theta$.

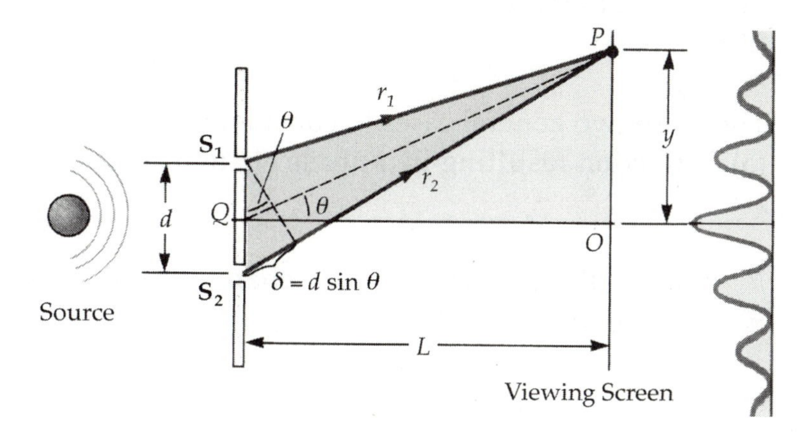


**Figure 27.1** Young's double slit experiment. The relative intensity of light on the screen is shown on the right.

## Section 27.4 Change of Phase Due to Reflection

Consider a light wave traveling in a medium with an index of refraction $n_1$. When partial reflection occurs at the surface of a medium with an index of refraction $n_2$:

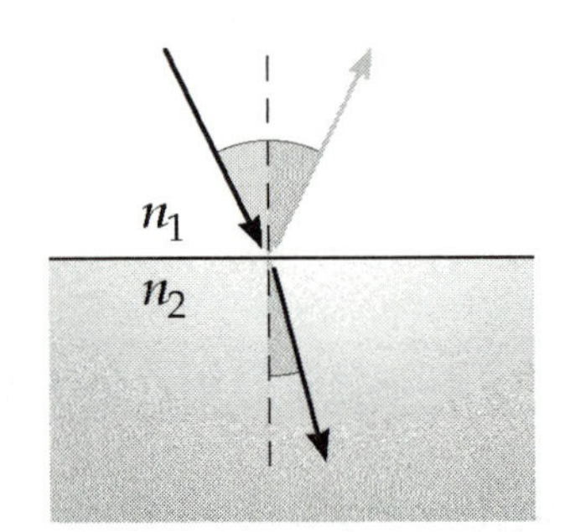


- if $n_1 < n_2$, the reflected ray experiences a phase change of 180°.
- if $n_1 > n_2$, there is no phase change in the reflected ray.
- there is no phase change in the transmitted ray regardless of the relative values of $n_1$ and $n_2$.

## Section 27.5 Interference in Thin Films

In order to predict constructive or destructive interference in thin films you must consider:

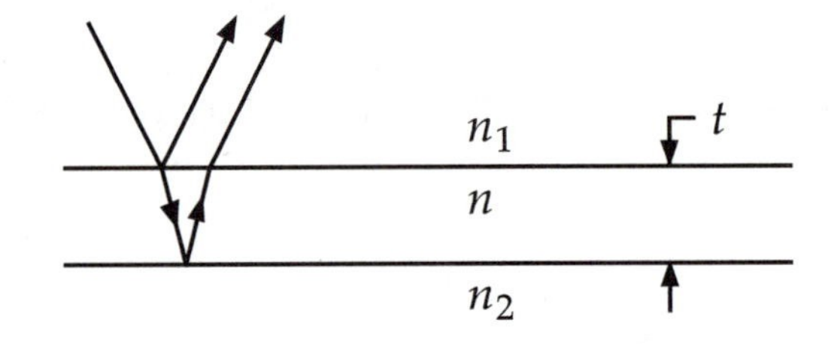


- the difference in path-length traveled by the two interfering waves;
- any expected changes in phase due to reflection;
- the change in wave length of the light as it enters the film.

There are two general cases to consider (see figure above):

**(a) Reflection resulting in a phase change at only one surface of the film.**

$n_1$ and $n_2$ both less than $n$ (phase change at the top surface)
$n_1$ and $n_2$ both greater than $n$ (phase change at the bottom surface)

**Constructive interference** will occur under these conditions, when the path difference (which equals $2t$) is an odd number of half wavelengths, $2t = (2m+1)\frac{\lambda_n}{2}$. Thus, the film thickness is $t = \left(\frac{m+1}{2}\right)\frac{\lambda_n}{2}$. *Here,* $\lambda_n = \frac{\lambda}{n}$ *is the wavelength as measured in the film (the wavelength in vacuum, divided by the index of refraction of the film).* The condition for constructive interference in this case then becomes $t = \left(\frac{m+1}{2}\right)\frac{\lambda}{2n}$.

**Destructive interference** will occur under these conditions, when the path difference, $2t$, equals an integer number of wavelengths so that $t = m\lambda/2n$.

**(b) Reflection resulting in phase changes at both top and bottom surfaces of the film (or at neither surface).**

$n_1 < n$ and $n_2 > n$ (phase change at both surfaces)
$n_1 > n$ and $n_2 < n$ (no phase change at either surface)

In this case, these two phase changes are offsetting and interference of the reflected rays depends only on the difference in distance traveled by the two reflected rays and the index of refraction of the film.

**Constructive interference** will occur when the path difference equals an integer number of wavelengths; the film thickness must be an integer number of half wavelengths, $t = \frac{m\lambda}{2n}$.

**Destructive interference** in this case will be observed when the path difference equals an odd number of half wavelengths; that is when $t = \left(m + \frac{1}{2}\right)\frac{\lambda}{2n}$.

## Section 27.6 Diffraction Patterns

Diffraction occurs when light waves deviate (or spread) from their initial direction of travel when passing through small openings, around obstacles, or by sharp edges.

The diffraction pattern produced by a single narrow slit consists of a broad, intense central band (the central maximum), flanked by a series of narrower and less intense secondary bands (called secondary maxima) alternating with a series of dark bands, or minima.

In the case of single slit diffraction, each portion of the slit acts as a source of waves; and light from one portion of the slit can interfere with light from another portion.

The resultant intensity on the screen depends on the angle $\theta$ between the perpendicular to the plane of the slit and the direction to a point on the screen.

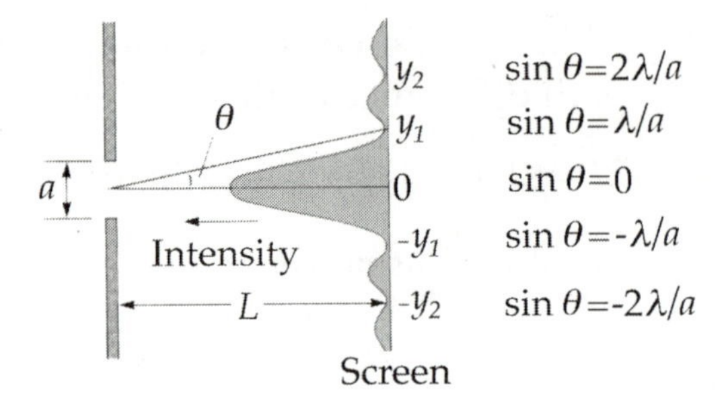


**FIG. 27.3** Single Slit Diffraction Pattern

One type of diffraction, called Fraunhofer diffraction, occurs when the rays reaching the observing screen are approximately parallel.

## Section 27.7 Resolution of Single-Slit and Circular Apertures

When the central maximum of one image falls on the first minimum of another image, the images are said to be just resolved. This limiting condition of resolution is known as **Rayleigh's criterion.**

## Section 27.8 The Diffraction Gratting

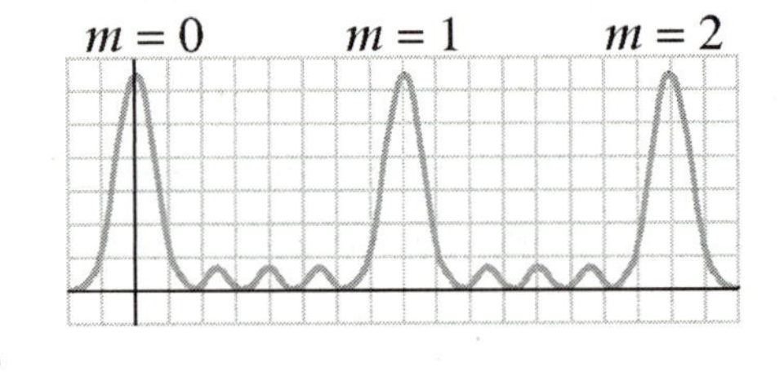


A diffraction grating, consisting of many equally spaced parallel slits, separated by a distance $d$, will produce an diffraction pattern. There will be a series of principle maxima (bright lines) for each wavelength component in the incident light. *The figure illustrates the case for which the incident light contains a single wavelength component*. Although not shown in the figure above, there will also be maxima corresponding to $m = -1, -2, \ldots$ These maxima occur at angles $\theta$, measured from the line perpendicular to the grating, where $m\lambda = d\sin\theta$.

If the incident light contains a second wavelength component, a second series of principle maxima (in general with a different intensity) will be present in the diffraction pattern. In a given spectral order, denoted by the number $m$, there will be principle maxima corresponding to each wavelength component incident on the grating.

# EQUATIONS AND CONCEPTS

In **Young's double-slit experiment**, two slits, $\mathbf{S}_1$ and $\mathbf{S}_2$, separated by a distance $d$, serve as monochromatic coherent sources. The light intensity at any point on the screen is the resultant of light reaching the screen from both slits. *As illustrated in the figure, a point P on the screen can be identified by the angle $\theta$ or by the distance y from the center of the screen.* When using $y = L\tan\theta$, $y$ is measured from the center of the interference pattern and $L$ is the distance from the double slit to the screen. Light from the two slits reaching any point on the screen (except the center) travels unequal path lengths. This difference in length of path, $\delta$, is called the path difference.

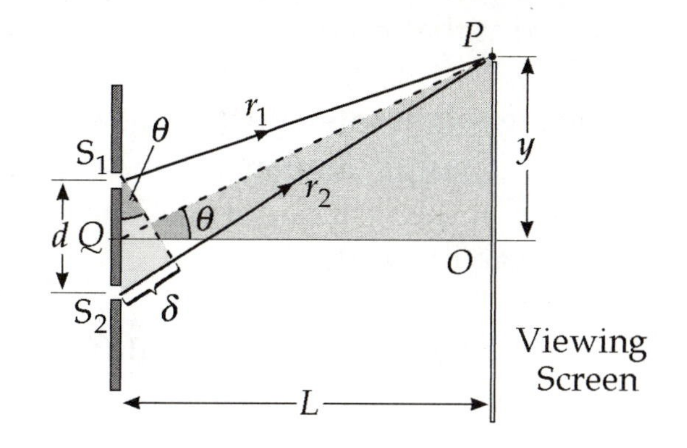


$$\delta = r_2 - r_1 = d\sin\theta \quad (27.1)$$

$$y = L\tan\theta \quad (27.4)$$

**Constructive interference** (bright fringes) will appear at points on the screen for which the path difference is equal to an integral multiple of the wavelength. The positions of bright fringes can also be located by calculating their distance from the center of the screen ($y$). In each case, the number $m$ is called the order number of the fringe. *The central bright fringe ($\theta = 0$, $m = 0$) is called the zeroth-order maximum.*

$$\delta = d\sin\theta_{\text{bright}} = m\lambda \quad (27.2)$$

for $m = 0, \pm 1, \pm 2, \ldots$

$$y_{\text{bright}} = L\tan\theta_{\text{bright}} \quad (27.5)$$

$$y_{\text{bright}} \cong m\frac{\lambda L}{d}$$

(for small $\theta$)

**Destructive interference** (dark fringes) will appear at points on the screen, which correspond to path differences of an odd multiple of half wavelengths. For these points of destructive interference, waves which leave the two slits in phase arrive at the screen 180° out of phase.

$$\delta = d\sin\theta_{\text{dark}} = \left(m + \frac{1}{2}\right)\lambda \quad (27.3)$$

for $m = 0, \pm 1, \pm 2, \ldots$

$$y_{\text{dark}} = L\tan\theta_{\text{dark}} \quad (27.6)$$

$$y_{\text{dark}} \cong \left(m + \frac{1}{2}\right)\frac{\lambda L}{d}$$

(for small $\theta$)

The **phase difference** $\phi$ between the two waves at any point on the screen depends on the path difference at that point.

$$\phi = \frac{2\pi}{\lambda}\delta = \frac{2\pi}{\lambda} d \sin\theta \qquad (27.7)$$

The **average light intensity** ($I_{av}$) at any point $P$ on the screen is proportional to the square of the amplitude of the resultant wave. The average intensity can be written:

- as a **function of phase difference** $\phi$,

$$I_{av} = I_{max} \cos^2\left(\frac{\phi}{2}\right)$$

- as a **function of the angle** ($\theta$) subtended by the screen point at the source midpoint; or

$$I_{av} = I_{max} \cos^2\left(\frac{\pi d \sin\theta}{\lambda}\right) \qquad (27.8)$$

- as a **function of the distance** ($y$) from the center of the screen.

$$I_{av} \cong I_{max} \cos^2\left(\frac{\pi d}{\lambda L} y\right)$$

(for small $\theta$)

***Increasing the number of equally spaced slits*** *will increase the number of secondary maxima; the principal maxima will become narrower but remain fixed in position. See Figure 27.21 of the textbook.*

The wavelength of light in a medium having an index of refraction $n$ is smaller than the wavelength in vacuum. In thin-film interference the wavelength of light within the film is $\lambda_n$.

$$\lambda_n = \frac{\lambda}{n} \qquad (27.9)$$

**Interference in thin films** depends on wavelength, film thickness, and the indices of refraction of the film and surrounding media. *Differences in phase may be due to path difference or phase change upon reflection.* There are two general cases.

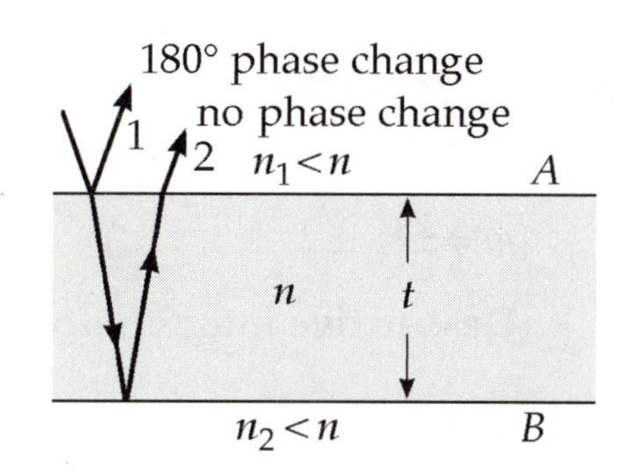


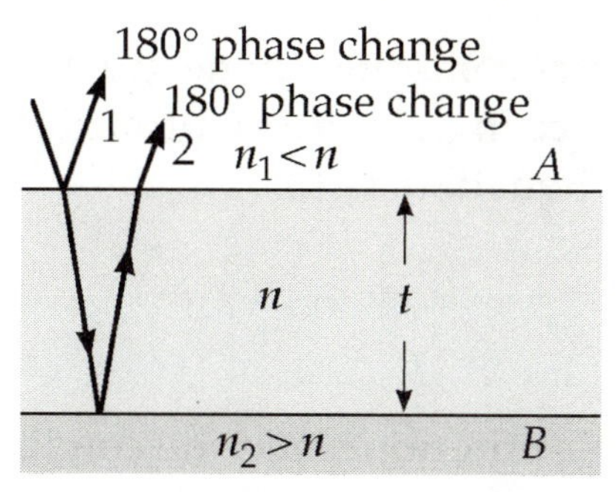


Case (I) **Phase change at only one film surface** $(n_1 < n$ and $n_2 < n)$ or $(n_1 > n$ and $n_2 > n)$. Indices of refraction of media on both sides of the film are less than that of the film (as shown in figure above left) or both greater than that of the film.

**Constructive interference (Case I)**

$$2nt = \left(m + \frac{1}{2}\right)\lambda \qquad (27.11)$$

$(m = 0, 1, 2, \ldots)$

**Destructive interference (Case I),**

$$2nt = m\lambda \qquad (24.10)$$

$(m = 0, 1, 2, \ldots)$

Case (II) **Phase changes at either both surfaces or at neither surface** $(n_1 < n < n_2)$or $(n_1 > n > n_2)$. Film is between two media either of which has an index of refraction greater than that of the film and the other a smaller index (as shown figure above right).

**Constructive interference (Case II)**

$$2nt = m\lambda$$

$(m = 0, 1, 2, \ldots)$

**Destructive interference(Case II)**

$$2nt = \left(m + \frac{1}{2}\right)\lambda \qquad (27.12)$$

$(m = 0, 1, 2, \ldots)$

A **single-slit diffraction pattern** consists of a central maximum with a series of alternating bright and dark bands (minima) on each side. The intensity at a given point on the screen depends on the angle between the direction to the central maximum and the direction to that point.

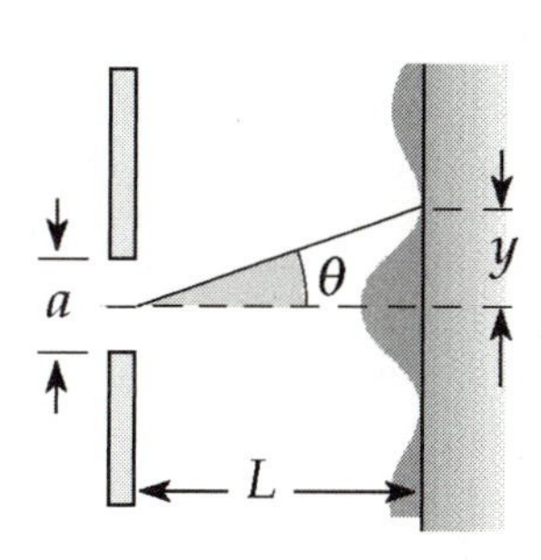

The **general condition for destructive interference** in a single slit diffraction pattern is stated in Equation 27.13.

$$\sin\theta_{\text{dark}} = m\frac{\lambda}{a} \qquad (27.13)$$

$(m = \pm 1,\ \pm 2,\ \pm 3,\ \ldots)$

**(Destructive interference)**

**Rayleigh's criterion** states the condition for the resolution of two images due to nearby sources.

For a **slit**, the angular separation between the sources must be greater than the ratio of the wavelength to slit width if the two sources are to be resolved.

$$\theta_{\min} = \frac{\lambda}{a} \qquad (27.14)$$

**(Limiting angle of resolution for a slit)**

For a **circular aperture,** the minimum angular separation depends on $D$, the diameter of the aperture (or lens).

$$\theta_{\min} = 1.22\frac{\lambda}{D} \qquad (27.15)$$

**(Limiting angle of resolution for a circular aperture)**

A **diffraction grating** (an array of a large number of parallel slits separated by a distance $d$) will produce an interference pattern in which there is a series of maxima for each wavelength. *Maxima due to wavelengths of different values comprise a spectral order denoted by order number m.*

$$d\sin\theta_{\text{bright}} = m\lambda \qquad (27.16)$$

$(m = 0,\ 1,\ 2,\ 3,\ \ldots)$

Each spectral order will contain a line characteristic of each wavelength. Equation 27.16 gives the angle of deviation for constructive interference (bright lines in the spectrum).

**Bragg's law** gives the conditions for **constructive interference** of x-rays reflected from the parallel planes of a crystalline solid separated by a distance $d$. *The angle $\theta$ is the angle between the incident beam and the surface.*

$$2d\sin\theta = m\lambda \qquad (27.17)$$

$(m = 1,\ 2,\ 3,\ \ldots)$

# SUGGESTIONS, SKILLS, AND STRATEGIES

## Thin-Film Interference Problems

- Identify the thin film from which interference effects are being observed.
- The type of interference that occurs in a specific problem is determined by the phase relationship between that portion of the wave reflected at the upper surface of the film and that portion reflected at the lower surface of the film.
- Phase differences between the two portions of the wave occur because of differences in the distances traveled by the two portions and by phase changes occurring upon reflection.

### Phase Difference due to Path Difference

The wave reflected from the lower surface of the film has to travel a distance equal to twice the thickness of the film before it returns to the upper surface of the film where it interferes with that portion of the wave reflected at the upper surface.

### Phase Change due to Reflection

When a wave traveling in a particular medium reflects off a surface having a higher index of refraction than the one it is in, a 180° phase shift occurs. This has the same effect as if the wave had traveled a lesser distance of $\frac{1}{2}\lambda$. This effect must be considered in addition to the phase difference related to the greater distance traveled by one of the waves.

- When distance and phase changes upon reflection are both taken into account:

Constructive interference will occur when the effective path difference is an integral multiple of $\lambda$—zero, $\lambda$, $2\lambda$, $3\lambda$ . . .

Destructive interference will occur when the effective path difference is an odd number of half wavelengths—$\frac{1}{2}\lambda, \frac{3}{2}\lambda, \frac{5}{2}\lambda, \ldots$

## REVIEW CHECKLIST

✓ Describe Young's double-slit experiment to demonstrate the wave nature of light. Account for the phase difference between light waves from the two sources as they arrive at a given point on the screen. State the conditions for constructive and destructive interference in terms of each of the following: path difference, phase difference, distance from the center of the screen, and angle subtended by the observation point at the source mid-point.

✓ Outline the manner in which the superposition principle leads to the correct expression for the intensity distribution on a distant screen due to two coherent sources of equal intensity.

✓ Account for the conditions of constructive and destructive interference in thin films considering both path difference and any expected phase changes due to reflection.

✓ Determine the positions of the minima in a single-slit diffraction pattern. Determine the positions of the principal maxima in the interference pattern of a diffraction grating.

✓ Determine whether or not two sources under a given set of conditions are resolvable as defined by Rayleigh's criterion.

## ANSWERS TO SELECTED QUESTIONS

**Q27.1** What is the necessary condition on the path length between two waves that interfere

(a) constructively and

(b) destructively?

**Answer** (a) Two waves interfere constructively if their path difference is either zero or some integral multiple of the wavelength; that is, if the path difference equals $m\lambda$.

(b) Two waves interfere destructively if their path difference is an odd multiple of one-half of a wavelength; that is, if the path difference equals $\left(m+\frac{1}{2}\right)\lambda$.

**Q27.3** In Young's double-slit experiment, why do we use monochromatic light? If white light is used, how would the pattern change?

**Answer** Every color produces its own pattern, with a spacing between the maxima that is characteristic of the wavelength. With several colors, the patterns are superimposed, and it can become difficult to pick out a single maximum. Using monochromatic light can eliminate this problem.

With white light, the central maximum is white. The first side maximum is a full spectrum, with violet on the inside and red on the outside. The second side maximum is a full spectrum also, but red in the second maximum overlaps the violet in the third maximum. At larger angles, the light soon starts mixing to white again, though it is often so faint that you would call it gray.

**Q27.11** Why can you hear around corners, but not see around corners?

**Answer** Audible sound has wavelengths on the order of meters or centimeters, while visible light has wavelengths on the order of half a micron. In this world of breadbox-size objects, $\lambda$ is comparable to the object size $a$ for sound, and sound diffracts around walls and through doorways. But $\frac{\lambda}{a}$ is much smaller for visible light passing ordinary-size objects or apertures, so light diffracts only through very small angles.

Another way of answering this question would be as follows. We can see by a small angle around a small obstacle or around the edge of a small opening. The side fringes in Figure 27.12 and the Arago spot in the center of Figure 27.13 (in the textbook) show this diffraction. Conversely, we cannot always hear around corners. Out-of-doors, away from reflecting surfaces, have someone a few meters distant face away from you and whisper. The high-frequency, short-wavelength, information-carrying components of the sound do not diffract around his head enough for you to understand his words.

**Q27.13** Describe the change in width of the central maximum of the single-slit diffraction pattern as the width of the slit is made narrower.

**Answer** Equation 27.13 describes the angles at which you get destructive interference; from it, we can obtain an estimate of the width of the central maximum. For small angles, the equation can be rewritten as

$$\theta_m = \sin^{-1}\left(\frac{m\lambda}{a}\right) \approx \frac{m\lambda}{a}.$$

Thus, as the width of the slit $a$ decreases, the angle of the first destructive interference $\theta_1$ grows, and the width of the central maximum grows as well.

## SOLUTIONS TO SELECTED PROBLEMS

**P27.3** Two radio antennas separated by 300 m as shown in Figure P27.3 simultaneously broadcast identical signals at the same wavelength. A radio in a car traveling due north receives the signals.

(a) If the car is at the position of the second maximum, what is the wavelength of the signals?

(b) How much farther must the car travel to encounter the next maximum in reception? (Note: Do not use the small-angle approximation in this problem.)

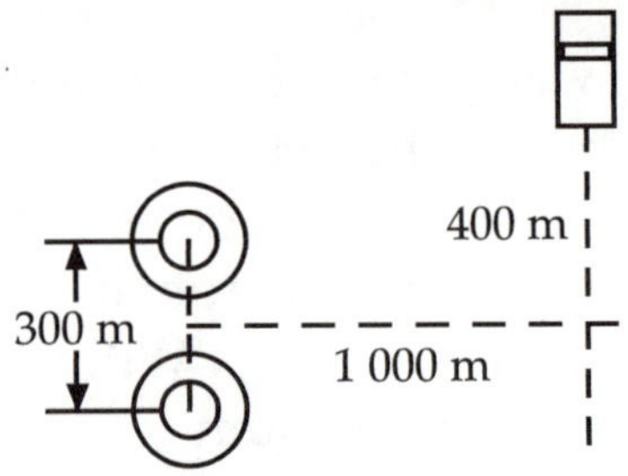

400 m
300 m
1 000 m

**FIG. P27.3**

**Solution** Note that with the conditions given, the small angle approximation **does not work well**. That is, $\sin\theta$, $\tan\theta$, and $\theta$ are significantly different. We use the Fraunhofer interference model, treating the waves from the two sources as moving along essentially parallel rays.

(a) At the $m = 2$ maximum, $\tan\theta = \dfrac{400\text{ m}}{1\,000\text{ m}} = 0.400$

and $\theta = 21.8°$ so $\lambda = \dfrac{d\sin\theta}{m} = \dfrac{(300\text{ m})(\sin 21.8°)}{2} = 55.7\text{ m}$. ◊

(b) The next maximum encountered is the $m = 2$ minimum.

At that point $d\sin\theta = \left[m + \dfrac{1}{2}\right]\lambda$: $d\sin\theta = \dfrac{5}{2}\lambda$

or $\sin\theta = \dfrac{5\lambda}{2d} = \dfrac{5(55.7\text{ m})}{2(300\text{ m})} = 0.464$

and $\theta = 27.7°$ so $y = (1\,000\text{ m})(\tan 27.7°) = 524\text{ m}$.

Therefore, the car must travel an additional 124 m. ◊

If we considered Fresnel interference, we would more precisely find

*continued on next page*

(a) $\lambda = \frac{1}{2}\left[\sqrt{550^2 + 1\,000^2} - \sqrt{250^2 + 1\,000^2}\right] = 55.2 \text{ m}.$

(b) $\Delta y = 123 \text{ m}$

**P27.5** Young's double-slit experiment is performed with 589-nm light and a distance of 2.00 m between the slits and the screen. The tenth interference minimum is observed 7.26 mm from the central maximum. Determine the spacing of the slits.

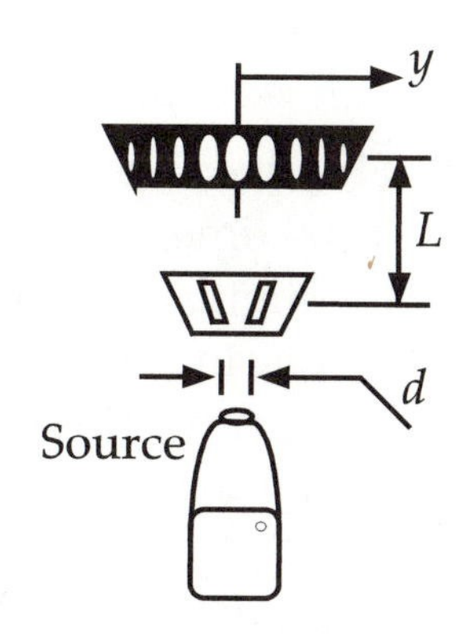


**FIG. P27.5**

**Solution** **Conceptualize:** For the situation described, the observed interference pattern is very narrow (the minima are less than 1 mm apart when the screen is 2 m away). In fact, the minima and maxima are so close together that it would probably be difficult to resolve adjacent maxima, so the pattern might look like a solid blur to the naked eye. Since the angular spacing of the pattern is inversely proportional to the slit width, we should expect that for this narrow pattern, the space between the slits will be larger than the typical fraction of a millimeter, and certainly much greater than the wavelength of the light ($d >> \lambda = 589$ nm).

**Categorize:** Since we are given the location of the tenth minimum for this interference pattern, we should use the equation for **destructive interference** from a double slit.

It can be confusing to keep track of four different symbols for distances. Three are shown in Figure P27.5. Note that:

$y$ is the unknown distance from the bright central maximum ($m = 0$) to another maximum or minimum on either side of the center of the interference pattern.

$\lambda$ is the wavelength of the light, determined by the source.

**Analyze:** In the equation $d\sin\theta = \left[m + \frac{1}{2}\right]\lambda$, the first minimum is described by $m = 0$ and the tenth by $m = 9$.

So, $\sin\theta = \frac{\lambda}{d}\left[9 + \frac{1}{2}\right]$.

*continued on next page*

Also, $\tan\theta = \frac{y}{L}$, but for small $\theta$, $\sin\theta \approx \tan\theta$. Thus, the distance between the slits is,

$$d = \frac{9.5\lambda}{\sin\theta} = \frac{9.5\lambda L}{y} = \frac{9.5\left(5\,890\times10^{-10}\text{ m}\right)(2.00\text{ m})}{7.26\times10^{-3}\text{ m}} = 1.54\times10^{-3}\text{ m} = 1.54\text{ mm}. \quad \Diamond$$

**Finalize:** The spacing between the slits is relatively large, as we expected (about 3 000 times greater than the wavelength of the light). In order to more clearly distinguish between maxima and minima, the pattern could be expanded by increasing the distance to the screen. However, as $L$ is increased, the overall pattern would be less bright as the light expands over a larger area, so that beyond some distance, the light would be too dim to see.

**P27.11** In Figure P27.11, let $L = 120$ cm and $d = 0.250$ cm. The slits are illuminated with coherent 600-nm light. Calculate the distance $y$ above the central maximum for which the average intensity on the screen is 75.0% of the maximum.

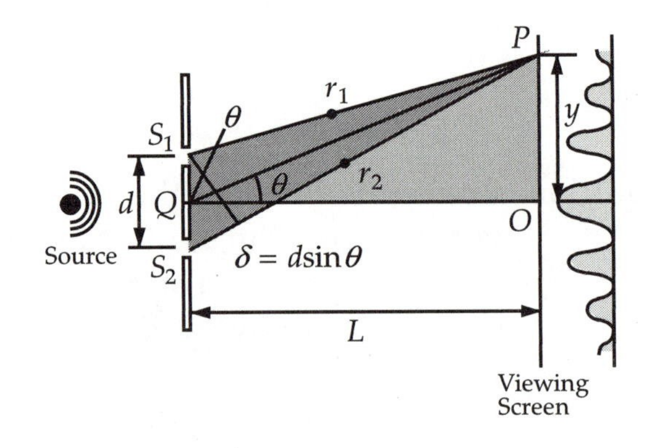


**FIG. P27.11**

**Solution** We use $$I = I_{\max}\cos^2\left[\frac{\pi d\sin\theta}{\lambda}\right].$$

From the drawing for small $\theta$,

$$\sin\theta \approx \frac{y}{L}: \qquad y = \frac{\lambda L}{\pi d}\cos^{-1}\sqrt{\frac{I}{I_{\max}}}.$$

In addition, since $I = 0.750 I_{\max}$, we can substitute a value for each variable:

$$y = \frac{\left(6.00\times10^{-7}\text{ m}\right)(1.20\text{ m})}{\pi\left(2.50\times10^{-3}\text{ m}\right)}\cos^{-1}\sqrt{0.750} = 0.048\,0\text{ mm}. \quad \Diamond$$

**P27.15** An oil film ($n = 1.45$) floating on water is illuminated by white light at normal incidence. The film is 280 nm thick. Find

(a) the color of the light in the visible spectrum most strongly reflected and

(b) the color of the light in the spectrum most strongly transmitted. Explain your reasoning.

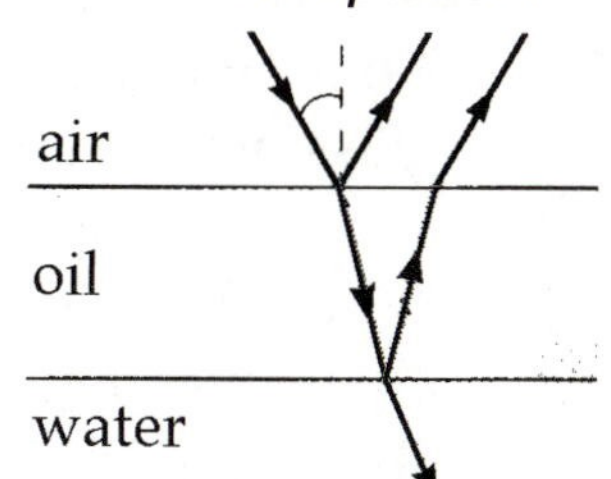


**FIG. P27.15**

**Solution** The light reflected from the top of the oil film undergoes phase reversal. Since $1.45 > 1.33$, the light reflected from the bottom undergoes no reversal. For constructive interference of reflected light, we then have

$$2nt = \left[m + \frac{1}{2}\right]\lambda \quad \text{or} \quad \lambda_m = \frac{2nt}{m + \frac{1}{2}} = \frac{2(1.45)(280 \text{ nm})}{m + \frac{1}{2}}.$$

(a) Substituting for $m$, we have

$m = 0$: $\lambda_0 = 1\,624$ nm (infrared)

$m = 1$: $\lambda_1 = 541$ nm (green)

$m = 2$: $\lambda_2 = 325$ nm (ultraviolet)

Both infrared and ultraviolet light are invisible to the human eye, so the dominant color is green. ◊

(b) Any light that is not reflected is transmitted. To find the color transmitted most strongly, we find the wavelengths reflected least strongly. According to the condition for destructive interference $2nt = m\lambda$.

Therefore,

$$\lambda = \frac{2nt}{m} = \frac{2(1.45)(280 \text{ nm})}{m}.$$

For

$m = 1$, $\lambda = 812$ nm (near infrared)

$m = 2$, $\lambda = 406$ nm (violet) ◊

$m = 3$, $\lambda = 271$ nm (ultraviolet)

Thus violet is the visible color least attenuated by reflection, and the dominant color in the transmitted light.

**P27.19** An air wedge is formed between two glass plates separated at one edge by a very fine wire, as shown in Figure P27.19. When the wedge is illuminated from above by 600-nm light and viewed from above, 30 dark fringes are observed. Calculate the radius of the wire.

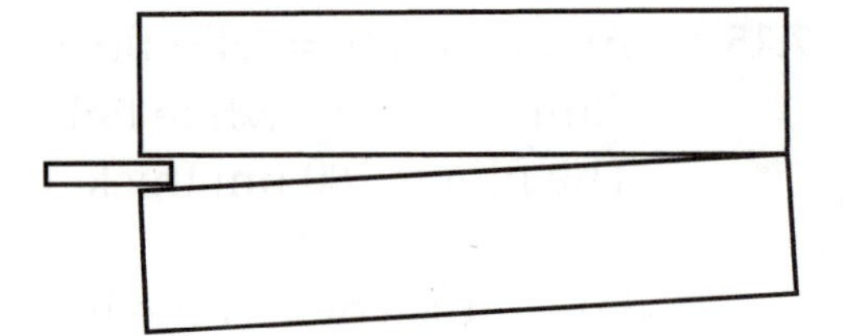

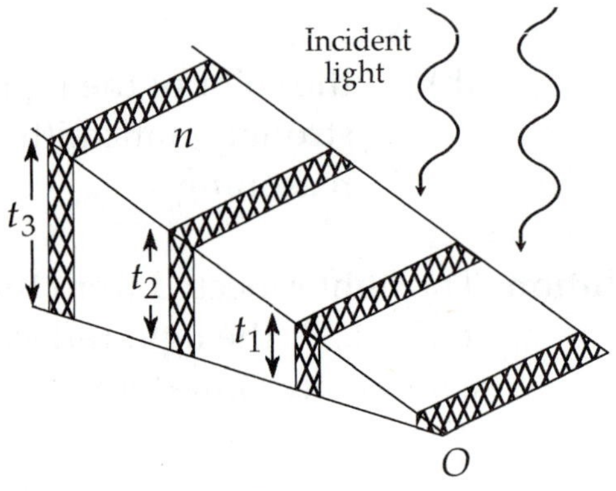


**FIG. P27.19**

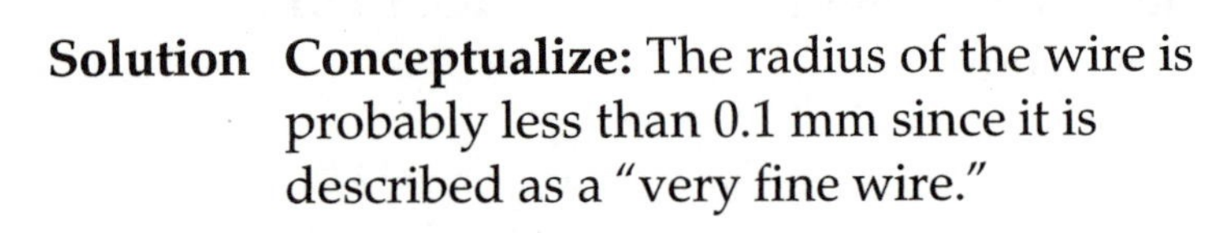

**Solution** **Conceptualize:** The radius of the wire is probably less than 0.1 mm since it is described as a "very fine wire."

**Categorize:** Light reflecting from the bottom surface of the top plate undergoes no phase shift, while light reflecting from the top surface of the bottom plate is shifted by $\pi$, and also has to travel an extra distance $2t$, where $t$ is the thickness of the air wedge.

**Analyze:** For destructive interference, $2t = m\lambda$ $(m = 0, 1, 2, 3, \ldots)$.

The first dark fringe appears where $m = 0$ at the line of contact between the plates. The 30th dark fringe gives for the diameter of the wire $2t = 29\lambda$, and $t = 14.5\lambda$.

$$r = \frac{t}{2} = 7.25\lambda = 7.25\left(600 \times 10^{-9} \text{ m}\right) = 4.35\ \mu\text{m} \qquad \Diamond$$

**Finalize:** This wire is not only less than 0.1 mm; it is even thinner than a typical human hair ($\sim 50\ \mu$m).

**P27.21** A screen is placed 50.0 cm from a single slit, which is illuminated with 690-nm light. If the distance between the first and third minima in the diffraction pattern is 3.00 mm, what is the width of the slit?

**Solution** In the equation for single-slit diffraction minima at small angles

$$\frac{y}{L} \approx \sin\theta_{\text{dark}} = \frac{m\lambda}{a}$$

take differences between the first and third minima,

*continued on next page*

to see that $\frac{\Delta y}{L} = \frac{\Delta m\lambda}{a}$ with $\Delta y = 3.00 \times 10^{-3}$ m and $\Delta m = 3 - 1 = 2$.

The width of the slit is then $a = \frac{\lambda L \Delta m}{\Delta y} = \frac{(690 \times 10^{-9}\ \text{m})(0.500\ \text{m})(2)}{3.00 \times 10^{-3}\ \text{m}}$

$$a = 2.30 \times 10^{-4}\ \text{m} \qquad \Diamond$$

**P27.27** A helium-neon laser emits light that has a wavelength of 632.8 nm. The circular aperture through which the beam emerges has a diameter of 0.500 cm. Estimate the diameter of the beam 10.0 km from the laser.

**Solution** **Conceptualize:** A typical laser pointer makes a spot about 5 cm in diameter at 100 m, so the spot size at 10 km would be about 100 times bigger, or about 5 m across. Assuming that this HeNe laser is similar, we could expect a comparable beam diameter.

**Categorize:** We assume that the light is parallel and not diverging as it passes through and fills the circular aperture. However, as the light passes through the circular aperture, it will spread from diffraction according to Equation 27.15.

**Analyze:** The beam spreads into a cone of half-angle

$$\theta_{\text{min}} = 1.22\frac{\lambda}{D} = 1.22\left[\frac{632.8 \times 10^{-9}\ \text{m}}{0.005\,00\ \text{m}}\right] = 1.54 \times 10^{-4}\ \text{rad}.$$

The radius of the beam ten kilometers away is, from the definition of radian measure

$$r_{\text{beam}} = \theta_{\text{min}}(1.00 \times 10^{4}\ \text{m}) = 1.54\ \text{m}$$

and its diameter is $d_{\text{beam}} = 2r_{\text{beam}} = 3.09$ m. $\Diamond$

**Finalize:** The beam is several meters across as expected, and is about 600 times larger than the laser aperture. Since most HeNe lasers are low power units in the mW range, the beam at this range would be so spread out that it be too dim to see on a screen.

**P27.29** The Impressionist painter Georges Seurat created paintings with an enormous number of dots of pure pigment, each of which was approximately 2.00 mm in diameter. The idea was to have colors such as red and green next to each other to form a scintillating canvas (Fig. P27.29). Outside what distance would one be unable to discern individual dots on the canvas? (Assume that $\lambda = 500$ nm and that the pupil diameter is 4.00 mm.)

**Solution** We will assume that the dots are just touching and do not overlap, so that the distance between their centers is 2.00 mm. By Rayleigh's criterion, two dots separated center to center by 2.00 mm would be seen to overlap when

$$\theta_{\min} = \frac{d}{L} = 1.22\frac{\lambda}{D} \qquad \text{with} \quad d = 2.00 \text{ mm}, \lambda = 500 \text{ nm, and } D = 4.00 \text{ mm}.$$

Thus,
$$L = \frac{Dd}{1.22\lambda} = \frac{(4.00\times10^{-3} \text{ m})(2.00\times10^{-3} \text{ m})}{1.22(500\times10^{-9} \text{ m})} = 13.1 \text{ m}. \qquad \Diamond$$

**P27.33** The hydrogen spectrum has a red line at 656 nm and a blue line at 434 nm. What are the angular separations between these two spectral lines obtained with a diffraction rating that has 4 500 grooves/cm?

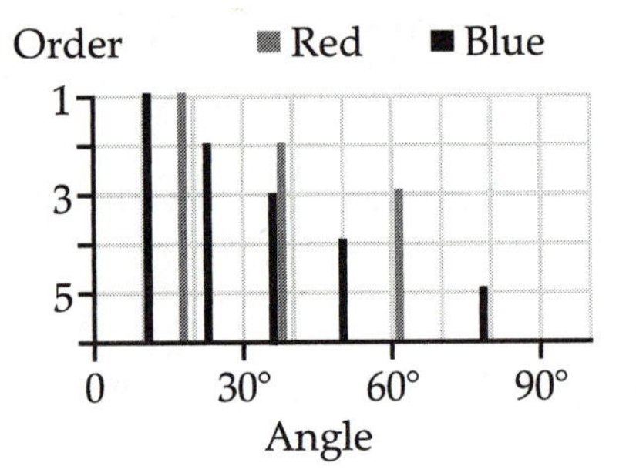


**FIG. P27.33**

**Solution** **Conceptualize:** Most diffraction gratings yield several spectral orders within the 180° viewing range, which means that the angle between red and blue lines is probably 10° to 30°.

**Categorize:** The angular separation is the difference between the angles corresponding to the red and blue wavelengths for each visible spectral order according to the diffraction grating equation, $d\sin\theta = m\lambda$.

**Analyze:** The grating spacing is

$$d = \frac{1.00\times10^{-2} \text{ m}}{4\,500 \text{ lines}} = 2.22\times10^{-6} \text{ m}.$$

In the first-order spectrum ($m = 1$), the angles of diffraction are given by $\sin\theta = \frac{\lambda}{d}$:

*continued on next page*

$$\sin\theta_{1r} = \frac{656\times10^{-9}\text{ m}}{2.22\times10^{-6}\text{ m}} = 0.295 \qquad \text{so} \qquad \theta_{1r} = 17.17°$$

$$\sin\theta_{1b} = \frac{434\times10^{-9}\text{ m}}{2.22\times10^{-6}\text{ m}} = 0.195 \qquad \text{so} \qquad \theta_{1b} = 11.26°$$

The angular separation is $\Delta\theta_1 = \theta_{1r} - \theta_{1b} = 17.17° - 11.26° = 5.91°$. ◊

In the 2nd-order ($m = 2$) $\Delta\theta_2 = \sin^{-1}\left(\frac{2\lambda_r}{d}\right) - \sin^{-1}\left(\frac{2\lambda_b}{d}\right) = 13.2°$. ◊

In the third order ($m = 3$), $\Delta\theta_3 = \sin^{-1}\left(\frac{3\lambda_r}{d}\right) - \sin^{-1}\left(\frac{3\lambda_b}{d}\right) = 26.5°$. ◊

Examining the fourth order, we find the red line is not visible:

$\theta_{4r} = \sin^{-1}\left(\frac{4\lambda_r}{d}\right) = \sin^{-1}(1.18)$ does not exist. ◊

**Finalize:** The full spectrum is visible in the first 3 orders with this diffraction grating, and the fourth is partially visible. We can also see that the pattern is dispersed more for higher spectral orders so that the angular separation between the red and blue lines increases as $m$ increases. It is also worth noting that the spectral orders can overlap (as is the case for the second and third order spectra above), which makes the pattern look confusing if you do not know what are you are looking for.

**P27.35** A grating with 250 grooves/mm is used with an incandescent light source. Assume the visible spectrum to range in wavelength from 400 to 700 nm. In how many orders can one see

(a) the entire visible spectrum?

(b) the short-wavelength region?

**Solution** The grating spacing is $d = \frac{1.00\text{ mm}}{250} = 4.00\times10^{-6}\text{ m}$.

(a) In each order of interference $m$, red light diffracts at a larger angle than the other colors with shorter wavelengths. We find the largest integer $m$ satisfying

$$d\sin\theta = m\lambda \qquad \text{with} \qquad \lambda = 700\text{ nm}.$$

*continued on next page*

With $\sin\theta$ having its largest value, $\left(4.00\times10^{-6}\text{ m}\right)(\sin 90°) = m\left(700\times10^{-9}\text{ m}\right)$

so that $\qquad m = 5.71$.

Thus the red light cannot be seen in the 6th order, and the full visible spectrum appears in only five orders. ◊

(b) Now consider light at the boundary between violet and ultraviolet.

$d\sin\theta = m\lambda \qquad$ becomes $\qquad \left(4.00\times10^{-6}\text{ m}\right)(\sin 90°) = m\left(400\times10^{-9}\text{ m}\right)$

and $m = 10$. ◊

**P27.39** If the interplanar spacing of NaCl is 0.281 nm, what is the predicted angle at which 0.140-nm x-rays are diffracted in a first-order maximum?

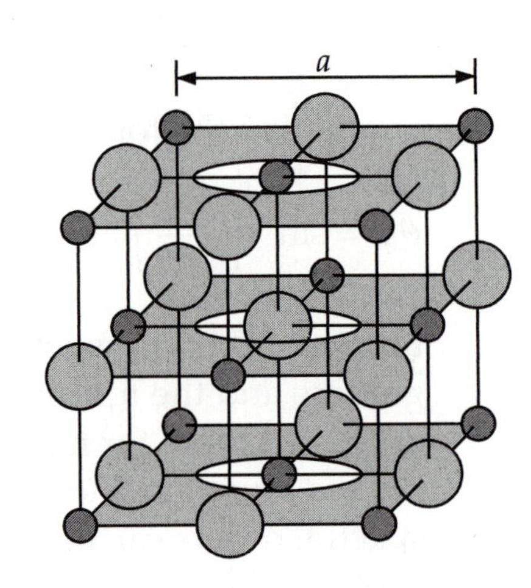


**FIG. P27.39**

**Solution** The atomic planes in this crystal are shown in Figure P27.39. The diffraction they produce is described by the Bragg condition,

that $2d\sin\theta = m\lambda$:

$$\sin\theta = \frac{m\lambda}{2d} = \frac{1\left(0.140\times10^{-9}\text{ m}\right)}{2\left(0.281\times10^{-9}\text{ m}\right)} = 0.249$$

$$\theta = 14.4°$$ ◊

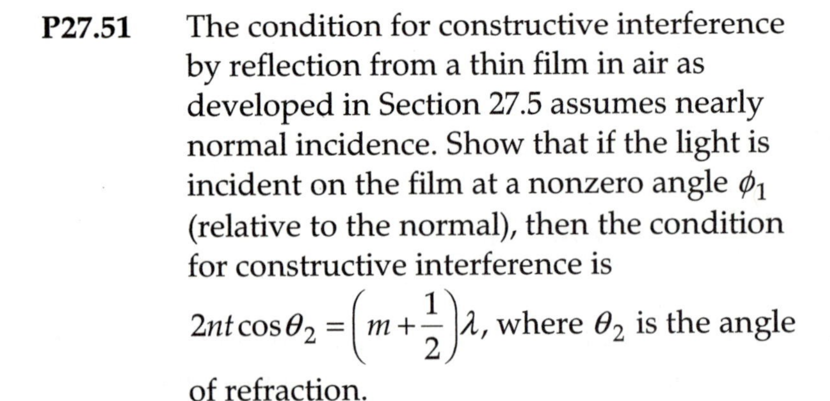

**P27.51** The condition for constructive interference by reflection from a thin film in air as developed in Section 27.5 assumes nearly normal incidence. Show that if the light is incident on the film at a nonzero angle $\phi_1$ (relative to the normal), then the condition for constructive interference is $2nt\cos\theta_2 = \left(m+\frac{1}{2}\right)\lambda$, where $\theta_2$ is the angle of refraction.

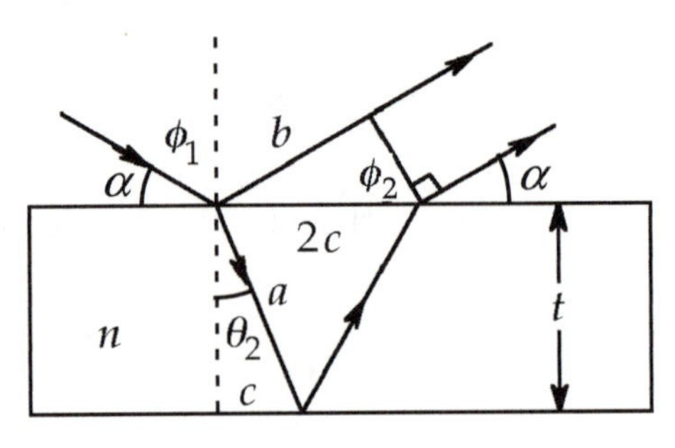


**FIG. P27.51**

*continued on next page*

**Solution** It may be helpful to draw a ray diagram of the interference. Because refraction out of the film is the opposite of refraction into the film, the refracted ray leaves the film parallel to the reflected ray. Then, since the angles marked $\alpha$ are equal,

$$\phi_2 + \alpha + 90° = 180°.$$

From the angle of incidence, $\phi_1 + \alpha = 90°.$

So $\phi_2 = \phi_1.$

Before they head off together along parallel paths, one beam travels a distance $b$ and undergoes phase reversal on reflection; the other beam travels a distance $2a$ inside the film, where its wavelength is $\frac{\lambda}{n}$. The number of cycles it completes along this path is

$$\frac{2a}{\lambda/n} = \frac{2na}{\lambda}.$$

The optical path length in the film is $2na$. Then the shift between the two outgoing rays is

$$\delta = 2na - b - \frac{\lambda}{2}$$

where $a$ and $b$ are shown in the ray diagram, $n$ is the index of refraction, and the term $\frac{\lambda}{2}$ is due to phase reversal at the top surface. For constructive interference, $\delta = m\lambda$ where $m$ has integer values.

This condition becomes $2na - b = \left[m + \frac{1}{2}\right]\lambda.$ (1)

From the figure's geometry, $a = \frac{t}{\cos\theta_2}$

and $c = a\sin\theta_2 = \frac{t\sin\theta_2}{\cos\theta_2}.$

Therefore, $b = 2c\sin\phi_1 = \frac{2t\sin\theta_2}{\cos\theta_2}\sin\phi_1.$

Also, from Snell's law, $\sin\phi_1 = n\sin\theta_2$

so $b = \frac{2nt\sin^2\theta_2}{\cos\theta_2}.$

*continued on next page*

With these results, the condition for constructive interference given in Equation (1) becomes:

$$2n\left(\frac{t}{\cos\theta_2}\right)-\frac{2nt\sin^2\theta_2}{\cos\theta_2}=\frac{2nt}{\cos\theta_2}\left(1-\sin^2\theta_2\right)=\left[m+\frac{1}{2}\right]\lambda.$$

$$2nt\cos\theta_2=\left[m+\frac{1}{2}\right]\lambda. \qquad \Diamond$$

For normal incidence the condition for constructive interference is $2t=\left(m+\frac{1}{2}\right)\lambda_n$ or $2nt=\left(m+\frac{1}{2}\right)\lambda$. As the angle of incidence $\phi_1$ goes to zero, the angle of refraction $\theta_2$ also becomes zero, and $\cos\theta_2$ is 1. Then the equation derived reduces to $2nt=\left(m+\frac{1}{2}\right)\lambda$, in agreement with the condition stated in the chapter.

# Chapter 28

# Quantum Physics

## NOTES FROM SELECTED CHAPTER SECTIONS

### Section 28.1 Blackbody Radiation and Planck's Theory

A **black body** is an ideal body that absorbs all radiation incident on it. Any body at some temperature $T$ emits thermal radiation, which is characterized by the properties of the body and its temperature. The spectral distribution of blackbody radiation at various temperatures is shown in the figure. As the temperature increases, the total power of the emitted radiation (area under the curve) increases, while the peak of the distribution shifts to shorter wavelengths.

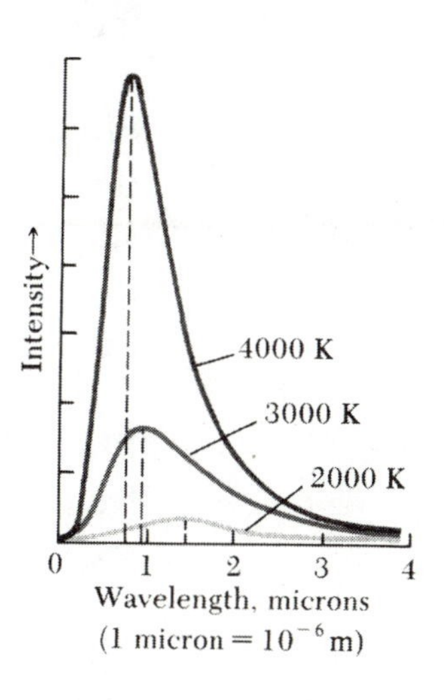


Classical theories failed to explain blackbody radiation. A new theory, proposed by Max Planck, is consistent with this distribution at all wavelengths. Planck made two basic assumptions in the development of this result:

- The oscillators emitting the radiation can only have **discrete energies** given by $E_n = nhf$, where $n$ is a quantum number ($n = 1, 2, 3, \ldots$), $f$ is the oscillator frequency, and $h$ is Planck's constant.

- These oscillators can emit or absorb energy in discrete units called **quanta** (or photons), where the energy of a light quantum obeys the relation $E = hf$.

Subsequent developments showed that the quantum concept was necessary in order to explain several phenomena at the atomic level, including the photoelectric effect, the Compton effect, and atomic spectra.

## Section 28.2 The Photoelectric Effect

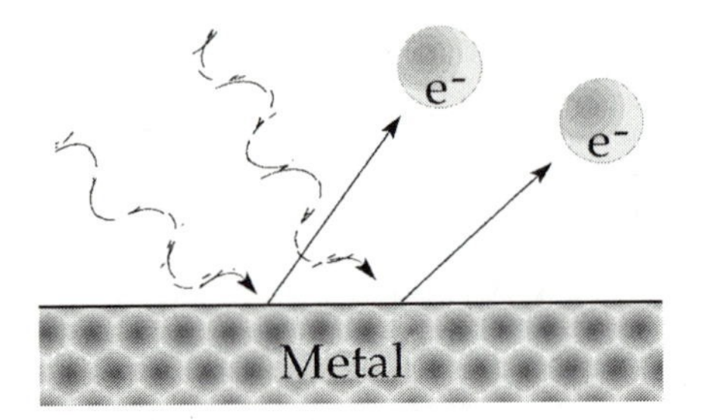


When light is incident on certain metallic surfaces, electrons can be emitted from the surfaces. This is called the photoelectric effect, discovered by Hertz. Several features of the photoelectric effect can not be explained with classical physics or with the wave theory of light. In 1905, Einstein provided a successful explanation of the photoelectric effect by extending Planck's quantum concept to include electromagnetic fields.

Observed features of the photoelectric effect can be explained and understood on the basis of the photon theory of light. These features include:

- **Photoelectrons have a maximum kinetic energy that is independent of light intensity** (number of photons incident per unit area of the photo surface per second). This is confirmed experimentally by showing that the photocurrent becomes zero at the same stopping potential for different light intensities.

- **Photoelectrons are emitted almost instantaneously** by photons with a frequency above the cutoff frequency (a threshold or minimum value characteristic of a particular material). This instantaneous emission occurs even at very low light intensity.

- **No photoelectrons are emitted when the frequency of incident light is below the cutoff frequency** characteristic of the material being illuminated. This is true regardless of the degree of light intensity.

- **The maximum kinetic energy of photoelectrons increases with increasing light frequency.** This is seen by examining the photoelectric equation (Eq. 28.5), which states that $K_{\text{max}} = hf - \phi$. The work function $\phi$ represents the minimum energy with which an electron is bound in the surface of a metal. The cutoff frequency is $f_c = \phi/h$.

## Section 28.3 The Compton Effect

The Compton effect involves the scattering of x-rays by electrons. The scattered x-ray undergoes a change in wavelength $\Delta\lambda$, called the Compton shift, which cannot be explained using classical concepts. By treating the x-ray as a photon (the quantum concept), the scattering process between the photon and electron predicts a shift in photon (x-ray) wavelength given by Equation 28.8, where $\theta$ is the angle between the incident and scattered x-ray and $m_e$ is the mass of the electron. The formula is in excellent agreement with experimental results.

The figure to the right represents the data for Compton scattering of x-rays from graphite at $\theta = 90.0°$. In this case, the Compton shift is $\Delta\lambda = 0.0024$ nm and $\lambda_0$ is the wavelength of the incident x-ray beam.

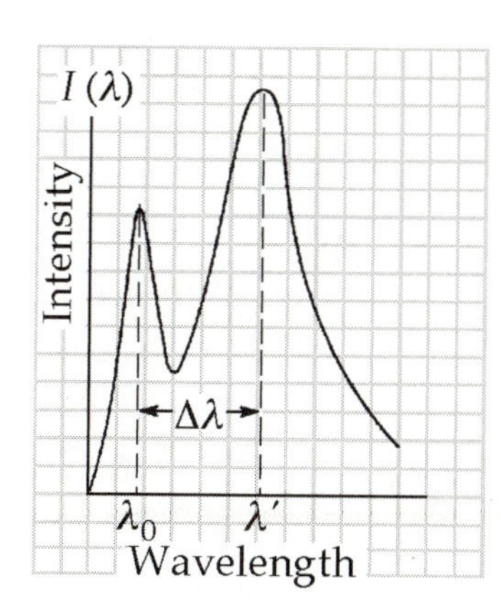


## Section 28.4 Photons and Electromagnetic Waves

The results of some experiments are better described based on the particle nature of light; other experimental outcomes are better described in terms of the wave properties of light. Both aspects of the dual nature of light can be described by the **quantum particle model**.

## Section 28.5 The Wave Properties of Particles

De Broglie postulated that a particle in motion has wave properties and a corresponding wavelength inversely proportional to the particle's momentum.

## Section 28.8 The Uncertainty Principle

It is physically impossible to simultaneously measure the exact position and exact momentum of a particle. The uncertainties in position and momentum are due to the quantum structure of matter.

## Section 28.9 An Interpretation of Quantum Mechanics

The wave function is a complex valued quantity. The square of the absolute value of the wave function gives the probability per volume of finding a particle at a given point in the volume. The wave function contains all the information that can be known about the particle. The **Schrödinger equation** describes the manner in which matter waves change in time and space. The average experimental value of a quantity such as position or energy is called the **expectation value** of the quantity.

## Section 28.10 A Particle in a Box

A particle confined to a line segment and represented by a well-defined de Broglie wave function is represented by a sinusoidal wave. The allowed states of the system are called **stationary states** since they represent **standing waves.** The minimum energy that the particle can have is called the zero-point energy.

## Section 28.12 The Schrödinger Equation

The basic problem in wave mechanics is to determine a solution to the Schrödinger equation. The solution will provide the allowed wave functions and energy levels of the system.

## Section 28.13 Tunneling Through a Potential Energy Barrier

When a particle is incident onto a potential energy barrier, the height of which is greater than the energy of the particle, there is a finite probability that the particle will penetrate the barrier. In this process, called **tunneling**, part of the incident wave is transmitted and part is reflected.

# EQUATIONS AND CONCEPTS

According to the **Wien displacement law,** as the temperature of a blackbody increases the radiation intensity increases and the peak of the distribution shifts to shorter wavelengths. *The total radiation emitted is the area under the curve and increases with temperature.*

$$\lambda_{max} T = 2.898 \times 10^{-3} \text{ m} \cdot \text{K} \tag{28.1}$$

**Discrete energy values** of an atomic oscillator are determined by a quantum number, $n$. Each discrete energy value corresponds to a quantum state.

$$E_n = nhf \quad (28.2)$$

$n = 1, 2, 3, \ldots$

$h = 6.626 \times 10^{-34}\ \text{J} \cdot \text{s}$

The **energy of a quantum** or photon (a discrete unit of energy) corresponds to the energy difference between initial and final quantum states. *An oscillator emits or absorbs energy only when there is a transition between quantum states.*

$$E = hf \quad (28.3)$$

The **maximum kinetic energy of a photoelectron** depends on the frequency of the incident light and the work function of the metal, $\phi$, which is typically a few eV. This model is in excellent agreement with experimental results.

$$K_{\text{max}} = hf - \phi \quad (28.5)$$

**The cutoff wavelength** (and corresponding frequency) depends on the value of the work function of a specific surface; for wavelengths greater than $\lambda_c$ no photoelectric effect will be observed. *The work function represents the minimum energy with which an electron is bound in a metal.*

$$\lambda_c = \frac{hc}{\phi} \quad (28.7)$$

The **Compton shift** is the change in wavelength of an x-ray when scattered from an electron. *The scattered x-ray makes an angle $\theta$ with the direction of the incident x-ray.*

$$\lambda' - \lambda_0 = \frac{h}{m_e c}(1 - \cos\theta) \quad (28.8)$$

$h/m_e c = 0.002\,43$ nm *is the Compton wavelength*

The **de Broglie wavelength** of a particle is inversely proportional to the momentum of the particle.

$$\lambda = \frac{h}{p} = \frac{h}{mv} \quad (28.10)$$

**Matter waves** can be represented by a sinusoidal wave function $\Psi$, which, in general, depends on both position and time. Equation 28.21 is the equation for $\psi$, the part of the wave function which depends only on position.

$$\psi(x) = Ae^{ikx} \quad (28.21)$$

$k = 2\pi / \lambda$ is the angular wave number

$A$ = constant amplitude

The **normalization condition** is a statement of the requirement that the particle exists at some point (along the $x$ axis in the one-dimensional case) at all times.

$$\int_{-\infty}^{\infty} |\psi|^2 \, dx = 1 \qquad (28.23)$$

**The probability, $P_{ab}$, of finding the particle within an arbitrary interval** $a \le x \le b$ equals the area under the curve of $|\psi|^2$ vs $x$, between the end points of the interval.

$$P_{ab} = \int_a^b |\psi|^2 \, dx \qquad (28.24)$$

*The **probability density**, $|\psi|^2$, is the relative probability of finding the particle at a given point along the interval.*

The **expectation value for the position** is the expected average value of the coordinate. *This is average of values of position if calculated from the known wave function. To find the expectation value of any function f(x), replace x in Equation 28.25 with f(x).*

$$\langle x \rangle \equiv \int_{-\infty}^{\infty} \psi^* x \psi \, dx \qquad (28.25)$$

$\psi^*$ = complex conjugate of $\psi$

The **Heisenberg uncertainty principle** can be stated in two forms:

- Simultaneous measurements of **position and momentum** with respective uncertainties $\Delta x$ and $\Delta p_x$.

$$\Delta x \Delta p_x \ge \frac{\hbar}{2} \qquad (28.17)$$

- Simultaneous measurements of **energy and lifetime** with uncertainties $\Delta E$ and $\Delta t$.

$$\Delta E \Delta t \ge \frac{\hbar}{2} \qquad (28.18)$$

For a **particle-in-a-box of width $L$** in one dimensional motion:

- The **wave function** can be represented as a real sinusoidal function.

$$\psi(x) = A \sin\left(\frac{n\pi x}{L}\right) \qquad (28.29)$$

$$(n = 1, 2, 3, \ldots)$$

- The **energy of the particle** is quantized. The ground state corresponds to $n = 1$ and is the lowest energy state.

$$E_n = \left(\frac{h^2}{8mL^2}\right)n^2 \quad (n = 1, 2, 3, \ldots) \tag{28.30}$$

The **time independent Schrödinger equation** for a particle confined to moving along the $x$-axis (total energy $E$ is constant) allows in principle the determination of the wave functions and energies of the allowed states if the potential energy function is known.

$$\frac{-\hbar^2}{2m}\frac{d^2\psi}{dx^2} + U\psi = E\psi \tag{28.31}$$

## REVIEW CHECKLIST

✓ Describe the account of blackbody radiation proposed by Planck.

✓ Describe the Einstein model for the photoelectric effect, and the predictions of the fundamental photoelectric effect equation for the maximum kinetic energy of photoelectrons. Recognize that Einstein's model of the photoelectric effect involves the photon concept ($E = hf$), and that the basic features of the photoelectric effect are consistent with this model.

✓ Describe the Compton effect (the scattering of x-rays by electrons) and be able to use the formula for the Compton shift. Recognize that the Compton effect can only be explained using the photon concept.

✓ Discuss the wave properties of particles, the de Broglie wavelength concept, and the dual nature of both matter and light.

✓ Describe the concept of wave function for the representation of matter waves and state in equation form the normalization condition and expectation value of the coordinate.

✓ Discuss the manner in which the uncertainty principle makes possible a better understanding of the dual wave-particle nature of light and matter.

## ANSWERS TO SELECTED QUESTIONS

**Q28.3** If the photoelectric effect is observed for one metal, can you conclude that the effect will also be observed for another metal under the same conditions? Explain.

**Answer** No. Suppose that the incident light frequency at which you first observed the photoelectric effect is above the cutoff frequency of the first metal, but less than the cutoff frequency of the second metal. In that case, the photoelectric effect would not be observed at all in the second metal.

**Q28.5** Why does the existence of a cutoff frequency in the photoelectric effect favor a particle theory for light over a wave theory?

**Answer** Wave theory predicts that the photoelectric effect should occur at any frequency, provided that the light intensity is high enough. As is implied by the question, this is in contradiction to experimental results.

**Q28.7** An x-ray photon is scattered by an electron. What happens to the frequency of the scattered photon relative to that of the incident photon?

**Answer** The x-ray photon transfers some of its energy to the electron. Thus, its energy, and therefore its frequency, must be decreased.

**Q28.11** If matter has a wave nature, why is this wave-like characteristic not observable in our daily experiences?

**Answer** For any object that we can perceive directly, the de Broglie wavelength $\lambda = \frac{h}{mv}$ is too small to be measured by any means; therefore, no wavelike characteristics can be observed. The object will not diffract noticeably when it goes through an aperture. It will not show resolvable interference maxima and minima when it goes through two openings. It will not show resolvable nodes and antinodes if it is in resonance.

## SOLUTIONS TO SELECTED PROBLEMS

**P28.5** An FM radio transmitter has a power output of 150 kW and operates at a frequency of 99.7 MHz. How many photons per second does the transmitter emit?

**Solution** Each photon has an energy

$$E = hf = \left(6.63\times10^{-34}\ \text{J}\cdot\text{s}\right)\left(99.7\times10^{-6}\ \text{s}^{-1}\right) = 6.61\times10^{-26}\ \text{J}.$$

The number of photons per second is the power divided by the energy per photon:

$$R = \frac{\mathcal{P}}{E} = \frac{150\times10^{3}\ \text{J/s}}{6.61\times10^{-26}\ \text{J}} = 2.27\times10^{30}\ \text{photons/s}. \quad \Diamond$$

**P28.11** Two light sources are used in a photoelectric experiment to determine the work function for a particular metal surface. When green light from a mercury lamp ($\lambda = 546.1$ nm) is used, a stopping potential of 0.376 V reduces the photocurrent to zero.

(a) Based on this measurement, what is the work function for this metal?

(b) What stopping potential would be observed when using the yellow light from a helium discharge tube ($\lambda = 587.5$ nm)?

**Solution** **Conceptualize:** According to Table 28.1, the work function for most metals is on the order of a few eV, so this metal is probably similar. We can expect the stopping potential for the yellow light to be slightly lower than 0.376 V since the yellow light has a longer wavelength (lower frequency) and therefore less energy than the green light.

**Categorize:** In this photoelectric experiment, the green light has sufficient energy $hf$ to overcome the work function of the metal $\phi$ so that the ejected electrons have a maximum kinetic energy of 0.376 eV. With this information, we can use the photoelectric effect equation to find the work function, which can then be used to find the stopping potential for the less energetic yellow light.

**Analyze:**

(a) Einstein's photoelectric effect equation is $K_{\text{max}} = hf - \phi$ and the energy required to raise an electron through a 1-V potential is 1 eV, so that

*continued on next page*

$$K_{\max} = e\Delta V_s = 0.376 \text{ eV}.$$

The energy of a photon from the mercury lamp is:

$$hf = \frac{hc}{\lambda} = \frac{(4.14\times10^{-15} \text{ eV}\cdot\text{s})(3.00\times10^{8} \text{ m/s})}{546.1\times10^{-9} \text{ m}} = 2.27 \text{ eV}.$$

Therefore, the work function for this metal is:

$$\phi = hf - K_{\max} = 2.27 \text{ eV} - 0.376 \text{ eV} = 1.90 \text{ eV}. \quad \Diamond$$

(b) For the yellow light, $\lambda = 587.5$ nm

and
$$hf = \frac{hc}{\lambda} = \frac{(4.14\times10^{-15} \text{ eV}\cdot\text{s})(3.00\times10^{8} \text{ m/s})}{587.5\times10^{-9} \text{ m}} = 2.11 \text{ eV}.$$

Therefore,
$$K_{\max} = hf - \phi = 2.11 \text{ eV} - 1.90 \text{ eV} = 0.216 \text{ eV}$$

so
$$\Delta V_s = 0.216 \text{ V}. \quad \Diamond$$

**Finalize:** The work function for this metal is lower than we expected, and does not correspond with any of the values in Table 28.1. Further examination in the **CRC Handbook of Chemistry and Physics** reveals that all of the metal elements have work functions between 2 and 6 eV. However, a single metals' work function may vary by about 1 eV depending on impurities in the metal, so it is just barely possible that a metal might have a work function of 1.90 eV.

The stopping potential for the yellow light is indeed lower than for the green light as we expected. An interesting calculation is to find the wavelength for the lowest energy light that will eject electrons from this metal. That threshold wavelength giving $K_{\max} = 0$ is 654 nm, which is red light in the visible electromagnetic spectrum.

**P28.17** A 0.001 60-nm photon scatters from a free electron. For what (photon) scattering angle does the recoiling electron have kinetic energy equal to the energy of the scattered photon?

**Solution** The energy of the incoming photon is

$$E_0 = \frac{hc}{\lambda_0} = \frac{(6.63\times10^{-34} \text{ J}\cdot\text{s})(3.00\times10^{8} \text{ m/s})}{0.001\,60\times10^{-9} \text{ m}} = 1.24\times10^{-13} \text{ J}.$$

*continued on next page*

The outgoing photon and the electron share equally in this energy. The kinetic energy of the electron and the energy of the scattered photon are each one half of $E_0$.

$$E' = 6.22\times10^{-14}\text{ J} \qquad \text{and} \qquad \lambda' = \frac{hc}{E'} = 3.20\times10^{-12}\text{ m}$$

The shift in wavelength is $\Delta\lambda = \lambda' - \lambda_0 = 1.60\times10^{-12}$ m.

But by the equation for the Compton shift $\Delta\lambda = \lambda_C(1-\cos\theta)$ where $\lambda_C$ is the Compton wavelength.

Then $$\cos\theta = 1 - \frac{\Delta\lambda}{\lambda_C} = 1 - \frac{1.60\times10^{-12}\text{ m}}{0.002\,43\times10^{-9}\text{ m}} = 0.342$$

and $\theta = 70.0°$. ◊

**P28.23** The nucleus of an atom is on the order of $10^{-14}$ m in diameter. For an electron to be confined to a nucleus, its de Broglie wavelength would have to be on this order of magnitude or smaller.

(a) What would be the kinetic energy of an electron confined to this region?

(b Make also an order-of-magnitude estimate of the electric potential energy of an electron inside an atomic nucleus. Would you expect to find an electron in a nucleus? Explain.

**Solution** **Conceptualize:** The de Broglie wavelength of a normal ground-state orbiting electron is on the order of $10^{-10}$ m (the diameter of a hydrogen atom) so with a shorter wavelength, the electron would have more kinetic energy if confined inside the nucleus. If the kinetic energy is much greater than the potential energy characterizing its attraction with the positive nucleus, then the electron will escape from its electrostatic potential well.

**Categorize:** If we try to calculate the velocity of the electron from the de Broglie wavelength,

we find that $$v = \frac{h}{m_e\lambda} = \frac{6.63\times10^{-34}\text{ J}\cdot\text{s}}{\left(9.11\times10^{-31}\text{ kg}\right)\left(10^{-14}\text{ m}\right)} = 7.27\times10^{10}\text{ m/s}$$

which is not possible since it exceeds the speed of light. Therefore, we must use the relativistic energy expression to find the kinetic energy of this fast-moving electron.

*continued on next page*

**Analyze:**

(a) We find the momentum of the particle:

$$p = \frac{h}{\lambda} = \frac{6.63\times10^{-34}\ \text{J}\cdot\text{s}}{10^{-14}\ \text{m}} = 6.63\times10^{-20}\ \text{N}\cdot\text{s}.$$

We find the particle's relativistic total energy from $E^2 = (pc)^2 + \left(mc^2\right)^2$:

$$E = \sqrt{\left(1.99\times10^{-11}\ \text{J}\right)^2 + \left(8.19\times10^{-14}\ \text{J}\right)^2} = 1.99\times10^{-11}\ \text{J}.$$

Its relativistic kinetic energy is $K = E - mc^2$:

$$K = \frac{1.99\times10^{-11}\ \text{J} - 8.19\times10^{-14}\ \text{J}}{1.60\times10^{-19}\ \text{J/eV}} = 124\ \text{MeV} \sim 100\ \text{MeV}.$$ ◊

(b) The electric potential energy of an electron-proton system with a separation of $10^{-14}$ m is:

$$U = -\frac{k_e e^2}{r} = -\frac{\left(8.99\times10^9\ \text{N}\cdot\text{m}^2/\text{C}^2\right)\left(1.60\times10^{-19}\ \text{C}\right)^2}{10^{-14}\ \text{m}}$$
$$= -2.30\times10^{-14}\ \text{J} \sim -0.1\ \text{MeV}.$$

Since the kinetic energy is nearly 1 000 times greater than the potential energy, the electron would immediately escape the proton's attraction and would not be confined to the nucleus. ◊

**Finalize:** It is also interesting to notice in the above calculations that the rest energy of the electron is negligible compared to the momentum contribution to the total energy.

**P28.25** The resolving power of a microscope depends on the wavelength used. If one wished to "see" an atom, a resolution of approximately $1.00\times10^{-11}$ m would be required.

(a) If electrons are used (in an electron microscope), what minimum kinetic energy is required for the electrons?

(b) If photons are used, what minimum photon energy is needed to obtain the required resolution?

*continued on next page*

**Solution** (a) Since the de Broglie wavelength is $\lambda = \frac{h}{p}$,

$$p_e = \frac{h}{\lambda} = \frac{6.63\times10^{-34}\ \text{J}\cdot\text{s}}{1.00\times10^{-11}\ \text{m}} = 6.63\times10^{-23}\ \text{kg}\cdot\text{m/s}$$

$$K_e = \frac{p_e^2}{2m_e} = \frac{\left(6.63\times10^{-23}\ \text{kg}\cdot\text{m/s}\right)^2}{2\left(9.11\times10^{-31}\ \text{kg}\right)} = 2.41\times10^{-15}\ \text{J} = 15.1\ \text{keV}. \quad \Diamond$$

For better accuracy, you can use the relativistic equation

$$\left(m_e c^2 + K\right)^2 = p_e^2 c^2 + m_e^2 c^4 \text{ to find that } K = 14.9\ \text{keV}. \quad \Diamond$$

(b) For photons:

$$E = hf = \frac{hc}{\lambda} = \frac{\left(6.63\times10^{-34}\ \text{J}\cdot\text{s}\right)\left(3.00\times10^{8}\ \text{m/s}\right)}{1.00\times10^{-11}\ \text{m}} = 1.99\times10^{-14}\ \text{J} = 124\ \text{keV}. \quad \Diamond$$

For the photon, this wavelength $\lambda = 10$ pm is in the x-ray range of the electromagnetic spectrum.

**P28.29** Neutrons traveling at 0.400 m/s are directed through a pair of slits having a 1.00-mm separation. An array of detectors is placed 10.0 m from the slits.

(a) What is the de Broglie wavelength of the neutrons?

(b) How far off axis is the first zero-intensity point on the detector array?

(c) When a neutron reaches a detector, can we say which slit the neutron passed through? Explain.

**Solution** We use the waves in interference model.

(a) $$\lambda = \frac{h}{mv} = \frac{6.63\times10^{-34}\ \text{J}\cdot\text{s}}{\left(1.67\times10^{-27}\ \text{kg}\right)\left(0.400\ \text{m/s}\right)} = 9.93\times10^{-7}\ \text{m} \quad \Diamond$$

(b) The condition for destructive interference in a multiple-slit experiment is

$$d\sin\theta = \left(m + \frac{1}{2}\right)\lambda \text{ with } m = 0 \text{ for the first minimum.}$$

*continued on next page*

Then $\theta = \sin^{-1}\left(\dfrac{\lambda}{2d}\right) = 0.028\ 4°$

$$\frac{y}{L} = \tan\theta\text{: } y = L\tan\theta = (10.0\ \text{m})\tan(0.028\ 4°) = 4.96\ \text{mm}. \qquad \Diamond$$

(c) We cannot say the neutron passed through one slit. We can only say it passed through the pair of slits, as a water wave does to produce an interference pattern. ◊

**P28.31** An electron $(m_e = 9.11\times10^{-31}\ \text{kg})$ and a bullet $(m = 0.020\ 0\ \text{kg})$ each have a velocity with a magnitude of 500 m/s, accurate to within $0.010\ 0\%$. Within what limits could we determine the position of the objects along the direction of the velocity?

**Solution** **Conceptualize:** It seems reasonable that a tiny particle like an electron could be located within a more narrow region than a bigger object like a bullet, but we often find that the realm of the very small does not obey common sense.

**Categorize:** Heisenberg's uncertainty principle can be used to find the uncertainty in position from the uncertainty in the momentum.

**Analyze:** The uncertainty principle states: $\Delta x\Delta p_x \ge \dfrac{\hbar}{2}$

where $\Delta p_x = m\Delta v$ and $\hbar = \dfrac{h}{2\pi}$.

Both the electron and bullet have a velocity uncertainty,

$$\Delta v = (0.000\ 100)(500\ \text{m/s}) = 0.050\ 0\ \text{m/s}.$$

For the electron, the minimum uncertainty in position is

$$\Delta x = \frac{h}{4\pi m\Delta v} = \frac{6.63\times10^{-34}\ \text{J}\cdot\text{s}}{4\pi\left(9.11\times10^{-31}\ \text{kg}\right)(0.050\ 0\ \text{m/s})} = 1.16\ \text{mm}. \qquad \Diamond$$

For the bullet
$$\Delta x = \frac{h}{4\pi m\Delta v} = \frac{6.63\times10^{-34}\ \text{J}\cdot\text{s}}{4\pi(0.020\ 0\ \text{kg})(0.050\ 0\ \text{m/s})} = 5.28\times10^{-32}\ \text{m}. \qquad \Diamond$$

**Finalize:** Our intuition did not serve us well here, since the position of the center of the larger bullet can be determined much more precisely than the electron. Quantum mechanics describes all objects, but the quantum fuzziness in position is too small to observe for the bullet, yet large for the small-mass electron.

**P28.37** A free electron has a wave function $\psi(x) = Ae^{i(5.00\times10^{10}x)}$ where $x$ is in meters. Find

(a) its de Broglie wavelength,

(b) its momentum, and

(c) its kinetic energy in electron volts.

**Solution** (a) The wave function, $\psi(x) = Ae^{i(5\times10^{10}x)} = A\cos(5\times10^{10}x) + iA\sin(5\times10^{10}x)$, will go through one full cycle between $x_1 = 0$ and $(5.00\times10^{10})x_2 = 2\pi$.

The wavelength is then $\lambda = x_2 - x_1 = \dfrac{2\pi}{5.00\times10^{10}\text{ m}^{-1}} = 1.26\times10^{-10}\text{ m}$. ◊

To say the same thing, we can inspect $Ae^{i(5\times10^{10}x)}$ to see that the wave number is $k = 5.00\times10^{10}\text{ m}^{-1}$.

(b) Since $\lambda = \dfrac{h}{p}$, the momentum is

$$p = \frac{h}{\lambda} = \frac{6.63\times10^{-34}\text{ J}\cdot\text{s}}{1.26\times10^{-10}\text{ m}} = 5.28\times10^{-24}\text{ kg}\cdot\text{m/s}.$$ ◊

(c) The electron's kinetic energy is

$$K = \frac{1}{2}mv^2 = \frac{p^2}{2m}: K = \frac{(5.28\times10^{-24}\text{ kg}\cdot\text{m/s})^2}{2(9.11\times10^{-31}\text{ kg})}\left(\frac{1\text{ eV}}{1.60\times10^{-19}\text{ J}}\right) = 95.5\text{ eV}.$$ ◊

Its relativistic total energy is $511\text{ keV} + 95.5\text{ eV}$.

**P28.39** An electron is contained in a one-dimensional box of length 0.100 nm.

(a) Draw an energy level diagram for the electron for levels up to $n = 4$.

(b) Find the wavelengths of all photons than can be emitted by the electron in making downward transitions that could eventually carry it from the $n = 4$ state to the $n = 1$ state.

*continued on next page*

**Solution** (a) We can draw a diagram that parallels our treatment of mechanical waves under boundary conditions. In each standing-wave state, we measure the distance $d$ from one node to another (N to N), and base our solution upon that distance:

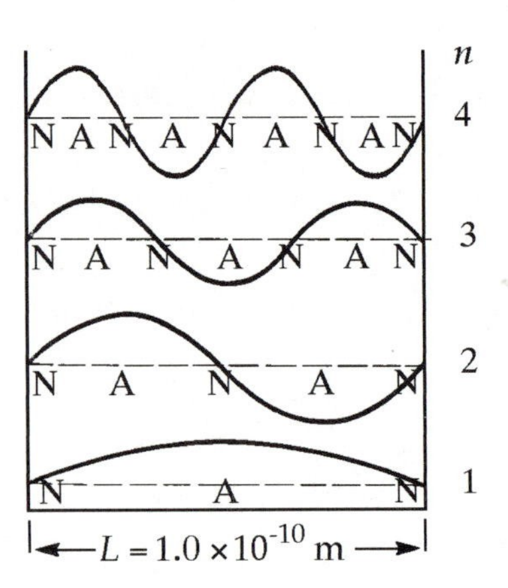


Since $d_{\text{N to N}} = \dfrac{\lambda}{2}$ and $\lambda = \dfrac{h}{p}$, $p = \dfrac{h}{\lambda} = \dfrac{h}{2d}$

Next,

$$K = \frac{p^2}{2m_e} = \frac{h^2}{8m_e d^2} = \frac{1}{d^2}\frac{\left(6.63\times10^{-34}\ \text{J}\cdot\text{s}\right)^2}{8\left(9.11\times10^{-31}\ \text{kg}\right)}$$

Evaluating,

$$K = \frac{6.03\times10^{-38}\ \text{J}\cdot\text{m}^2}{d^2} = \frac{3.77\times10^{-19}\ \text{eV}\cdot\text{m}^2}{d^2}$$

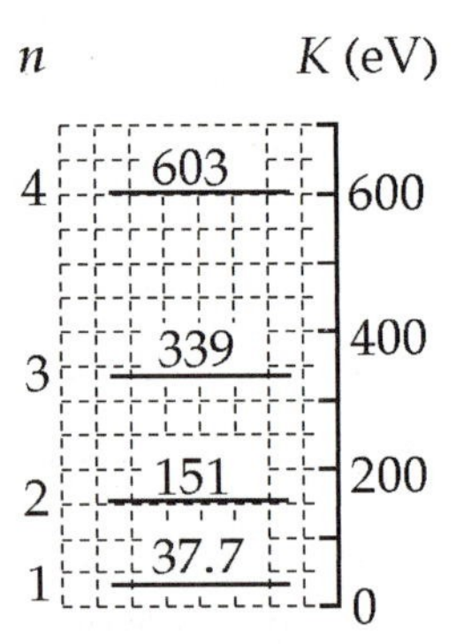


**FIG. P28.39(a)**

In state 1, $d = 1.00\times10^{-10}$ m $\quad K_1 = 37.7$ eV

In state 2, $d = 5.00\times10^{-11}$ m $\quad K_2 = 151$ eV

In state 3, $d = 3.33\times10^{-11}$ m $\quad K_3 = 339$ eV

In state 4, $d = 2.50\times10^{-11}$ m $\quad K_4 = 603$ eV

These energy levels are shown in the diagram to the right. ◊

(b) When the charged, massive electron inside the box makes a downward transition from one energy level to another, a chargeless, massless photon comes out of the box, carrying the difference in energy, $\Delta E$. Its wavelength is

$$\lambda = \frac{c}{f} = \frac{hc}{\Delta E} = \frac{\left(6.63\times10^{-34}\ \text{J}\cdot\text{s}\right)\left(3.00\times10^{8}\ \text{m/s}\right)}{\Delta E\left(1.602\times10^{-19}\ \text{J/eV}\right)} = \frac{1.24\times10^{-6}\ \text{eV}\cdot\text{m}}{\Delta E}.$$

| Transition | $4\to3$ | $4\to2$ | $4\to1$ | $3\to2$ | $3\to1$ | $2\to1$ |
|---|---|---|---|---|---|---|
| $\Delta E$ (eV) | 264 | 452 | 565 | 188 | 302 | 113 |
| Wavelength (nm) | 4.71 | 2.75 | 2.20 | 6.60 | 4.12 | 11.0 |

The wavelengths of light radiated in each transition are given in the table above. ◊

**P28.43** Show that the wave function $\psi = Ae^{i(kx-\omega t)}$ is a solution to the Schrödinger equation (Eq. 28.31) where $k = \frac{2\pi}{\lambda}$ and $U = 0$.

**Solution** From $$\psi = Ae^{i(kx-\omega t)} \quad (1)$$

we evaluate $$\frac{d\psi}{dx} = ikAe^{i(kx-\omega t)}$$

and $$\frac{d^2\psi}{dx^2} = -k^2 Ae^{i(kx-\omega t)}. \quad (2)$$

We substitute Equations (1) and (2) into the Schrödinger equation, so that Eq. 28.31,

$$\frac{d^2\psi}{dx^2} = -\frac{2m}{\hbar^2}(E-U)\psi$$

becomes $$-k^2 Ae^{i(kx-\omega t)} = \left(-\frac{2m}{\hbar^2}K\right)Ae^{i(kx-\omega t)} \quad (3)$$

where $K$ is the kinetic energy. The wave function $\psi = Ae^{i(kx-\omega t)}$ is a solution to the Schrödinger equation if Eq. (3) is true. Both sides depend on $A$, $x$, and $t$ in the same way, so we can divide out several factors and determine that we have a solution if

$$k^2 = \frac{2m}{\hbar^2}K.$$

But this is true for a nonrelativistic particle with mass,

since $$\frac{2m}{\hbar^2}K = \frac{2m}{(h/2\pi)^2}\left(\frac{1}{2}mv^2\right) = \frac{4\pi^2 m^2 v^2}{h^2} = \left(\frac{2\pi p}{h}\right)^2 = \left(\frac{2\pi}{\lambda}\right)^2 = k^2.$$

Therefore, the given wave function does satisfy Equation 28.31. ◊

**P28.49** An electron with kinetic energy $E = 5.00$ eV is incident on a barrier with thickness $L = 0.200$ nm and height $U = 10.0$ eV (Fig. P28.49). What is the probability that the electron

(a) will tunnel through the barrier and

(b) will be reflected?

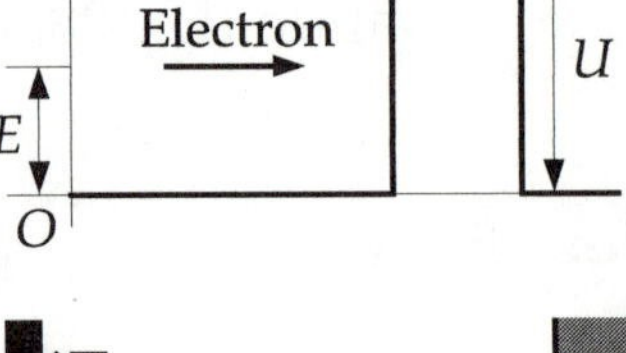


**FIG. P28.49**

**Solution** **Conceptualize:** Since the barrier energy is higher than the kinetic energy of the electron, transmission is not likely, but should be possible since the barrier is not infinitely high or thick.

**Categorize:** The probability of transmission is found from Equations 28.37 and 28.38.

**Analyze:** The decay constant for the wave function inside the barrier is:

$$C = \frac{\sqrt{2m(U-E)}}{\hbar} = \frac{\sqrt{2(9.11\times10^{-31}\text{ kg})(10.0\text{ eV}-5.00\text{ eV})(1.60\times10^{-19}\text{ J/eV})}}{6.63\times10^{-34}\text{ J}\cdot\text{s}/2\pi}$$

$$C = 1.14\times10^{10}\text{ m}^{-1}$$

(a) The probability of transmission is

$$T \approx e^{-2CL} = e^{-2(1.14\times10^{10}\text{ m}^{-1})(2.00\times10^{-10}\text{ m})} = e^{-4.58} = 0.010\,3. \qquad \lozenge$$

(b) If the electron does not tunnel, it is reflected, with probability $1 - 0.010\,3 = 0.990.$ $\lozenge$

**Finalize:** Our expectation was correct; there is only a 1% chance that the electron will penetrate the barrier. This tunneling probability would be greater if the barrier were thinner, shorter, or if the kinetic energy of the electron were greater.

*continued on next page*

**Related Comment:** A typical scanning-tunneling electron microscope (STM) can be built for less than $5 000, and can be tied directly to a home computer.

As shown in the figure above, the STM essentially consists of a needle (1) that is mounted on a few piezo-electric crystals (2). When a voltage is applied across the piezo-electric crystals, the crystals change their shape, and the needle moves.

The STM charges the needle, and moves the tip of the needle to within a few tenths of a nanometer of the test sample (3); the remaining gap provides the energy barrier that is required for tunneling. When the electron cloud of one of the needle's atoms is close to the electron cloud of one of the sample's atoms, a relatively large number of electrons tunnel from one to the other, constituting a current. The current is then measured by a computer. As the needle moves across the surface of the sample, an image of the atoms is generated.

**P28.57** The following table shows data obtained in a photoelectric experiment.

(a) Using these data, make a graph similar to Active Figure 28.9 that plots as a straight line. From the graph, determine

(b) an experimental value for Planck's constant (in joule-seconds) and

(c) the work function (in electron volts) for the surface. (Two significant figures for each answer are sufficient.)

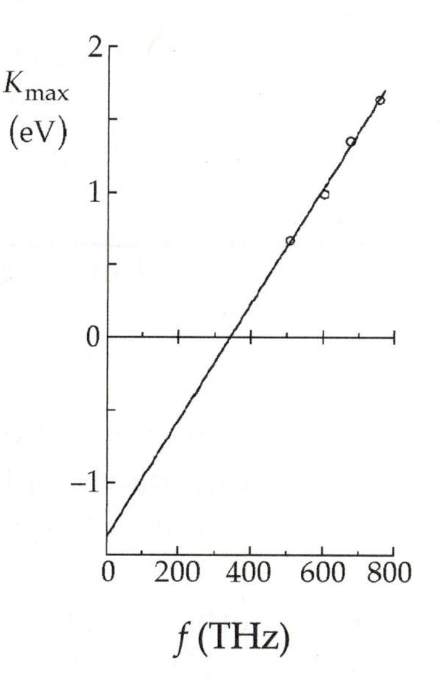


**FIG. P28.57**

| Wavelength (nm) | Maximum Kinetic Energy of Photoelectrons (eV) |
|---|---|
| 588 | 0.67 |
| 505 | 0.98 |
| 445 | 1.35 |
| 399 | 1.63 |

*continued on next page*

**Solution** Convert each wavelength to a frequency using the relation $\lambda f = c$, where $c$ is the speed of light:

$$\lambda_1 = 588 \times 10^{-9}\ \text{m} \qquad f_1 = 5.10 \times 10^{14}\ \text{Hz}$$

$$\lambda_2 = 505 \times 10^{-9}\ \text{m} \qquad f_2 = 5.94 \times 10^{14}\ \text{Hz}$$

$$\lambda_3 = 445 \times 10^{-9}\ \text{m} \qquad f_3 = 6.74 \times 10^{14}\ \text{Hz}$$

$$\lambda_4 = 399 \times 10^{-9}\ \text{m} \qquad f_4 = 7.52 \times 10^{14}\ \text{Hz}$$

(a) Plot each point on an energy vs. frequency graph, as shown at the right. Extend a line through the set of 4 points, as far as the $-y$ intercept. ◊

(b) Our basic equation is $K_{\text{max}} = hf - \phi$. Therefore, Planck's constant should be equal to the slope of the linear $K$-$f$ graph, which can be found from a least-squares fit or from reading the graph as:

$$h_{\text{exp}} = \frac{\text{Rise}}{\text{Run}} = \frac{1.25\ \text{eV} - 0.25\ \text{eV}}{6.5 \times 10^{14}\ \text{Hz} - 4.0 \times 10^{14}\ \text{Hz}} = 4.0 \times 10^{-15}\ \text{eV}\cdot\text{s}$$

$$= 6.4 \times 10^{-34}\ \text{J}\cdot\text{s}$$ ◊

From the scatter of the data points on the graph, we estimate the uncertainty of the slope to be about 3%. Thus we choose to show two significant figures in writing the experimental value of Planck's constant.

(c) From the linear equation $K_{\text{max}} = hf - \phi$, the work function for the metal surface is the negative of the $y$-intercept of the graph, so $\phi_{\text{exp}} = -(-1.4\ \text{eV}) = 1.4\ \text{eV}$. ◊

Based on the range of slopes that appear to fit the data, the estimated uncertainty of the work function is 5%.

**P28.61** An electron is represented by the time-independent wave function

$$\psi(x) = \begin{cases} Ae^{-\alpha x} & \text{for } x > 0 \\ Ae^{+\alpha x} & \text{for } x < 0 \end{cases}$$

(a) Sketch the wave function as a function of $x$.

(b) Sketch the probability density representing the likelihood that the electron is found between $x$ and $x + dx$.

(c) Only an infinite value of potential energy could produce the discontinuity in the derivative of the wave function at $x = 0$. Aside from this feature, argue that $\psi(x)$ can be a physically reasonable wave function.

(d) Normalize the wave function.

(e) Determine the probability of finding the electron somewhere in the range

$$x_1 = -\frac{1}{2\alpha} \qquad \text{to} \qquad x_2 = \frac{1}{2\alpha}.$$

**Solution** (a)

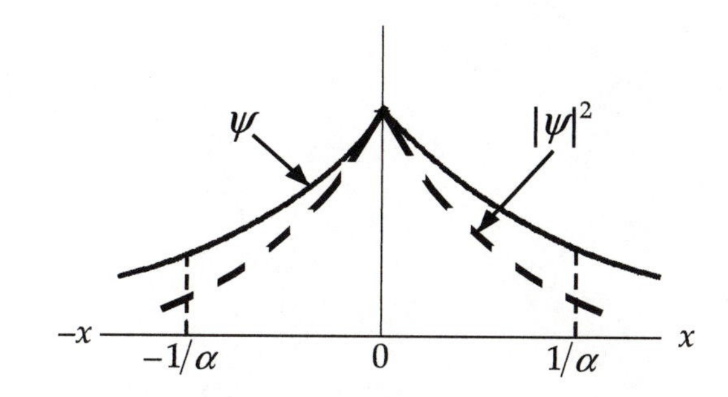


**FIG. P28.61(a)**

(b) For $x > 0$:

$$P|_x^{x+dx} = |\psi|^2 dx = A^2 e^{-2\alpha x} dx. \qquad \Diamond$$

For $x < 0$:

$$P|_x^{x+dx} = |\psi|^2 dx = A^2 e^{+2\alpha x} dx. \qquad \Diamond$$

*continued on next page*

(c) This might be reasonable since (1) $\psi$ is continuous, (2) $\psi \to 0$ as $x \to \pm\infty$, and (3) the wave function can represent an electron bound by an infinitely deep, infinitely narrow potential well at $x = 0$. The wave function's derivative has no single value at the origin, but the wave function need not be differentiable at a point where the potential is infinite. (4) The wave function is integrable and can be normalized, as we show in part (d). ◊

(d) As $\psi$ is symmetric, $\int_{-\infty}^{\infty} |\psi|^2 dx = 2\int_0^{\infty} |\psi|^2 dx = 1$ or $2A^2 \int_0^{\infty} e^{-2\alpha x} dx = 1$.

Integrating, $\left[\frac{2A^2}{-2\alpha}\right]\left[e^{-\infty} - e^0\right] = 1$ gives $A = \sqrt{\alpha}$. ◊

(e) $P\Big|_{-1/2\alpha}^{1/2\alpha} = 2\int_0^{1/2\alpha} \left(\sqrt{\alpha}\right)^2 e^{-2\alpha x} dx = \left[\frac{2\alpha}{-2\alpha}\right]\left[e^{-2\alpha/2\alpha} - 1\right] = \left[1 - e^{-1}\right] = 0.632$ ◊

# Chapter 29

# Atomic Physics

## NOTES FROM SELECTED CHAPTER SECTIONS

### Section 29.1 Early Structural Models of the Atom

A need for modification of the Bohr theory became apparent when improved spectroscopic techniques were used to examine the spectral lines of hydrogen. It was found that many of the lines in the Balmer and other series were not single lines at all. Instead, each was a group of lines spaced very close together. An additional difficulty arose when it was observed that, in some situations, certain single spectral lines were split into three closely spaced lines when the atoms were placed in a strong magnetic field.

### Section 29.2 The Hydrogen Atom Revisited

### Section 29.4 Physical Interpretation of the Quantum Numbers

The possible stationary energy states of an electron in an atom are determined by the values of four quantum numbers.

- Principle quantum number, $n$; integer values from 1 to $\infty$

  The principal quantum number follows from the concept of quantization of angular momentum and specifies the radii of the non-radiating orbits. Electron energy states with the same principal quantum number form a shell identified by the letters K, L, M . . . corresponding to $n = 1, 2, 3 \ldots$

- Orbital quantum number, $\ell$; integer values from 0 to $(n-1)$

  An electron in a given allowed energy state may exist in different elliptical orbits determined by the value of $\ell$. For each value of $n$ there are $n$ possible orbits corresponding to different values of $\ell$. Energy states with given values of $n$ and $\ell$ form a subshell identified by the letters s, p, d, f, . . . corresponding to $\ell = 0, 1, 2, 3, \ldots$. The maximum number of electrons allowed in any subshell is $2(2\ell + 1)$.

- Orbital magnetic quantum number, $m_\ell$, integer values from –1 to +1

  The orbital magnetic quantum number accounts for the observed Zeeman effect; when a gas is placed in an external magnetic field, single spectral lines are split into several lines.

- Spin magnetic quantum number, $m_s$; values of –1/2 and +1/2

  The spin magnetic quantum number $m_s$ accounts for the two closely spaced energy states in spectral lines ("doublets") corresponding to the two possible orientations of electron spin (+1/2 is "up" spin and –1/2 is "down" spin).

## Section 29.5 The Exclusion Principle and the Periodic Table

The **Pauli exclusion principle** states that no two electrons can exist in identical quantum states. This means that no two electrons in a given atom can be characterized by the same set of quantum numbers at the same time.

**Hund's rule** states that when an atom has orbitals of equal energy, the order in which they are filled by electrons is such that a maximum number of electrons will have unpaired spins.

## Section 29.6 More on Atomic Spectra: Visible and X-Ray

An atom will emit electromagnetic radiation if the atom in an excited state makes a transition to a lower energy state. The set of wavelengths observed for a species by such processes is called an **emission spectrum**. Likewise, atoms in the ground-state configuration can also absorb electromagnetic radiation at specific wavelengths, giving rise to an **absorption spectrum.** Such spectra can be used to identify the elements in gases.

Since the orbital angular momentum of an atom changes when a photon is emitted or absorbed (that is, as a result of a transition) and since angular momentum must be conserved, we conclude that **the photon involved in the process must carry angular momentum.**

The x-ray spectrum of a metal target consists of a broad continuous spectrum on which are superimposed a series of sharp lines. **Characteristic x-rays** are emitted by atoms when an electron undergoes a transition from an outer shell into an electron vacancy in one of the inner shells. Transitions into a vacant state in the K shell give rise to the K series of spectral lines, transitions into a vacant state in the L shell create the L series of lines, and so on. **Lines within a series are given a notation to designate the shell from which the transition originates. Examples from the K and L series (in order of increasing energy and decreasing intensity) are:**

- $L_\alpha$ line; an electron drops from the L shell to the K shell
- $L_\beta$ line; an electron drops from the M shell to the K shell
- $K_\alpha$ line; an electron drops from the M shell to the L shell
- $K_\beta$ line; an electron drops from the N shell to the L shell

## EQUATIONS AND CONCEPTS

The **potential energy** of the hydrogen atom (electron-proton system) is negative and depends only on the distance of the electron from the nucleus.

$$U(r) = -k_e \frac{e^2}{r} \qquad (29.1)$$

**Quantized energy level values** can be expressed in units of electron volts (eV). *The lowest allowed energy state or ground state corresponds to the principal quantum number $n = 1$. The absolute value of the ground state energy is equal to the ionization energy of the atom. The energy level approaches $E = 0$ as the electron radius, r approaches infinity.*

$$E_n = -\left(\frac{ke^2}{2a_0}\right)\frac{1}{n^2} = -\frac{13.606}{n^2}\text{eV} \qquad (29.2)$$

$$(n = 1,\ 2,\ 3\ldots)$$

The **ground state wave function for hydrogen** (1s state) depends only on the radial distance $r$. The parameter $a_0$ is the Bohr radius. *All s states have spherical symmetry.*

$$\psi_{1s}(r) = \frac{1}{\sqrt{\pi a_0{}^3}} e^{-r/a_o} \qquad (29.3)$$

$$a_0 = \frac{\hbar^2}{m_e k_e e^2} = 0.052\ 9\ \text{nm}$$

The **radial probability density** for the 1s state of hydrogen is defined as the probability per unit radial distance of finding the electron in a spherical shell of radius $r$ and thickness $dr$. As shown in the figure, the largest value of $P_{1s}(r)$ corresponds to $r = a_0$.

$$P_{1s}(r) = \left(\frac{4r^2}{a_0{}^3}\right) e^{-2r/a_0} \tag{29.7}$$

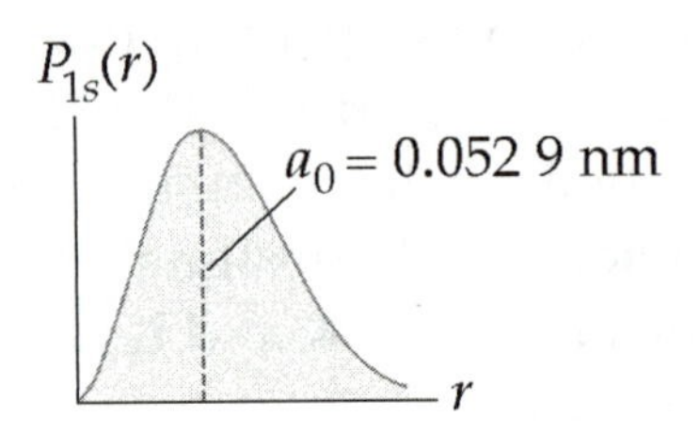


**Quantum numbers** determine the allowed wave functions and energy levels for the hydrogen atom.

The **orbital quantum number** $\ell$ determines discrete values of the magnitude of the angular momentum, $L$. *There are n different possible values of* $\ell$. *All states having the same values of n and* $\ell$ *form a sub-shell.*

$$L = \sqrt{\ell(\ell+1)}\,\hbar \tag{29.9}$$

$$(\ell = 0, 1, 2, 3, \ldots, n-1)$$

The **orbital magnetic quantum number** $m_\ell$ specifies the allowed values of the z component of the orbital momentum vector $\vec{\mathbf{L}}$. *When an atom is placed in an external magnetic field, the vector* $\vec{\mathbf{L}}$ *lies on the surface of a cone and precesses about the z axis.* This effect is referred to as **space quantization**.

$$L_z = m_\ell \hbar \tag{29.10}$$

$$(m_\ell = -\ell, -\ell+1, -\ell+2, \ldots, \ell)$$

The **Zeeman effect** ("splitting" of spectral lines) is due to the quantization of the angle $\theta$ that $\vec{\mathbf{L}}$ makes with the direction of a weak external magnetic field. $\vec{\mathbf{L}}$ can never be parallel or antiparallel to the $z$-axis ($\theta$ cannot be zero).

$$\cos\theta = \frac{L_z}{|\vec{\mathbf{L}}|} = \frac{m_\ell}{\sqrt{\ell(\ell+1)}} \tag{29.11}$$

The **spin quantum number,** $s$, is related to the intrinsic angular momentum or spin angular momentum, $S$, of the electron spinning on its axis.

$$S = \sqrt{s(s+1)}\,\hbar = \frac{\sqrt{3}}{2}\hbar \tag{29.12}$$

$$s = \frac{1}{2}$$

The **spin magnetic quantum number** $m_s$ specifies the space orientation, relative to a $z$ axis, of the spin angular momentum vector.

$$S_z = m_s\hbar = \pm\frac{1}{2}\hbar \tag{29.13}$$

$$m_s = \pm\frac{1}{2}$$

The **spin magnetic moment** $\vec{\mu}_s$ is related to the spin angular momentum.

$$\vec{\mu}_s \triangleq \left(-\frac{e}{m_e}\right)\vec{\mathbf{S}} \tag{29.14}$$

The *z*-**component of the spin magnetic moment** is quantized; there are two allowed values.

$$\mu_{Sz} = \pm\frac{e\hbar}{2m_e} \tag{29.15}$$

**Bohr magneton,** $\mu_B = \dfrac{e\hbar}{2m_e}$

$$\mu_B = 9.274\times10^{-24}\ \text{J/T}$$

The **exclusion principle** dictates that no two electrons in the same atom can have the exact same set of quantum numbers.

**Selection rules** govern the allowed changes in the orbital quantum number and the orbital magnetic quantum number when an electron undergoes a transition between stationary states.

$$\Delta\ell = \pm1 \text{ and } \Delta m_\ell = 0 \text{ or } \pm1 \tag{29.16}$$

**Energy levels of multielectron atoms** are calculated by taking into account the shielding effect of the nuclear charge by inner-core electrons. *The atomic number is replaced by an effective atomic number* $Z_{eff}$*, which depends on the value of* $n$ *and* $\ell$.

$$E_n = -\frac{(13.606\ \text{eV})Z_{\text{eff}}^2}{n^2} \tag{29.18}$$

## SUGGESTIONS, SKILLS, AND STRATEGIES

Before reading this chapter, it may be helpful to review Section 5 of Chapter 11. This section deals with the Bohr model of the atom, and makes it easier to understand the modern concept of the atomic structure.

After reading the chapter in your text, review the significance of each of the quantum numbers that is used to describe the various electronic states of electrons in an atom. Next, review the set of allowed values for each of the quantum numbers.

In addition to the principal quantum number $n$ (which can range from 1 to $\infty$), other quantum numbers are necessary to specify completely the possible energy levels in the hydrogen atom and also in more complex atoms.

All energy states with the same principal quantum number, $n$, form a shell. These shells are identified by the spectroscopic notation K, L, M, . . . corresponding to $n = 1, 2, 3, \ldots$.

The orbital quantum number $\ell$ [which can range from 0 to $(n-1)$], determines the allowed value of orbital angular momentum. All energy states having the same values of $n$ and $\ell$ form a subshell. The letter designations $s, p, d, f, \ldots$ correspond to values of $\ell = 0, 1, 2, 3, \ldots$.

The magnetic orbital quantum number $m_\ell$ (which can range from $-\ell$ to $\ell$) determines the possible orientations of the electron's orbital angular momentum vector in the presence of an external magnetic field.

The spin magnetic quantum number $m_s$, can have only two values, $m_s = -\frac{1}{2}$ and $m_s = \frac{1}{2}$, which in turn correspond to the two possible directions of the electron's intrinsic spin.

## REVIEW CHECKLIST

✓ Understand the significance of the wave function and the associated radial probability density for the ground state of hydrogen.

✓ For each of the quantum numbers, $n$ (the principal quantum number), $\ell$ (the orbital quantum number), $m_\ell$ (the orbital magnetic quantum number), and $m_s$ (the spin magnetic quantum number):

- qualitatively describe what each implies concerning atomic structure;
- state the allowed values which may be assigned to each, and the number of allowed states that may exist in a particular atom corresponding to each quantum number.

✓ Associate the customary shell and subshell spectroscopic notations with allowed combinations of quantum numbers $n$ and $\ell$. Calculate the possible values of the orbital angular momentum, $L$, corresponding to a given value of the principal quantum number.

✓ Describe how allowed values of the magnetic orbital quantum number, $m_\ell$, may lead to a restriction on the orientation of the orbital angular momentum vector in an external magnetic field. Find the allowed values for $L_Z$ (the component of the angular momentum along the direction of an external magnetic field) for a given value of $L$.

✓ State the Pauli exclusion principle and describe its relevance to the periodic table of the elements. Show how the exclusion principle leads to the known electronic ground state configuration of the light elements.

## ANSWERS TO SELECTED QUESTIONS

**Q29.7** Could the Stern-Gerlach experiment be performed with ions rather than neutral atoms? Explain.

**Answer** Practically speaking, the answer would be no. Since ions have a net charge, the magnetic force $q\vec{\mathbf{v}} \times \vec{\mathbf{B}}$ would deflect the beam, making it very difficult to separate ions with different magnetic-moment orientations.

**Q29.9** Discuss some of the consequences of the exclusion principle.

**Answer** If the Pauli exclusion principle were not valid, the elements and their chemical behavior would be grossly different because every electron would end up in the lowest energy level of the atom. All matter would therefore be nearly alike in its chemistry and composition, since the shell structures of each element would be identical. Most materials would have a much higher density, and the spectra of atoms and molecules would be very simple, resulting in the existence of less color in the world.

**Q29.11** Why do lithium, potassium, and sodium exhibit similar chemical properties?

**Answer** The three elements have similar electronic configurations, with filled inner shells plus one electron in an $s$ orbital. Since atoms typically interact through their unfilled outer shells, and the outer shell of each of these atoms is similar, the chemical interactions of the three atoms are also similar.

## SOLUTIONS TO SELECTED PROBLEMS

**P29.1** According to classical physics, a charge $e$ moving with an acceleration $a$ radiates at a rate

$$\frac{dE}{dt} = -\frac{1}{6\pi \in_0}\frac{e^2a^2}{c^3}.$$

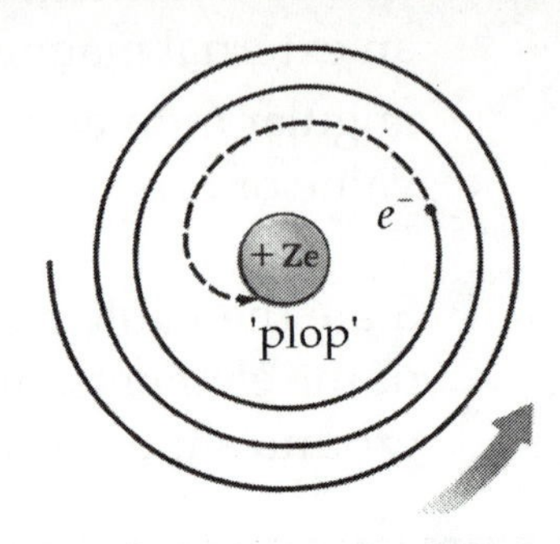


(a) Show that an electron in a classical hydrogen atom (see Figure P29.1) spirals into the nucleus at a rate

$$\frac{dr}{dt} = \frac{-e^4}{12\pi^2 \in_0^2 r^2 m_e^2 c^3}.$$

(b) Find the time interval over which the electron will reach $r = 0$, starting from $r_0 = 2.00 \times 10^{-10}$ m.

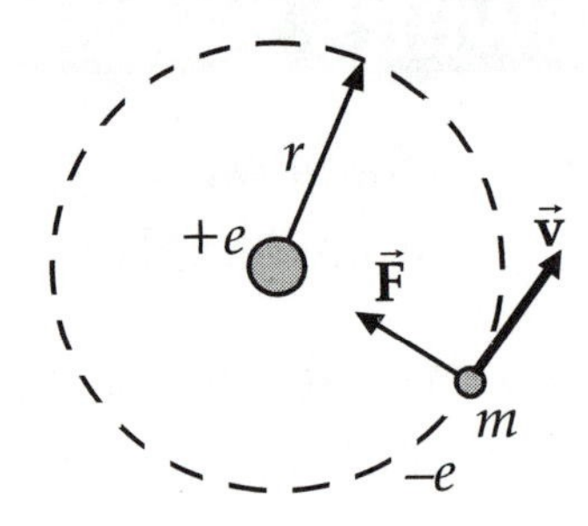


FIG. P29.1

**Solution** According to a classical model, the electron moving as a particle in uniform circular motion about the proton in the hydrogen atom experiences a force $\frac{k_e e^2}{r^2}$; and from Newton's second law, $F = ma$, its acceleration is $\frac{k_e e^2}{m_e r^2}$.

(a) Using the fact that the Coulomb constant

$$k_e = \frac{1}{4\pi \in_0}, \qquad a = \frac{v^2}{r} = \frac{k_e e^2}{m_e r^2} = \frac{e^2}{4\pi \in_0 m_e r^2} \tag{1}$$

From the Bohr structural model of the atom, we can write the total energy of the atom as

$$E = -\frac{k_e e^2}{2r} = -\frac{e^2}{8\pi \in_0 r} \quad \text{so} \quad \frac{dE}{dt} = \frac{e^2}{8\pi \in_0 r^2}\frac{dr}{dt} = -\frac{1}{6\pi \in_0}\frac{e^2a^2}{c^3} \tag{2}$$

continued on next page

Substituting (1) into (2) for $a$, solving for $\frac{dr}{dt}$, and simplifying gives

$$\frac{dr}{dt}=-\frac{4r^2}{3c^3}\left(\frac{e^2}{4\pi\epsilon_0 m_e r^2}\right)^2=-\frac{e^4}{12\pi^2\epsilon_0^2 r^2 m_e^2 c^3}. \quad \Diamond$$

(b) We can express $\frac{dr}{dt}$ in the simpler form:

$$\frac{dr}{dt}=-\frac{A}{r^2}=-\frac{3.15\times10^{-21}}{r^2} \qquad \text{thus} \qquad -\int_{2.00\times10^{-10}\text{ m}}^{0} r^2 dr=3.15\times10^{-21}\int_0^T dt$$

and $$T=\left(3.17\times10^{20}\right)\left.\frac{r^3}{3}\right|_0^{2.00\times10^{-10}\text{ m}}=8.46\times10^{-10}\text{ s}=0.846\text{ ns}. \quad \Diamond$$

We know that atoms last much longer than 0.8 ns; thus, classical physics does not hold (fortunately) for atomic systems.

**P29.7** A general expression for the energy levels of one-electron atoms and ions is

$$E_n=-\frac{\mu k_e^2 q_1^2 q_2^2}{2\hbar^2 n^2}$$

where $k_e$ is the Coulomb constant, $q_1$ and $q_2$ are the charges of the electron and the nucleus, and $\mu$ is the reduced mass of the atom, given by $\mu=\frac{m_1 m_2}{m_1+m_2}$ where $m_1$ is the mass of the electron and $m_2$ is the mass of the nucleus. The wavelength for the $n=3$ to $n=2$ transition of the hydrogen atom is 656.3 nm (visible red light). What are the wavelengths for this same transition in

(a) positronium, which consists of an electron and a positron (*Note*: A positron is a positively charged electron.), and

(b) singly ionized helium?

*continued on next page*

**Solution** **Conceptualize:** The reduced mass of positronium is **less** than hydrogen, so the energy will be less for positronium than for hydrogen. This means that the wavelength of the emitted photon will be longer than 656.3 nm. On the other hand, helium has about the same reduced mass but more charge than hydrogen, so its transition energy will be **larger**, corresponding to a wavelength shorter than 656.3 nm.

**Categorize:** All the factors in the above equation are constant for this problem except for the reduced mass and the nuclear charge. Therefore, the wavelength corresponding to the energy difference for the transition can be found simply from the ratio of mass and charge variables.

**Analyze:** For hydrogen, $\mu = \dfrac{m_p m_e}{m_p + m_e} \approx m_e$.

The photon energy is $\Delta E = E_3 - E_2$.

Its wavelength is $\lambda = 656.3\text{ nm}$, where $\lambda = \dfrac{c}{f} = \dfrac{hc}{\Delta E}$.

(a) For positronium, $\mu = \dfrac{m_e m_e}{m_e + m_e} = \dfrac{m_e}{2}$

so the energy of each level is one half as large as in hydrogen, which we could call "protonium." The photon energy is inversely proportional to its wavelength, so for positronium,

$$\lambda_{32} = 2(656.3\text{ nm}) = 1.31\ \mu\text{m} \qquad \text{(in the infrared region)} \qquad \Diamond$$

(b) For $He^+$, $\mu \approx m_e$, $q_1 = e$, and $q_2 = 2e$, so the transition energy is $2^2 = 4$ times larger than hydrogen.

Then, $$\lambda_{32} = \left(\frac{656}{4}\right)\text{nm} = 164\text{ nm} \qquad \text{(in the ultraviolet region)} \qquad \Diamond$$

**Finalize:** As expected, the wavelengths for positronium and helium are respectively larger and smaller than for hydrogen. Other energy transitions should have wavelength and shifts consistent with this pattern. It is important to remember that the reduced mass is not the total mass, but is generally close in magnitude to the smaller mass of the system (hence the name reduced mass).

**P29.13** For a spherically symmetric state of a hydrogen atom, the Schrödinger equation in spherical coordinates is

$$-\frac{\hbar^2}{2m}\left(\frac{d^2\psi}{dr^2}+\frac{2}{r}\frac{d\psi}{dr}\right)-\frac{k_e e^2}{r}\psi=E\psi.$$

Show that the 1$s$ wave function for hydrogen,

$$\psi_{1s}(r)=\frac{1}{\sqrt{\pi a_0^3}}e^{-r/a_0}$$

satisfies the Schrödinger equation.

**Solution** We use the quantum particle under boundary conditions model. We substitute the wave function and its derivatives into the Schrödinger equation, and then start simplifying the resulting equation. If we find that the resulting equation is true, then we know that the Schrödinger equation is satisfied.

$$\frac{d\psi}{dr}=\frac{1}{\sqrt{\pi a_0^3}}\frac{d}{dr}\left(e^{-r/a_0}\right)=-\frac{1}{\sqrt{\pi a_0^3}}\left(\frac{1}{a_0}\right)e^{-r/a_0}=-\frac{\psi}{a_0} \quad (1)$$

Likewise,
$$\frac{d^2\psi}{dr^2}=-\frac{1}{\sqrt{\pi a_0^5}}\frac{d}{dr}e^{-r/a_0}=\frac{1}{\sqrt{\pi a_0^7}}e^{-r/a_0}=\frac{1}{a_0^2}\psi \quad (2)$$

Substituting (1) and (2) into the Schrödinger equation, and noting that $m$ is the electron's mass $m_e$,

$$-\frac{\hbar^2}{2m_e}\left(\frac{1}{a_0^2}-\frac{2}{a_0 r}\right)\psi-\frac{k_e e^2}{r}\psi=E\psi \quad (3)$$

Substituting $\hbar^2=m_e k_e e^2 a_0$

and canceling $m_e$ and $\psi$, $$-\frac{k_e e^2 a_0}{2}\left(\frac{1}{a_0^2}-\frac{2}{a_0 r}\right)-\frac{k_e e^2}{r}=E.$$

We find that the second and third terms add to zero, leaving the equation that was given for the ground state energy of hydrogen:

$$E=-\frac{k_e e^2}{2a_0}.$$

This is true, so the Schrödinger equation is satisfied. ◊

**P29.19** How many sets of quantum numbers are possible for a hydrogen atom for which

(a) $n = 1$,

(b) $n = 2$,

(c) $n = 3$,

(d) $n = 4$, and

(e) $n = 5$?

Check your results to show that they agree with the general rule that the number of sets of quantum numbers for a shell is equal to $2n^2$.

**Solution** (a) For $n = 1$,

we have $\ell = 0$, $m_\ell = 0$, $m_s = \pm\frac{1}{2}$.

| $n$ | $\ell$ | $m_\ell$ | $m_s$ |
|---|---|---|---|
| 1 | 0 | 0 | −1/2 |
| 1 | 0 | 0 | +1/2 |

This yields $2n^2 = 2(1)^2 = 2$ sets. ◊

(b) For $n = 2$, we have

| $n$ | $\ell$ | $m_\ell$ | $m_s$ |
|---|---|---|---|
| 2 | 0 | 0 | ±1/2 |
| 2 | 1 | −1 | ±1/2 |
| 2 | 1 | 0 | ±1/2 |
| 2 | 1 | +1 | ±1/2 |

This yields $2n^2 = 2(2)^2 = 8$ sets ◊

*continued on next page*

Note that the number is twice the number of $m_\ell$ values. Also, for each $\ell$ there are $(2\ell+1)$ different $m_\ell$ values. Finally, $\ell$ can take on values ranging from 0 to $n-1$.

So the general expression is number $= \sum_0^{n-1} 2(2\ell+1)$.

The series is the sum of an arithmetic progression like $2+6+10+14$.

The sum is $$\sum_0^{n-1} 4\ell + \sum_0^{n-1} 2 = 4\left[\frac{n^2-n}{2}\right] + 2n = 2n^2.$$

(c) $n=3$, $2(1)+2(3)+2(5)=2+6+10=18$: $2n^2=2(3)^2=18$ ◊

(d) $n=4$, $2(1)+2(3)+2(5)+2(7)=32$: $2n^2=2(4)^2=32$ ◊

(e) $n=5$, $32+2(9)=32+18=50$: $2n^2=2(5)^2=50$ ◊

**P29.23** The $\rho^-$ meson has a charge of $-e$, a spin quantum number of 1, and a mass 1 507 times that of the electron. The possible values for its spin magnetic quantum number are –1, 0, and 1. Imagine that the electrons in atoms were replaced by $\rho^-$ mesons. List the possible sets of quantum numbers for $\rho^-$ mesons in the 3*d* subshell.

**Solution** The 3*d* subshell has $\ell=2$, and $n=3$. Also, we have $s=1$. Therefore, we can have $n=3$, $\ell=2$, $m_\ell=-2,-1,0,1,2$, $s=1$, and $m_s=-1,0,1$, leading to the following table:

| $n$ | 3 | 3 | 3 | 3 | 3 | 3 | 3 | 3 | 3 | 3 | 3 | 3 | 3 | 3 | 3 |
|---|---|---|---|---|---|---|---|---|---|---|---|---|---|---|---|
| $\ell$ | 2 | 2 | 2 | 2 | 2 | 2 | 2 | 2 | 2 | 2 | 2 | 2 | 2 | 2 | 2 |
| $m_\ell$ | –2 | –2 | –2 | –1 | –1 | –1 | 0 | 0 | 0 | 1 | 1 | 1 | 2 | 2 | 2 |
| $s$ | 1 | 1 | 1 | 1 | 1 | 1 | 1 | 1 | 1 | 1 | 1 | 1 | 1 | 1 | 1 |
| $m_s$ | –1 | 0 | 1 | –1 | 0 | 1 | –1 | 0 | 1 | –1 | 0 | 1 | –1 | 0 | 1 |

◊

**P29.29** (a) Scanning through Table 29.4 in order of increasing atomic number, note that the electrons fill the subshells in such a way that those subshells with the lowest values of $n+\ell$ are filled first. If two subshells have the same value of $n+\ell$, the one with the lower value of $n$ is filled first. Using these two rules, write the order in which the subshells are filled through $n+\ell=7$.

(b) Predict the chemical valence for the elements that have atomic numbers 15, 47, and 86 and compare your predictions with the actual valences (which may be found in a chemistry text).

**Solution** (a)

| $n+\ell$ | 1 | 2 | 3 | 4 | 5 | 6 | 7 |
|---|---|---|---|---|---|---|---|
| subshell | 1*s* | 2*s* | $2p$, $3s$ | $3p$, $4s$ | $3d$, $4p$, $5s$ | $4d$, $5p$, $6s$ | $4f$, $5d$, $6p$, $7s$ |

◊

(b) The valence of an atom, also called its oxidation number, is the number of electrons it "loses" when the atom is in a molecule. A negative number indicates electron "gain." In stable molecules, atoms tend to be in states with full subshells. Full shells are even more preferred, but this is balanced by a preference for the net charge to be not too great.

$Z=15$: Inner subshells: 1*s*, 2*s*, 2*p* (10 electrons)
Outer subshells: 2 electrons in the 3*s* subshell, and 3 electrons in the 3*p* subshell
Prediction: Valence +5 or –3
Element is phosphorus: Valence –3, +3, or +5 (Prediction works) ◊

$Z=47$: Filled subshells: 1*s*, 2*s*, 2*p*, 3*s*, 3*p*, 4*s*, 3*d*, 4*p*, 5*s* (38 electrons)
Outer subshell: 9 electrons in the 4*d* subshell
Prediction: Valence –1
Element is silver: Valence +1 (Prediction fails) ◊

Note that the ground-state electron configuration of silver disobeys the second rule stated in the problem, by having 10 electrons in the 4*d* subshell and only 1 in the 5*s* subshell.

$Z=86$: Filled subshells: 1*s*, 2*s*, 2*p*, 3*s*, 3*p*, 4*s*, 3*d*, 4*p*, 5*s*, 4*d*, 5*p*, 6*s*, 4*f*, 5*d*, 6*p*
Outer subshell: Full
Prediction: Inert gas with valence 0
Element is radon: Inert gas (prediction works) ◊

**P29.35** Use the method illustrated in Example 29.8 to calculate the wavelength of the x-ray emitted from a molybdenum target ($Z = 42$) when an electron moves from the L shell ($n = 2$) to the K shell ($n = 1$).

**Solution** Following Example 29.8, suppose the electron is originally in the L shell with just one other electron in the K shell between it and the nucleus, so it moves in a field of effective charge $(42-1)e$.

Its energy is then $E_L = -(42-1)^2\left(\frac{13.6\text{ eV}}{4}\right)$.

In its final state we estimate the screened charge holding it in orbit as again $(42-1)e$, so its energy is

$$E_K = -(42-1)^2(13.6\text{ eV}).$$

The photon energy emitted is the difference.

$$E_\gamma = \frac{3}{4}(42-1)^2(13.6\text{ eV}) = 1.71\times10^4\text{ eV} = 2.74\times10^{-15}\text{ J}$$

Then $f = \frac{E}{h} = 4.14\times10^{18}\text{ Hz}$ and $\lambda = \frac{c}{f} = 72.5\text{ pm}$. ◊

**P29.41** A distant quasar is moving away from Earth at such high speed that the blue 434-nm $H_\gamma$ line of hydrogen is observed at 510 nm, in the green portion of the spectrum (Fig. P29.41).

(a) How fast is the quasar receding? You may use the result of Problem 29.39.

(b) Edwin Hubble discovered that all objects outside the local group of galaxies are moving away from us, with speeds proportional to their distances. Hubble's law is expressed as $v = HR$, where Hubble's constant has the approximate value $H = 17\times10^{-3}\text{ m/s}\cdot\text{ly}$. Determine the distance from the Earth to this quasar.

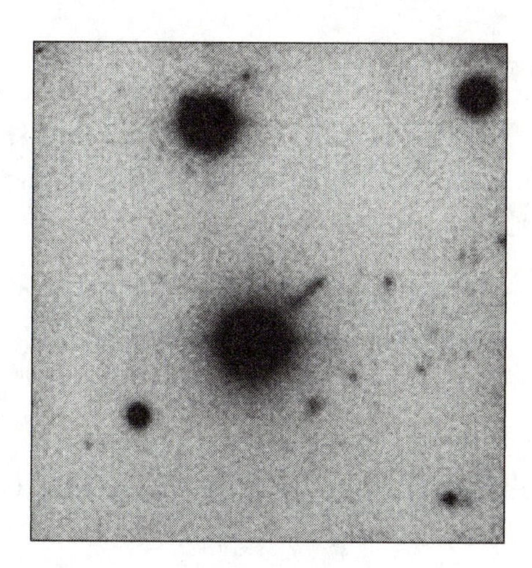

**FIG. P29.41**

*continued on next page*

**Solution** **Conceptualize:** The problem states that the quasar is moving very fast, and since there is a significant red shift of the light, the quasar must be moving away from the Earth at a relativistic speed ($v > 0.1c$). Quasars are very distant astronomical objects, and since our universe is estimated to be about 15 billion years old, we should expect this quasar to be $\sim 10^9$ light-years away.

**Categorize:** As suggested, we can use the equation in Problem 39 to find the speed of the quasar from the Doppler red shift, and this speed can then be used to find the distance using Hubble's law.

**Analyze:**

(a) $$\frac{\lambda'}{\lambda} = \frac{510\text{ nm}}{434\text{ nm}} = 1.18 = \sqrt{\frac{1+v/c}{1-v/c}}$$

Squared this becomes $\frac{1+v/c}{1-v/c} = 1.38$ or $2.38\frac{v}{c} = 0.381$.

Therefore, $v = 0.160c$ (or 16.0% of the speed of light). ◊

(b) Hubble's law asserts that the universe is expanding at a constant rate so that the speeds of galaxies are proportional to their distance $R$ from Earth, $v = HR$.

So, $$R = \frac{v}{H} = \frac{0.160\left(3.00\times10^8\text{ m/s}\right)}{1.70\times10^{-2}\text{ m/s}\cdot\text{ly}} = 2.82\times10^9\text{ ly}.$$ ◊

**Finalize:** The speed and distance of this quasar are consistent with our predictions. It appears that this quasar is quite far from Earth but not the most distant objects in the visible universe.

**P29.45** Show that the average value of $r$ for the 1s state of hydrogen has the value $\frac{3}{2}a_0$. (Suggestion: Use Eq. 29.7.)

**Solution** The average value (expectation value) of $r$ is $$\langle r\rangle = \int_0^\infty rP_{1s}(r)dr$$

where $P_{1s}(r) = \left(\frac{4r^2}{a_0^3}\right)e^{-2r/a_0}$: $$\langle r\rangle = \frac{4}{a_0^3}\int_0^\infty r^3e^{-2r/a_0}dr$$

Letting $x = \frac{2r}{a_0}$ we find: $$\langle r\rangle = \frac{1}{4}a_0\int_0^\infty x^3e^{-x}dx$$

*continued on next page*

In Appendix Table B.6 the integral is tabulated as $\frac{3!}{1^4}$, so we have $\langle r\rangle=\frac{1}{4}a_0 6=\frac{3}{2}a_0$. Alternatively, we can perform the integration by parts. Let $u=x^3$ and $dv=e^{-x}dx$, with $du=3x^2dx$ and $v=-e^{-x}$ to obtain $\int x^3e^{-x}dx=-x^3e^{-x}+3\int x^2e^{-x}dx$. Now let $u=x^2$ and $dv=e^{-x}dx$, with $du=2xdx$ and $v=-e^{-x}$ to get to

$$\int x^3e^{-x}dx=-x^3e^{-x}+3\left[-x^2e^{-x}+2\int xe^{-x}dx\right]=-x^3e^{-x}-3x^2e^{-x}+6\int xe^{-x}dx.$$

Next let $u=x$ and $dv=e^{-x}dx$ to proceed to

$$\begin{aligned}\int x^3e^{-x}dx&=-x^3e^{-x}-3x^2e^{-x}+6\left[-xe^{-x}+\int e^{-x}dx\right]\\&=-x^3e^{-x}-3x^2e^{-x}-6xe^{-x}-6e^{-x}+\text{constant.}\end{aligned}$$

Then $\int_0^\infty x^3e^{-x}dx=-x^3e^{-x}-3x^2e^{-x}-6xe^{-x}-6e^{-x}\Big|_0^\infty$. The exponential function with a growing negative exponent goes to zero so strongly that $\lim_{x\to\infty}x^ne^{-x}=0$, for $n=3, 2, 1, 0$. Also, $e^{-0}=1$, so $\int_0^\infty x^3e^{-x}dx=0+0+0+0-[0+0+0-6]=6$. So again the expectation value of the separation distance between electron and proton in the hydrogen ground state is $\langle r\rangle=\frac{1}{4}a_0 6=\frac{3}{2}a_0$. ◊

**P29.49** Suppose a hydrogen atom is in the 2*s* state, with its wave function given by Equation 29.8. Taking $r=a_0$, calculate values for

(a) $\psi_{2s}(a_0)$,

(b) $|\psi_{2s}(a_0)|^2$, and

(c) $P_{2s}(a_0)$.

**Solution** The wave function for the 2*s* state is given by

$$\psi_{2s}(r)=\frac{1}{4\sqrt{2\pi}}\left(\frac{1}{a_0}\right)^{3/2}\left(2-\frac{r}{a_0}\right)e^{-r/2a_0}.$$

*continued on next page*

(a) Taking $r = a_0 = 0.529 \times 10^{-10}$ m,

we find $$\psi_{2s}(a_0) = \frac{1}{4\sqrt{2\pi}}\left(\frac{1}{0.529 \times 10^{-10}\ \text{m}}\right)^{3/2}(2-1)e^{-1/2} = 1.57 \times 10^{14}\ \text{m}^{-3/2}$$ ◊

(b) $$|\psi_{2s}(a_0)| = \left(1.57 \times 10^{14}\ \text{m}^{-3/2}\right)^2 = 2.47 \times 10^{28}\ \text{m}^{-3}$$ ◊

(c) Using Equation 29.6 and the results to (b) gives

$$P_{2s}(a_0) = 4\pi a_0^2 |\psi_{2s}(a_0)|^2 = 8.69 \times 10^8\ \text{m}^{-1}$$ ◊

**P29.53** (a) Show that the most probable radial position for an electron in the 2*s* state of hydrogen is $r = 5.236a_0$.

(b) Show that the wave function given by Equation 29.8 is normalized.

**Solution** We are using the quantum particle model. We use Equation 29.8:

$$\psi_{2s}(r) = \frac{1}{4\sqrt{2\pi}}\left(\frac{1}{a_0}\right)^{3/2}\left[2 - \frac{r}{a_0}\right]e^{-r/2a_0}.$$

By Equation 29.6, the radial probability distribution function is

$$P(r) = 4\pi r^2 \psi^2 = \frac{1}{8}\left(\frac{r^2}{a_0^3}\right)\left(2 - \frac{r}{a_0}\right)^2 e^{-r/a_0}.$$

(a) Its extremes are given by

$$\frac{dP(r)}{dr} = \frac{1}{8}\left[\frac{2r}{a_0^3}\left(2 - \frac{r}{a_0}\right)^2 - \frac{2r^2}{a_0^3}\left(\frac{1}{a_0}\right)\left(2 - \frac{r}{a_0}\right) - \frac{r^2}{a_0^3}\left(2 - \frac{r}{a_0}\right)^2\left(\frac{1}{a_0}\right)\right]e^{-r/a_0} = 0.$$

We factor in the following manner:

$$\frac{1}{8}\left(\frac{r}{a_0^3}\right)\left(2 - \frac{r}{a_0}\right)\left[2\left(2 - \frac{r}{a_0}\right) - \frac{2r}{a_0} - \frac{r}{a_0}\left(2 - \frac{r}{a_0}\right)\right]e^{-r/a_0} = 0.$$

*continued on next page*

The roots of $\frac{dP}{dr}=0$ at $r=0$, $r=2a_0$, and $r=\infty$ are minima, because $P(r)=0$. Therefore, we focus on the roots given by

$$2\left(2-\frac{r}{a_0}\right)-2\frac{r}{a_0}-\left(\frac{r}{a_0}\right)\left(2-\frac{r}{a_0}\right)=4-\frac{6r}{a_0}+\left(\frac{r}{a_0}\right)^2=0$$

which has solutions $r=\left(3\pm\sqrt{5}\right)a_0$. We substitute the two roots into $P(r)$:

When $\quad r=\left(3-\sqrt{5}\right)a_0=0.764a_0 \quad$ then $\quad P(r)=\frac{0.051\,9}{a_0}$.

When $\quad r=\left(3+\sqrt{5}\right)a_0=5.24a_0 \quad$ then $\quad P(r)=\frac{0.191}{a_0}$.

Therefore, the most probable value of $r$ is $\left(3+\sqrt{5}\right)a_0=5.24a_0$. ◊

(b) The total probability of finding the electron at any distance from the nucleus is

$$\int_0^\infty P(r)dr=\int_0^\infty \frac{1}{8}\left(\frac{r^2}{a_0^3}\right)\left(2-\frac{r}{a_0}\right)^2 e^{-r/a_0}dr.$$

Let $\quad u=\frac{r}{a_0}$ and $dr=a_0du$

so that $\quad \int_0^\infty P(r)dr=\int_0^\infty \frac{1}{8}u^2\left(4-4u+u^2\right)e^{-u}du=\int_0^\infty \frac{1}{8}\left(u^4-4u^3+4u^2\right)e^{-u}du.$

Using a table of integrals, or integrating by parts repeatedly,

$$\int_0^\infty P(r)dr=-\frac{1}{8}\left(u^4+4u^2+8u+8\right)e^{-u}\Big|_0^\infty=1 \text{ as desired.}$$ ◊

**P29.55** An electron in chromium moves from the $n = 2$ state to the $n = 1$ state without emitting a photon. Instead, the excess energy is transferred to an outer electron (one in the $n = 4$ state), which is then ejected by the atom. This phenomenon is called an Auger (pronounced "ohjay") process, and the ejected electron is referred to as an Auger electron. Use the Bohr theory to find the kinetic energy of the Auger electron.

**Solution** The chromium atom with nuclear charge $Z = 24$ starts with one vacancy in the $n = 1$ shell, perhaps produced by the absorption of an x-ray, which ionized the atom. An electron from the $n = 2$ shell tumbles down to fill the vacancy. We suppose that the electron is shielded from the electric field of the full nuclear charge by the one K-shell electron originally below it.

Its change in energy is $\Delta E = -(Z-1)^2(13.6\text{ eV})\left(\frac{1}{1^2} - \frac{1}{2^2}\right) = -5.40\text{ keV}$.

Then $+5.40$ keV can be transferred to the single 4*s* electron. Suppose that it is shielded by the 22 electrons in the K, L, and M shells. To break the outermost electron out of the atom, producing a $Cr^{2+}$ ion, requires an energy investment of

$$E_{\text{ionize}} = \frac{(Z-22)^2(13.6\text{ eV})}{4^2} = \frac{2^2(13.6\text{ eV})}{16} = 3.40\text{ eV}.$$

As evidence that this relatively tiny amount of energy can still be the right order of magnitude, note that the (first) ionization energy for neutral chromium is tabulated as 6.76 eV. Then the remaining energy that can appear as kinetic energy is

$$K = \Delta E - E_{\text{ionize}} = 5\,395.8\text{ eV} - 3.4\text{ eV} = 5.39\text{ keV}. \qquad \lozenge$$

Because of conservation of momentum for the ion-electron system and the tiny mass of the electron compared to that of the $Cr^{2+}$ ion, almost all of this kinetic energy will belong to the electron.

# Chapter 30

# Nuclear Physics

## NOTES FROM SELECTED CHAPTER SECTIONS

### Section 30.1 Some Properties of Nuclei

Important quantities in the description of nuclear properties are:

- The **atomic number**, $Z$, which equals the number of protons in the nucleus.
- The **neutron number**, $N$, which equals the number of neutrons in the nucleus.
- The **mass number**, $A$, which equals the number of nucleons (neutrons plus protons) in the nucleus.

Nuclei can be represented symbolically as ${}_Z^AX$, where X designates the chemical symbol for a specific chemical element. The nuclei of all atoms of a particular element contain the same number of protons but may contain different numbers of neutrons. Nuclei that are related in this way are called isotopes. **The isotopes of an element have the same $Z$ value but different $N$ and $A$ values.** The **unified mass unit, u, is defined such that the mass of the isotope ${}^{12}C$ is exactly 12 u** $(1\text{ u} = 1.660\,559 \times 10^{-27}\text{ kg})$.

Based on nuclear scattering and other experiments, it is possible to conclude that:

- Most nuclei are approximately spherical and all have nearly the same density.
- Nucleons combine to form a nucleus as though they were tightly packed spheres.

- The stability of nuclei is due to a short-range attractive nuclear force.

- The nuclear force is approximately the same for an interaction between any pair of nucleons (p-p, n-n, or p-n).

Nuclei have **intrinsic angular momentum** which is quantized by the **nuclear spin quantum number**, which may be integer or half integer.

The **magnetic moment of the nucleus** is measured in terms of the **nuclear magneton.** When placed in an external magnetic field, nuclear magnetic moments precess with a frequency called the **Larmor precessional frequency.**

## Section 30.2 Binding Energy

The total mass of a nucleus is always less than the sum of the masses of its individual nucleons. This difference in mass is the origin of the nuclear **binding energy** and represents the energy that would be required to separate the nucleus into neutrons and protons

## Section 30.3 Radioactivity

The decay of radioactive substances can be accompanied by the emission of three forms of radiation that vary in ability to penetrate shielding materials:

- **alpha** ($\alpha$) **particles** — Helium nuclei ($^4_2\text{He}$) which can barely penetrate a sheet of paper.

- **beta** ($\beta$) **particles** — Either electrons ($e^-$) or positrons ($e^+$) able to penetrate a few millimeters of aluminum.

- **gamma** ($\gamma$) **rays** — High-energy photons ($\gamma$) capable of penetrating several centimeters of lead.

A positron is the antiparticle of the electron; it has a mass equal to $m_e$ and a charge of $+q_e$.

The decay rate, or activity, $R$, of a sample of a radioactive substance is defined as the number of decays occurring per second. The half life ($T_{1/2}$) of a radioactive substance is the time during which half of an initial number of radioactive nuclei in a sample will decay. The customary unit of radioactivity is the Curie (Ci); the SI unit of activity is the **becquerel** (Bq), where 1 Bq = 1 decay/s and $1 \text{ Ci} = 3.7 \times 10^{10}$ Bq.

## Section 30.4 The Radioactive Decay Processes

The overall decay process can be represented in equation form as:

$$\underset{\text{Parent nucleus}}{X} \rightarrow \underset{\text{Daughter nucleus}}{Y} + \text{emitted radiation}$$

In an equation representing a specific radioactive decay process, the sum of the mass numbers on each side of the equation must be equal and the sum of the atomic numbers on each side of the equation must be equal.

**Alpha decay** can occur because, according to quantum mechanics, some nuclei have energy barriers that can be penetrated by the alpha particles (the tunneling process). Beta decay is energetically more favorable for those nuclei having a large excess of neutrons.

**Beta decay** is accompanied by the emission of either an electron ($e^-$) and an antineutrino ($\bar{\nu}$), or a positron ($e^+$) and a neutrino ($\nu$). The neutrino and antineutrino shown in the $e^-$ and $e^+$ decay processes in the table below are required for conservation of energy and momentum.

The neutrino has the following properties:

- zero electric charge
- rest mass much smaller than that of the electron (recent experiments suggest that the neutrino mass may not be zero).
- spin of $\frac{1}{2}$ (satisfying the law of conservation of angular momentum)
- very weak interaction with matter (therefore very difficult to detect)

In the **electron-capture process**, the nucleus of an atom absorbs one of its own electrons (usually from the K shell) and emits a neutrino.

In **gamma decay**, a nucleus in an excited state decays to its ground state and emits a gamma ray. The disintegration energy is the energy released as a result of the decay process.

The general characteristics of alpha, beta, and gamma decay are summarized below.

| Decay | can be written as | Process |
|---|---|---|
| Alpha decay | ${}^{A}_{Z}\mathrm{X} \rightarrow {}^{A-4}_{Z-2}\mathrm{Y} + {}^{4}_{2}\mathrm{He}$ | The parent nucleus emits an alpha particle and loses two protons and two neutrons. |
| Beta decay | ${}^{A}_{Z}\mathrm{X} \rightarrow {}^{A}_{Z+1}\mathrm{Y} + \mathrm{e}^{-} + \bar{\nu}$ (electron emission) ${}^{A}_{Z}\mathrm{X} \rightarrow {}^{A}_{Z-1}\mathrm{Y} + \mathrm{e}^{+} + \nu$ (positron emission) | A nucleus can undergo beta decay in two ways. The parent nucleus can emit an electron ($\mathrm{e}^{-}$) and an antineutrino ($\bar{\nu}$) or emit a positron ($\mathrm{e}^{+}$) and a neutrino ($\nu$). |
| Gamma decay | ${}^{A}_{Z}\mathrm{X}^{*} \rightarrow {}^{A}_{Z}\mathrm{X} + \gamma$ * parent nucleus in an excited state | A nucleus in an excited state decays to a lower energy state (often the ground state) and emits a gamma ray. |

## Section 30.5 Nuclear Reactions

Nuclear reactions are events in which collisions between nuclei change the structure or properties of nuclei. The **Q-value** is the energy required to balance the overall reaction equation.

**Exothermic reactions release energy and have positive Q-values.**

**The Q-values of endothermic reactions are negative** and the minimum energy of the incoming particle for which an endothermic reaction will occur is called the **threshold energy.**

# EQUATIONS AND CONCEPTS

The **nuclear radius** is proportional to the cube root of the mass number (the total number of nucleons). *Most nuclei are nearly spherical in shape and all nuclei have nearly the same density.*

$$r = r_0 A^{1/3} \quad (30.1)$$

$$r_0 = 1.2 \times 10^{-15}\ \mathrm{m} = 1.2\ \mathrm{fm}$$

The femtometer (fm) sometimes called the **Fermi** is a convenient unit in which to express nuclear dimensions.

$$1\ \mathrm{fm} = 10^{-15}\ \mathrm{m}$$

The **nuclear magneton** $\mu_n$ is the unit of magnetic moment used to express the nuclear magnetic moment; a property associated with the angular momentum of the nucleus.

$$\mu_n \equiv \frac{e\hbar}{2m_p} = 5.05 \times 10^{-27}\ \mathrm{J/T} \quad (30.3)$$

The **binding energy** of any nucleus can be calculated in terms of the mass of a neutral hydrogen atom, $M(\mathrm{H})$, the mass of a neutron, $m_n$ and the atomic mass of an atom of the isotope associated with the compound nucleus, $M\left({}^{A}_{Z}\mathrm{X}\right)$.

$$E_b(\mathrm{MeV}) = \left(ZM(\mathrm{H}) + Nm_n - M\left({}^{A}_{Z}\mathrm{X}\right)\right) \times \left(931.494\ \frac{\mathrm{MeV}}{\mathrm{u}}\right)$$

masses are expressed in atomic mass units, u (30.4)

The **decay rate or activity** (number of decays per second) of a radioactive sample is defined as the number of decays per second. The decay rate is proportional to the number of nuclei present and depends on the **decay constant** which is characteristic of a particular radioactive species.

$$\frac{dN}{dt} = -\lambda N \quad (30.5)$$

$\lambda$ = decay constant

$$R = \left|\frac{dN}{dt}\right| = N_0 \lambda e^{-\lambda t} = R_0 e^{-\lambda t} \quad (30.7)$$

$R_0$ is the decay rate at $t = 0$.

The **exponential decay law** is illustrated in the graph ($N$ vs $t$). The decay constant $\lambda$ is characteristic of a specific isotope and is the probability of decay per nucleus per second. $N_0$ is the number of radioactive nuclei present at time $t = 0$.

$$N = N_0 e^{-\lambda t} \quad (30.6)$$

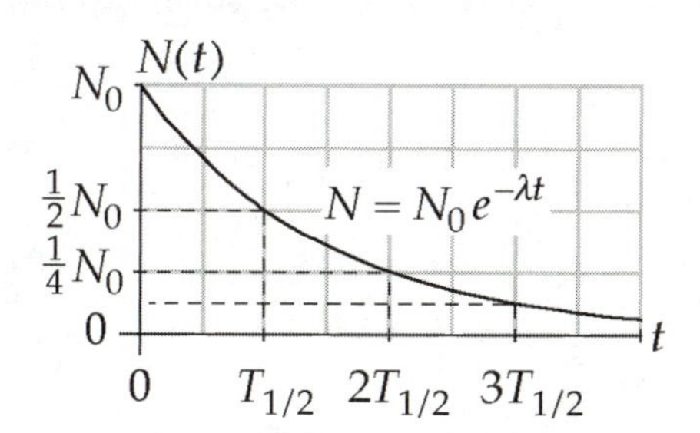


**Units of activity** are the becquerel (Bq) and the curie (Ci). *The becquerel is the SI unit and the curie is approximately the activity of 1 gram of radium.*

$$1\ \mathrm{Ci} \equiv 3.7 \times 10^{10}\ \mathrm{decays/s}$$

$$1\ \mathrm{Bq} = 1\ \mathrm{decay/s}$$

The **half-life** of a radioactive substance, $T_{1/2}$, is the time required for half of a given number of radioactive nuclei in a sample of the substance to decay. *After an elapsed time of n half-lives, the number of radioactive nuclei remaining will be $N_0/2^n$.*

$$T_{1/2} = \frac{\ln 2}{\lambda} = \frac{0.693}{\lambda} \quad (30.8)$$

**Spontaneous decay** of radioactive nuclei proceeds by one of the following processes: alpha decay, beta decay, electron capture, or gamma decay. The overall decay process can be represented as a decay equation. *The sum of the mass numbers on each side of the equation must be equal.*

$$\underset{\text{Parent nucleus}}{X} \rightarrow \underset{\text{Daughter nucleus}}{Y} + \text{Emitted radiation}$$

**Alpha Decay**

When a nucleus decays by alpha emission, the parent nucleus loses two neutrons and two protons. For alpha emission to occur, the mass of the parent nucleus ($^{A}_{Z}X$) must be greater than the combined masses of the daughter nuclei ($^{A-4}_{Z-2}Y$) and the emitted alpha particle. Two examples of alpha decay are shown.

$$^{A}_{Z}\text{X} \rightarrow {}^{A-4}_{Z-2}\text{Y} + {}^{4}_{2}\text{He} \quad (30.9)$$

$$^{238}_{92}\text{U} \rightarrow {}^{234}_{90}\text{Th} + {}^{4}_{2}\text{He} \quad (30.10)$$

$$^{226}_{88}\text{Ra} \rightarrow {}^{222}_{86}\text{Rn} + {}^{4}_{2}\text{He} \quad (30.11)$$

The **disintegration energy** $Q$ of the system (parent nucleus, daughter nucleus, and alpha particle) is required in order to conserve energy during spontaneous decay of the isolated parent nucleus. $Q$ appears in the form of kinetic energy of the daughter nucleus and the alpha particle.

$$Q = (M_X - M_Y - M_\alpha)(931.494 \text{ MeV/u}) \quad (30.13)$$

masses in atomic mass units, u

**Beta decay**

When a radioactive nucleus undergoes beta decay, the process is accompanied by a third particle required to conserve energy and momentum: antineutrino ($\bar{\nu}$) with $e^-$ emission and neutrino ($\nu$) with $e^+$ emission. For each mode of decay, the equations at right show: (i) The equation representing the process, (ii) an example of the decay mode, and (iii) an equation representing the origin of the electron or positron.

**Electron Emission**

(i) $^{A}_{Z}\text{X} \rightarrow {}^{A}_{Z+1}\text{Y} + e^- + \bar{\nu}$ (30.16)

(ii) $^{14}_{6}\text{C} \rightarrow {}^{14}_{7}\text{N} + e^- + \bar{\nu}$ (30.18)

(iii) $n \rightarrow p + e^- + \bar{\nu}$

**Positron Emission**

(i) $^{A}_{Z}\text{X} \rightarrow {}^{A}_{Z-1}\text{Y} + e^+ + \nu$ (30.17)

(ii) $^{12}_{7}\text{N} \rightarrow {}^{12}_{6}\text{C} + e^+ + \nu$ (30.19)

(iii) $p \rightarrow n + e^+ + \nu$

**Electron Capture**
This decay process competes with positron beta decay and occurs when a parent nucleus captures an orbital electron and emits a neutrino. *It is typically the K shell electron that is captured by the parent nucleus.*

$$ {}^{A}_{Z}X + e^{-} \rightarrow {}^{A}_{Z-1}Y + \nu \qquad (30.20)$$

**Gamma Decay**
Nuclei which undergo alpha or beta decay are often left in an excited energy state (indicated by the * symbol). The nucleus returns to the ground state by emission of one or more photons. The excited state of ${}^{12}C^{*}$ is an example of decay by gamma emission. *Gamma decay results in no change in mass number or atomic number.*

$$ {}^{A}_{Z}X^{*} \rightarrow {}^{A}_{Z}X + \gamma \qquad (30.21)$$

$$ {}^{12}_{6}C^{*} \rightarrow {}^{12}_{6}C + \gamma \qquad (30.23)$$

**Nuclear reactions** can occur when target nuclei (X) are bombarded with energetic particles (a) resulting in a daughter or product nucleus (Y) and an outgoing particle (b). The reaction (Eq. 30.32) can be represented in a compact form as shown. *In these reactions the structure, identity, or properties of the target nuclei are changed.*

$$ a + X \rightarrow Y + b \qquad (30.24)$$

$$ X(a, b)Y $$

The **reaction energy** associated with a nuclear reaction is the total change in mass-energy resulting from the reaction.

$$ Q = (M_a + M_X - M_Y - M_b)c^2 \qquad (30.25)$$

The **threshold energy** is the minimum energy required for an endothermic reaction to occur.

$Q > 0$, exothermic reaction
$Q < 0$, endothermic reaction

## SUGGESTIONS, SKILLS, AND STRATEGIES

The rest energy of a particle is given by $E = mc^2$. It is therefore often convenient to express the unified mass unit in terms of its equivalent energy, $1\ \text{u} = 1.660559 \times 10^{-27}$ kg or $1\ \text{u} = 931.494\ \text{MeV}/c^2$. When masses are expressed in units of u, energy values are then $E = m(931.494\ \text{MeV/u})$.

Equation 30.6 can be solved for the particular time $t$ after which the number of remaining nuclei will be some specified fraction of the original number $N_0$. This can be done by taking the natural log of each side of Equation 30.6 to find

$$ t = \frac{1}{\lambda}\ln\left(\frac{N_0}{N}\right). $$

## REVIEW CHECKLIST

✓ Use the appropriate nomenclature in describing the static properties of nuclei.

✓ Discuss nuclear stability in terms of the strong nuclear force and a plot of $N$ vs. $Z$.

✓ Account for nuclear binding energy in terms of the Einstein mass-energy relationship. Describe the basis for energy released by fission and fusion in terms of the shape of the curve of binding energy per nucleon vs. mass number.

✓ Identify each of the components of radiation that are emitted by the nucleus through natural radioactive decay and describe the basic properties of each. Write out generic equations (e.g. Eqs. 30.9, 30.16, 30.20, and 30.19) that illustrate the conservation of nucleon number and charge in the processes of radioactive decay by alpha, beta and gamma emission and by electron capture. Explain why the neutrino must be considered in the analysis of beta decay.

✓ State and apply to the solution of related problems, the formula which expresses decay rate as a function of the decay constant and the number of radioactive nuclei. Describe the process of carbon dating as a means of determining the age of ancient objects.

✓ Calculate the $Q$ value of given nuclear reactions and determine the threshold energy of endothermic reactions.

## ANSWERS TO SELECTED QUESTIONS

**Q30.3** Why do nearly all the naturally occurring isotopes lie above the $N = Z$ line in Figure 30.4?

**Answer** As $Z$ increases, extra neutrons are required to overcome the increasing electrostatic repulsion of the protons.

**Q30.7** Two samples of the same radioactive nuclide are prepared. Sample A has twice the initial activity of sample B. How does the half-life of A compare with the half-life of B? After each has passed through five half-lives, what is the ratio of their activities?

**Answer** Since the two samples are of the same radioactive isotope, they have the same half-life; the 2:1 ratio in activity is due to a 2:1 ratio in the number of radioactive nuclei in each sample. After 5 half lives, each will have decreased in number by a factor of $2^5 = 32$. However, since this simply means that the number of parent nuclei will still be $\left(\frac{2}{32}\right):\left(\frac{1}{32}\right)$, or 2:1. Therefore, the ratio of their activities will **always** be 2:1.

**Q30.9** If a nucleus such as $^{226}Ra$ initially at rest undergoes alpha decay, which has more kinetic energy after the decay, the alpha particle or the daughter nucleus?

**Answer** The alpha particle and the daughter nucleus carry momenta of equal magnitudes in opposite directions. Since kinetic energy can be written as $\frac{p^2}{2m}$, the less massive particle has much more of the decay energy than the recoiling nucleus.

**Q30.11** Suppose it could be shown that the cosmic-ray intensity at the Earth's surface was much greater 10 000 years ago. How would this difference affect what we accept as valid carbon-dated values of the age of ancient samples of once-living matter?

**Answer** If the cosmic ray intensity at the Earth's surface was much greater 10 000 years ago, the fraction of the Earth's carbon dioxide with the heavy nuclide $^{14}C$ would also be greater at that time, than now. Thus, there would initially be a greater fraction of $^{14}C$ in the organic artifacts, and we would believe that the artifact was more recent than it actually was.

For example, suppose that the actual ratio of atmospheric $^{14}C$ to $^{12}C$, two half-lives (11 460 years) ago was $2.6\times10^{-12}$. The current ratio of isotopes in a sample from that time would be

$$\left(\frac{1}{2}\right)\left(\frac{1}{2}\right)\left(2.6\times10^{-12}\right)=0.65\times10^{-12}.$$

We would still measure this ratio today; but, believing the initial ratio to be $1.3\times10^{-12}$, we would think that the creature had died only one half-life (5 730 years) ago.

$$\frac{1}{2}\left(1.3\times10^{-12}\right)=0.65\times10^{-12}.$$

## SOLUTIONS TO SELECTED PROBLEMS

**P30.5** A star ending its life with a mass of two times the mass of the Sun is expected to collapse, combining its protons and electrons to form a neutron star. Such a star could be thought of as a gigantic atomic nucleus. If a star of mass $2\times1.99\times10^{30}$ kg collapsed into neutrons $\left(m_n=1.67\times10^{-27}\text{ kg}\right)$, what would its radius be? (Assume that $r=r_0A^{1/3}$.)

*continued on next page*

**Solution** The number of nucleons in the star is

$$A = \frac{2\left(1.99 \times 10^{30} \text{ kg}\right)}{1.67 \times 10^{-27} \text{ kg}} = 2.38 \times 10^{57}.$$

Therefore, $r = r_0 A^{1/3} = \left(1.20 \times 10^{-15} \text{ m}\right)\left(2.38 \times 10^{57}\right)^{1/3} = 16.0 \text{ km}.$ ◊

**P30.9** A pair of nuclei for which $Z_1 = N_2$ and $Z_2 = N_1$ are called *mirror isobars* (the atomic and neutron numbers are interchanged). Binding-energy measurements on these nuclei can be used to obtain evidence of the charge independence of nuclear forces (that is, proton-proton, proton-neutron, and neutron-neutron nuclear forces are equal). Calculate the difference in binding energy for the two mirror isobars ${}^{15}_{8}\text{O}$ and ${}^{15}_{7}\text{N}$. The electric repulsion among eight protons rather than seven accounts for the difference.

**Solution** For ${}^{15}_{8}\text{O}$ we have (using Equation 30.4)

$$E_b = \left[8(1.007\,825) \text{ u} + 7(1.008\,665) \text{ u} - (15.003\,065) \text{ u}\right](931.494 \text{ MeV/u}) = 111.96 \text{ MeV}$$

For ${}^{15}_{7}\text{N}$ we have

$$E_b = \left[7(1.007\,825) \text{ u} + 8(1.008\,665) \text{ u} - (15.000\,109) \text{ u}\right](931.5 \text{ MeV/u}) = 115.49 \text{ MeV}.$$

Therefore, the difference in the two binding energies is

$\Delta E_b = 3.54 \text{ MeV}.$ ◊

**P30.11** Using the graph in Figure P30.11, estimate how much energy is released when a nucleus of mass number 200 fissions into nuclei each of mass number 100.

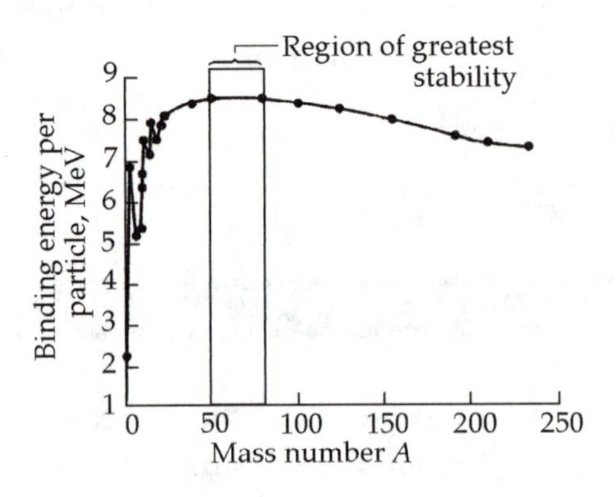


**FIG. P30.11**

**Solution** The curve of binding energy shows that a heavy nucleus of mass number $A = 200$ has total binding energy about

$$(7.8 \text{ MeV/nucleon})(200 \text{ nucleons}) \approx 1.56 \text{ GeV}.$$

*continued on next page*

Thus, it is less stable than its potential fission products, two middleweight nuclei of $A = 100$, having total binding energy

$$2(8.7\text{ MeV/nucleon})(100\text{ nucleons}) \approx 1.74\text{ GeV}.$$

Fission then releases about $1.74\text{ GeV} - 1.56\text{ GeV} \approx 200\text{ MeV}$. ◊

This is the energy source of uranium bombs and of nuclear electric-generating plants.

**P30.13** A freshly prepared sample of a certain radioactive isotope has an activity of 10.0 mCi. After 4.00 h, its activity is 8.00 mCi.

(a) Find the decay constant and half-life.

(b) How many atoms of the isotope were contained in the freshly prepared sample?

(c) What is the sample's activity 30.0 h after it is prepared?

**Solution** **Conceptualize:** Over the course of 4 hours, this isotope lost 20% of its activity, so its half-life appears to be around 10 hours, which means that its activity after 30 hours (~3 half-lives) will be about 1 mCi. The decay constant and number of atoms are not so easy to estimate.

**Categorize:** From the rate equation, $R = R_0 e^{-\lambda t}$, we can find the decay constant $\lambda$, which can then be used to find the half life, the original number of atoms, and the activity at any other time, $t$.

**Analyze:**

(a) $$\lambda = \frac{1}{t}\ln\left(\frac{R_0}{R}\right) = \left(\frac{1}{(4.00\text{ h})(3\,600\text{ s/h})}\right)\ln\left(\frac{10.0\text{ mCi}}{8.00\text{ mCi}}\right) = 1.55\times10^{-5}\text{ s}^{-1}$$ ◊

$$T_{1/2} = \frac{\ln 2}{\lambda} = \frac{0.693}{0.055\,8\text{ h}^{-1}} = 12.4\text{ h}$$ ◊

(b) The number of original atoms can be found if we convert the initial activity from curies into becquerels (decays per second):
$1\text{ Ci} \equiv 3.7\times10^{10}\text{ Bq}$

$$R_0 = 10.0\text{ mCi} = \left(10.0\times10^{-3}\text{ Ci}\right)\left(3.70\times10^{10}\text{ Bq/Ci}\right) = 3.70\times10^{8}\text{ Bq}.$$

*continued on next page*

Since $R_0 = \lambda N_0$, $N_0 = \dfrac{R_0}{\lambda} = \dfrac{3.70\times10^8 \text{ decays/s}}{1.55\times10^{-5}/\text{s}} = 2.39\times10^{13}$ atoms ◊

(c) $R = R_0 e^{-\lambda t} = (10.0 \text{ mCi})e^{-\left(5.58\times10^{-2}\text{ h}^{-1}\right)(30.0\text{ h})} = 1.88 \text{ mCi}$. ◊

**Finalize:** Our estimate of the half life was about 20% short because we did not account for the non-linearity of the decay rate. Consequently, our estimate of the final activity also fell short, but both of these calculated results are close enough to be reasonable.

The number of atoms is much less than one mole, so this appears to be a very small sample. To get a sense of how small, we can assume that the molar mass is about 100 g/mol, so the sample has a mass of only

$$m \approx \frac{\left(2.4\times10^{13}\text{ atoms}\right)(100 \text{ g/mol})}{6.02\times10^{23}\text{ atoms/mol}} \approx 0.004\ \mu\text{g}.$$

This sample is so small it cannot be measured by a commercial mass balance!

The problem states that this sample was "freshly prepared," from which we assumed that **all** the atoms within the sample are initially radioactive. Generally this is not the case, so that $N_0$ only accounts for the formerly radioactive atoms, and does not include additional atoms in the sample that were not radioactive. Realistically then, the sample mass should be significantly greater than our previous estimate.

**P30.19** Find the energy released in the alpha decay ${}^{238}_{92}\text{U} \to {}^{234}_{90}\text{Th} + {}^{4}_{2}\text{He}$. You will find Table A.3 useful.

**Solution** Table A.3 contains the following values:

$$M\left({}^{238}_{92}\text{U}\right) = 238.050\,783 \text{ u}$$

$$M\left({}^{234}_{90}\text{Th}\right) = 234.043\,596 \text{ u}$$

$$M\left({}^{4}_{2}\text{He}\right) = 4.002\,603 \text{ u}$$

We apply to the decay a relativistic energy version of the isolated system model. Some of the rest energy of the parent nucleus is converted into kinetic energy of the decay products according to $Q = \Delta mc^2$.

We calculate $Q = (M_{\text{U}} - M_{\text{Th}} - M_{\text{He}})(931.5 \text{ MeV/u})$

$$Q = (238.050\,783 - 234.043\,596 - 4.002\,603)(931.5) = 4.27 \text{ MeV}$$ ◊

**P30.23** The nucleus $^{15}_{8}\text{O}$ decays by electron capture. The nuclear reaction is written $^{15}_{8}\text{O} + e^- \rightarrow {}^{15}_{7}\text{N} + \nu$.

(a) Write the process going on for a single particle within the nucleus.

(b) Write the decay process referring to neutral atoms.

(c) Determine the energy of the neutrino. Disregard the daughter's recoil.

**Solution** (a) $e^- + p \rightarrow n + \nu$ ◊

(b) Add 7 protons, 7 neutrons, and 7 electrons to each side to give

$$^{15}\text{O atom} \rightarrow {}^{15}\text{N atom} + \nu$$ ◊

(c) From Table A.3,

$$M\left({}^{15}\text{O}\right) = M\left({}^{15}\text{N}\right) + \frac{Q}{c^2}$$

$$\Delta M = 15.003\,065 - 15.000\,109 = 0.002\,956 \text{ u}$$

$$Q = (931.5 \text{ MeV/u})(0.002\,956 \text{ u}) = 2.75 \text{ MeV}$$ ◊

**P30.27** Natural gold has only one isotope, $^{197}_{79}\text{Au}$. If natural gold is irradiated by a flux of slow neutrons, electrons are emitted.

(a) Write the reaction equation.

(b) Calculate the maximum energy of the emitted electrons.

**Solution** The $^{197}_{79}\text{Au}$ will absorb a neutron to become $^{198}_{79}\text{Au}$, which emits an $e^-$ to become $^{198}_{80}\text{Hg}$.

(a) For nuclei the net reaction is:
$^{197}_{79}\text{Au nucleus} + {}^{1}_{0}\text{n} \rightarrow {}^{198}_{80}\text{Hg nucleus} + {}^{0}_{-1}e^- + \bar{\nu}$ ◊

(b) Adding 79 electrons to both sides:

$$^{197}_{79}\text{Au atom} + {}^{1}_{0}\text{n} \rightarrow {}^{198}_{80}\text{Hg atom} + \bar{\nu}$$

From Table A.3, $196.966\,552 + 1.008\,665 = 197.966\,752 + 0 + \frac{Q}{c^2}$

$Q = \Delta mc^2$: $Q = (0.008\,465 \text{ u})(931.5 \text{ MeV/u}) = 7.89 \text{ MeV}$ ◊

**P30.29** **Review problem.** Suppose enriched uranium containing 3.40% of the fissionable isotope $^{235}_{92}\text{U}$ is used as fuel for a ship. The water exerts an average friction force of magnitude $1.00 \times 10^5$ N on the ship. How far can the ship travel per kilogram of fuel? Assume that the energy released per fission event is 208 MeV and that the ship's engine has an efficiency of 20.0%.

**Solution** **Conceptualize:** Nuclear fission is much more efficient for converting mass to energy than burning fossil fuels. However, without knowing the rate of diesel fuel consumption for a comparable ship, it is difficult to estimate the nuclear fuel rate. It seems plausible that a ship could cross the Atlantic ocean with only a few kilograms of nuclear fuel, so a reasonable range of uranium fuel consumption might be 10 km/kg to 10 000 km/kg.

**Categorize:** The fuel consumption rate can be found from the energy released by the nuclear fuel and the work required to push the ship through the water. We use the particle in equilibrium model. The thrust force $P$ exerted on the propeller by the water around it must be equal in magnitude to the backward frictional (drag) force on the hull.

**Analyze:** One kg of enriched uranium contains 3.40% $^{235}_{92}\text{U}$, so

$$m_{235} = 0.034\,0(1\,000 \text{ g}) = 34.0 \text{ g}.$$

In terms of number of nuclei, this is equivalent to

$$N_{235} = (34.0 \text{ g})\left(\frac{1}{235 \text{ g/mol}}\right)\left(6.02 \times 10^{23} \text{ atoms/mol}\right) = 8.71 \times 10^{22} \text{ nuclei}.$$

If all these nuclei fission, the thermal energy released is equal to

$$\left(8.71 \times 10^{22} \text{ nuclei}\right)(208 \text{ MeV/nucleus})\left(1.602 \times 10^{-19} \text{ J/eV}\right) = 2.90 \times 10^{12} \text{ J}.$$

Now, the engine,

$$\text{efficiency} = \frac{\text{work output}}{\text{heat input}} \quad \text{or} \quad e = \frac{P\Delta r \cos\theta}{Q_h}.$$

So the distance the ship can travel per kilogram of uranium fuel is

$$\Delta r = \frac{eQ_h}{P\cos(0)} = \frac{0.200\left(2.90 \times 10^{12} \text{ J}\right)}{1.00 \times 10^5 \text{ N}} = 5.80 \times 10^6 \text{ m}. \quad \Diamond$$

**Finalize:** The ship can travel 5 800 km/kg of uranium fuel, which is on the high end of our prediction range. The distance between New York and Paris is 5 851 km, so this ship could cross the Atlantic ocean on just one kilogram of uranium fuel.

**P30.31** It has been estimated that on the order of $10^9$ tons of natural uranium is available at concentrations exceeding 100 parts per million, of which 0.7% is the fissionable isotope $^{235}\text{U}$. Assume that all the world's energy use $\left(7\times10^{12}\ \text{J/s}\right)$ were supplied by $^{235}\text{U}$ fission in conventional nuclear reactors, releasing 208 MeV for each reaction. How long would the supply last? The estimate of uranium supply is taken from K. S. Deffeyes and I. D. MacGregor, "World Uranium Resources," *Scientific American* **242**(1):66, 1980.

**Solution** The mass of natural uranium reserves is

$$m_{\text{U, total}} = \left(10^9\ \text{metric tons}\right)\left(\frac{10^3\ \text{kg}}{1\ \text{metric ton}}\right) = 1\times10^{12}\ \text{kg}.$$

For fissionable U-235, $m_{\text{U-235}} = 0.007\left(m_{\text{U, total}}\right) = 7\times10^9\ \text{kg} = 7\times10^{12}\ \text{g}.$

In terms of U-235 nuclei, $N_{235} = \left(7\times10^{12}\ \text{g}\right)\left(\dfrac{6.02\times10^{23}\ \text{atoms}}{235\ \text{g}}\right) = 1.79\times10^{34}\ \text{nuclei}.$

Following the statement of the problem, we take the fission energy as 208 MeV/fission:

$$E = \left(1.79\times10^{34}\ \text{nuclei}\right)\left(208\ \text{MeV/fission}\right)\left(1.60\times10^{-19}\ \text{J/eV}\right) = 5.97\times10^{23}\ \text{J}.$$

Now, $\Delta t = \dfrac{E}{\mathcal{P}} = \dfrac{5.97\times10^{23}\ \text{J}}{7\times10^{12}\ \text{J/s}} = 8.53\times10^{10}\ \text{s} \sim 3\,000\ \text{yr}.$ ◊

**P30.35** Consider the two nuclear reactions

$$\text{(I)}\qquad \text{A}+\text{B}\rightarrow\text{C}+\text{E} \qquad \text{(II)}\qquad \text{C}+\text{D}\rightarrow\text{F}+\text{G}$$

(a) Show that the net disintegration energy for these two reactions $\left(Q_{\text{net}} = Q_{\text{I}} + Q_{\text{II}}\right)$ is identical to the disintegration energy for the net reaction.

$$\text{A}+\text{B}+\text{D}\rightarrow\text{E}+\text{F}+\text{G}$$

*continued on next page*

(b) One chain of reactions in the proton-proton cycle in the Sun's core is

$${}^{1}_{1}\mathrm{H} + {}^{1}_{1}\mathrm{H} \rightarrow {}^{2}_{1}\mathrm{H} + {}^{0}_{1}\mathrm{e} + \nu \qquad {}^{0}_{1}\mathrm{e} + {}^{0}_{-1}\mathrm{e} \rightarrow 2\gamma$$

$${}^{1}_{1}\mathrm{H} + {}^{2}_{1}\mathrm{H} \rightarrow {}^{3}_{2}\mathrm{He} + \gamma \qquad {}^{1}_{1}\mathrm{H} + {}^{3}_{2}\mathrm{He} \rightarrow {}^{4}_{2}\mathrm{He} + {}^{0}_{1}\mathrm{e} + \nu$$

$${}^{0}_{1}\mathrm{e} + {}^{0}_{-1}\mathrm{e} \rightarrow 2\gamma$$

Based on part (a), what is $Q_{\text{net}}$ for this sequence?

**Solution** $Q_I = (M_A + M_B - M_C - M_E)c^2 \qquad Q_{\text{II}} = (M_C + M_D - M_F - M_G)c^2$

$$Q_{\text{net}} = (M_A + M_B - M_C - M_E + M_C + M_D - M_F - M_G)c^2$$
$$= (M_A + M_B + M_D - M_E - M_F - M_G)c^2$$

(a) This value is identical to $Q$ for the reaction $\mathrm{A} + \mathrm{B} + \mathrm{D} \rightarrow \mathrm{E} + \mathrm{F} + \mathrm{G}$. Thus, any product (e.g., "C") that is a reactant in a subsequent reaction disappears from the energy balance. ◊

(b) Adding all five reactions, we have $4\left({}^{1}_{1}\mathrm{H}\right) + 2\left({}^{0}_{-1}\mathrm{e}\right) \rightarrow {}^{4}_{2}\mathrm{He} + 2\nu$

Here the symbol ${}^{1}_{1}\mathrm{H}$ represents a proton, the nucleus of a hydrogen-1 atom. So that we may use the tabulated masses of neutral atoms for calculation, we add two electrons to each side of the reaction. The four electrons on the initial side are enough to make four hydrogen atoms and the two electrons on the final side constitute, with the alpha particle, a neutral helium atom. The atomic-electronic binding energies are negligible compared to $Q_{\text{net}}$.

We use the arbitrary symbol "${}^{1}_{1}\mathrm{H}$ atom" to represent a neutral atom. Then we have

$$4\left({}^{1}_{1}\mathrm{H}\ \text{atom}\right) \rightarrow {}^{4}_{2}\mathrm{He}\ \text{atom} + 2\nu$$

$$4(1.007\,825\ \mathrm{u}) = 4.002\,603\ \mathrm{u} + \frac{Q_{\text{net}}}{c^2}$$

$$Q_{\text{net}} = \left[4(1.007\,825\ \mathrm{u}) - 4.002\,603\ \mathrm{u}\right](931.5\ \mathrm{MeV/u}) = 26.7\ \mathrm{MeV}$$ ◊

**P30.49** The decay of an unstable nucleus by alpha emission is represented by Equation 30.9. The disintegration energy $Q$ given by Equation 30.12 must be shared by the alpha particle and the daughter nucleus to conserve both energy and momentum in the decay process.

(a) Show that $Q$ and $K_\alpha$, the kinetic energy of the alpha particle, are related by the expression

$$Q = K_\alpha\left(1 + \frac{M_\alpha}{M}\right)$$

where $M$ is the mass of the daughter nucleus.

(b) Use the result of part (a) to find the energy of the alpha particle emitted in the decay of ${}^{226}\text{Ra}$. (See Example 30.5 for the calculation of $Q$.)

**Solution** For clarity, we will let the subscript $d$ denote the daughter nucleus.

(a) Let us assume that the parent nucleus (mass $M_p$) is initially at rest. We denote the masses of the daughter nucleus and alpha particle by $M_d$ and $M_\alpha$, respectively. Applying the equations of conservation of momentum and energy for the alpha decay process gives

$$M_d v_d = M_\alpha v_\alpha \tag{1}$$

$$M_p c^2 = M_d c^2 + M_\alpha c^2 + \frac{1}{2}M_\alpha v_\alpha^2 + \frac{1}{2}M_d v_d^2. \tag{2}$$

The disintegration energy $Q$ is given by

$$Q = \left(M_p - M_d - M_\alpha\right)c^2 = \frac{1}{2}M_\alpha v_\alpha^2 + \frac{1}{2}M_d v_d^2. \tag{3}$$

Eliminating $v_d$ from Equations (1) and (3) gives

$$Q = \frac{1}{2}M_\alpha v_\alpha^2 + \frac{1}{2}M_d\left(\frac{M_\alpha}{M_d}v_\alpha\right)^2 = \frac{1}{2}M_\alpha v_\alpha^2 + \frac{1}{2}\frac{M_\alpha^2}{M_d}v_\alpha^2$$

$$Q = \frac{1}{2}M_\alpha v_\alpha^2\left(1 + \frac{M_\alpha}{M_d}\right) = K_\alpha\left(1 + \frac{M_\alpha}{M_d}\right)$$ ◊

*continued on next page*

(b) $$K_\alpha = \frac{Q}{1+\frac{M_\alpha}{M_d}} = \frac{4.87 \text{ MeV}}{1+\frac{4}{222}} = 4.78 \text{ MeV}$$ ◊

**P30.51** A small building has become accidentally contaminated with radioactivity. The longest-lived material in the building is strontium-90. ($^{90}_{38}$Sr has an atomic mass 89.907 7 u, and its half-life is 29.1 yr. It is particularly dangerous because it substitutes for calcium in bones.) Assume that the building initially contained 5.00 kg of this substance uniformly distributed throughout the building and that the safe level is defined as less than 10.0 decays/min (to be small in comparison to background radiation). How long will the building be unsafe?

**Solution** The number of nuclei in the original sample is

$$N_0 = \frac{\text{mass present}}{\text{mass of nucleus}} = \frac{5.00 \text{ kg}}{(89.907\,7 \text{ u})(1.66\times10^{-27} \text{ kg/u})} = 3.35\times10^{25} \text{ nuclei}$$

$$\lambda = \frac{\ln 2}{T_{1/2}} = \frac{0.693}{29.1 \text{ yr}} = 2.38\times10^{-2} \text{ yr}^{-1} = 4.53\times10^{-8} \text{ min}^{-1}$$

$$R_0 = \lambda N_0 = \left(4.53\times10^{-8} \text{ min}^{-1}\right)\left(3.35\times10^{25} \text{ nuclei}\right) = 1.52\times10^{18} \text{ counts/min}$$

$$\frac{R}{R_0} = \frac{10.0 \text{ counts/min}}{1.52\times10^{18} \text{ counts/min}} = 6.59\times10^{-18} = e^{-\lambda t}$$

$$t = \frac{-\ln(R/R_0)}{\lambda} = \frac{-\ln\left(6.59\times10^{-18}\right)}{2.38\times10^{-2} \text{ yr}^{-1}} = 1\,660 \text{ yr}$$ ◊

# Chapter 31

# Particle Physics

## NOTES FROM SELECTED CHAPTER SECTIONS

### Section 31.1 The Fundamental Forces in Nature

There are four fundamental forces in nature: **strong** (hadronic), **electromagnetic**, **weak**, and **gravitational**. A residual effect of the strong force is the nuclear force, which acts between nucleons to keep the nucleus together. The weak force is responsible for beta decay. The electromagnetic and weak forces are now considered to be manifestations of a single force called the **electroweak** force.

### Section 31.2 Positrons and Other Antiparticles
### Section 31.3 Mesons and the Beginning of Particle Physics

An antiparticle and a particle have the same mass, but opposite charge. Furthermore, other properties may have opposite values such as lepton number and baryon number. It is possible to produce particle-antiparticle pairs in nuclear reactions if the available energy is greater than $2mc^2$, where $m$ is the mass of the particle.

**Pair production** is a process in which a gamma ray with an energy of at least 1.02 MeV interacts with a nucleus, and an electron-positron pair is created.

**Pair annihilation** is an event in which an electron and a positron can annihilate to produce two gamma rays, each with an energy of at least 0.511 MeV.

A **Feynman diagram** can be used to represent the interaction between two particles.

### Section 31.4 Classification of Particles

**We model the universe as composed of field particles and matter particles.** The exchange of field particles mediates the interactions of the matter particles.

- **Gluons** are the field particles for the strong force.
- **Photons** are exchanged by charged particles in electromagnetic interactions.
- **$W^+$, $W^-$, and Z particles** mediate the weak force.
- **Gravitons** are hypothetical quantum particles of the gravitational field.

The **matter particles** fall into two broad classifications: **hadrons** and **leptons**.

**Hadrons** are particles that interact via all fundamental forces. They are composed of quarks. There are two types of hadrons grouped according to their masses and spins.

- **Mesons** — decay finally into electrons, positrons, neutrinos, and photons; they have spin quantum number 0 or 1.
- **Baryons** — (with the exception of the proton) decay in a manner leading to end products which include a proton; they have spin quantum number $\frac{1}{2}$ or $\frac{3}{2}$.

**Leptons** ($e^-$, $\mu^-$, $\tau^-$, $\nu_e$, $\nu_\mu$, $\nu_\tau$ and their antiparticles) interact by the weak interaction, the electromagnetic interaction (if charged), and (presumably) the gravitational interaction, but not by the strong force. All leptons have spin quantum number $\frac{1}{2}$.

All particles have a corresponding antiparticle. A particle and its antiparticle have the same mass and spin, and have opposites of all other "charges" (electric charge, baryon number, strangeness, etc.). For a particle composed of quarks, its antiparticle is composed of the corresponding antiquarks. There are a few neutral mesons which are their own antiparticle.

## Section 31.5 Conservation Laws
## Section 31.6 Strange Particles and Strangeness

Conservation laws of elementary particles are based on empirical evidence:

- **Baryon number**—Whenever a nuclear reaction or decay occurs, the sum of the baryon numbers before the process must equal the sum of the baryon numbers after the process. This requirement is quantified by assigning a baryon number, $B$ ($B = +1$ for baryons, $B = -1$ for antibaryons, $B = 0$ for all other particles).

- **Lepton number**—In this case there are three conservation laws, quantified by assigning three lepton numbers, one for each variety of lepton: electron-lepton number $L_e$; muon-lepton number, $L_\mu$; and tau-lepton number,
. In each case the sum of the lepton numbers before a reaction or decay must equal the sum of the lepton numbers after the reaction or decay.

- **Strangeness**—Strange particles are always produced in pairs by the strong interaction and decay very slowly, as is characteristic of the weak interaction. Strange particles are assigned a strangeness quantum number, $S$; quantitatively $S = \pm 1, \pm 2, \pm 3$ for strange particles and $S = 0$ for nonstrange particles. Whenever a nuclear reaction or decay occurs via the strong or electromagnetic interactions, the sum of the strangeness numbers before the process must equal the sum of the strangeness numbers after the process ($\Delta S = 0$). Decays occurring via the weak interaction are accompanied by the loss of one strange particle; the process proceeds slowly via the weak interaction and violates the conservation of strangeness.

- **Relativistic total energy**—This quantity is conserved in all reactions, but an apparent fluctuation in energy $\Delta E$ can occur in the creation of a virtual particle provided that the particle exists for a time no longer than $\Delta t = \dfrac{\hbar}{2\Delta E}$.

- **Electric charge**—This scalar quantity is conserved in all reactions.

- **Linear momentum and angular momentum**—These vector quantities are conserved in all reactions.

Integer quantum numbers for strangeness, charm, topness, and bottomness are assigned to baryons. These properties are conserved in interactions involving the strong force, but not necessarily in processes occurring via the weak interaction. These quantum numbers are accounted for in the quark model of hadrons.

## Section 31.9 Quarks

**Quarks and antiquarks** (of six possible types or flavors) make up baryons and mesons; the identity of each particle is determined by the particular combination of quarks. *Each baryon contains three quarks and each meson contains one quark and one antiquark.* The table at right lists the charge and baryon number for each of the six quarks and antiquarks.

| Quark, antiquark | | Charge | Baryon number |
|---|---|---|---|
| Up: | $u, \bar{u}$ | $+\frac{2}{3}e, -\frac{2}{3}e$ | $\frac{1}{3}, -\frac{1}{3}$ |
| Down: | $d, \bar{d}$ | $-\frac{1}{3}e, +\frac{1}{3}e$ | $\frac{1}{3}, -\frac{1}{3}$ |
| Strange: | $s, \bar{s}$ | $-\frac{1}{3}e, +\frac{1}{3}e$ | $\frac{1}{3}, -\frac{1}{3}$ |
| Charm: | $c, \bar{c}$ | $+\frac{2}{3}e, -\frac{2}{3}e$ | $\frac{1}{3}, -\frac{1}{3}$ |
| Bottom: | $b, \bar{b}$ | $-\frac{1}{3}e, +\frac{1}{3}e$ | $\frac{1}{3}, -\frac{1}{3}$ |
| Top: | $t, \bar{t}$ | $+\frac{2}{3}e, -\frac{2}{3}e$ | $\frac{1}{3}, -\frac{1}{3}$ |

The strong force between quarks is referred to as the **color force**, and is mediated by massless particles called **gluons**. **Color charge** (red, green , blue) is a property assigned to quarks, which allows combinations of quarks to satisfy the exclusion principle.

## EQUATIONS AND CONCEPTS

**Pions** are in three varieties corresponding to three charge states: $\pi^+$, $\pi^-$, and $\pi^0$. Pions are very unstable; the equations at right show example decays for each type.

$$\pi^+ \to \pi^0 + e^+ + \nu_e$$

$$\pi^- \to \mu^- + \bar{\nu}_\mu \qquad (31.1)$$

$$\pi^0 \to \gamma + \gamma$$

**Muons** are in two varieties: $\mu^-$ and $\mu^+$ (the antiparticle). Example decay processes are shown.

$$\mu^+ \to e^+ + \nu_e + \bar{\nu}_\mu$$

$$\mu^- \to e^- + \nu_\mu + \bar{\nu}_e$$

**Hubble's law** states a linear relationship between the velocity of a galaxy and its distance $R$ from the Earth. The constant $H$ is called the **Hubble parameter**.

$$v = HR \qquad (31.7)$$

$$H = 17 \times 10^{-3}\ \text{m/(s light-year)}$$

## REVIEW CHECKLIST

✓ Be aware of the four fundamental forces in nature and the corresponding field particles or quanta via which these forces are mediated.

✓ Understand the concepts of the antiparticle, pair production, and pair annihilation.

✓ Know the broad classification of particles and the characteristic properties of the several classes (relative mass value, spin, decay mode).

## ANSWERS TO SELECTED QUESTIONS

**Q31.4** Describe the properties of baryons and mesons and the important differences between them.

**Answer** There are two types of hadrons, called baryons and mesons. Hadrons interact primarily through the strong force and are not elementary particles, being composed of either three quarks (baryons), or a quark and an antiquark (mesons). Baryons have a nonzero baryon number with a spin of either 1/2 or 3/2. Mesons have a baryon number of zero, and a spin of either 0 or 1.

**Q31.8** The $\Xi^0$ particle decays by the weak interaction according to the decay mode $\Xi^0 \rightarrow \Lambda^0 + \pi^0$. Would you expect this decay to be fast or slow? Explain.

**Answer** This decay should be slow, since decays which occur via the weak interaction typically take $10^{-10}$ s or longer to occur.

**Q31.10** Identify the particle decays in Table 31.2 that occur by the electromagnetic interaction. Justify your answers.

**Answer** The decays of the neutral pion, eta, and neutral sigma occur by the electromagnetic interaction. These are three of the shortest lifetimes in Table 31.2. All produce photons, which are the quanta of the electromagnetic force. All conserve strangeness.

**Q31.14** How many quarks are in each of the following:

(a) a baryon,

(b) an antibaryon,

(c) a meson,

(d) an antimeson?

How do you account for the fact that baryons have half-integral spins while mesons have spins of 0 or 1? (*Note*: Quarks have spin 1/2.)

**Answer** All baryons and antibaryons consist of three quarks. All mesons and antimesons consist of two quarks. When one quantum particle with spin 1/2 combines with another particle to form a new particle, the spin of the new particle must be either 1/2 greater or 1/2 less than that of the original. Since quarks have spins of 1/2, it follows that all baryons (which consist of three quarks) must have half-integral spins, and all mesons (which consist of two quarks) must have spins of 0 or 1.

## SOLUTIONS TO SELECTED PROBLEMS

**P31.5** A photon with an energy $E_\gamma = 2.09$ GeV creates a proton-antiproton pair in which the proton has a kinetic energy of 95.0 MeV. What is the kinetic energy of the antiproton? $\left(m_p c^2 = 938.3 \text{ MeV.}\right)$

**Solution** **Conceptualize:** An antiproton has the same mass as a proton, so it seems reasonable to expect that both particles will have similar kinetic energies.

**Categorize:** The total energy of each particle is the sum of its rest energy and its kinetic energy. Conservation of energy requires that the total energy before this pair production event equal the total energy after.

**Analyze:** $$E_\gamma = \left(E_{Rp} + K_p\right) + \left(E_{R\bar{p}} + K_{\bar{p}}\right).$$

The energy of the photon is given as $E_\gamma = 2.09 \text{ GeV} = 2.09 \times 10^3 \text{ MeV}$.

From Table 31.2 or from the problem statement, we see that the rest energy of both the proton and the antiproton is

$$E_{Rp} = E_{R\bar{p}} = m_p c^2 = 938.3 \text{ MeV}.$$

*continued on next page*

If the kinetic energy of the proton is observed to be 95.0 MeV, the kinetic energy of the antiproton is

$$K_{\bar{p}} = E_\gamma - E_{R\bar{p}} - E_{Rp} - K_p = 2.09\times10^3 \text{ MeV} - 2(938.5 \text{ MeV}) - 95.0 \text{ MeV}$$

or $$K_{\bar{p}} = 118 \text{ MeV}.$$ ◊

**Finalize:** The kinetic energy of the antiproton is slightly (~20%) greater than the proton. The magnitude of the momentum of each product particle is less than 1/4 of the gamma ray's momentum. Thus, another particle must have been involved to satisfy momentum conservation. It could be a pre-existing heavy nucleus with which the photon collided. This means that the nucleus must have also carried away some of the energy. For a heavy nucleus this could have been small—as little as 3 MeV for a favorable geometry. The actual value cannot be determined without further information. This extra particle also explains why the energy need not be shared equally between the proton and antiproton.

**P31.7** One of the mediators of the weak interaction is the $Z^0$ boson, with mass $91.2 \text{ GeV}/c^2$. Use this information to find the order of magnitude of the range of the weak interaction.

**Solution** The rest energy of the $Z^0$ boson is $E_0 = 91.2 \text{ GeV}$. The maximum time a virtual $Z^0$ boson can exist is found from

$$\Delta E \Delta t \geq \frac{1}{2}\hbar \quad \text{or} \quad \Delta t \approx \frac{\hbar}{2\Delta E} = \frac{1.055\times10^{-34} \text{ J}\cdot\text{s}}{2(91.2 \text{ GeV})\left(1.60\times10^{-10} \text{ J/GeV}\right)} = 3.61\times10^{-27} \text{ s}.$$

The maximum distance it can travel in this time is

$$d = c\Delta t = \left(3.00\times10^8 \text{ m/s}\right)\left(3.61\times10^{-27} \text{ s}\right) \sim 10^{-18} \text{ m}.$$ ◊

The distance $d$ is an approximate value for the range of the weak interaction.

**P31.9** A neutral pion at rest decays into two photons according to $\pi^0 \to \gamma + \gamma$. Find the energy, momentum, and frequency of each photon.

**Solution** We use the energy and momentum versions of the isolated system model. Since the pion is at rest and momentum is conserved in the decay, the two gamma-rays must have equal amounts of momentum in opposite directions. So, they must share equally in the energy of the pion.

$$m_{\pi^0} = 135.0 \text{ MeV}/c^2 \qquad \text{(Table 31.2)}$$

*continued on next page*

Therefore, $E_\gamma = 67.5 \text{ MeV} = 1.08\times10^{-11} \text{ J}$ ◊

$$p = \frac{E_\gamma}{c} = \frac{67.5 \text{ MeV}}{3.00\times10^8 \text{ m/s}} = 3.60\times10^{-20} \text{ kg}\cdot\text{m/s}$$ ◊

$$f = \frac{E_\gamma}{h} = 1.63\times10^{22} \text{ Hz}$$ ◊

**P31.13** Name one possible decay mode (see Table 31.2) for $\Omega^+$, $\overline{\text{K}}{}^0_\text{S}$, $\overline{\Lambda}{}^0$, and $\overline{\text{n}}$.

**Solution** The particles in this problem are the antiparticles of those listed in Table 31.2. Therefore, the decay modes include the antiparticles of those shown in the decay modes in Table 31.2.

$$\Omega^+ \to \overline{\Lambda}{}^0 + \text{K}^+ \qquad \overline{\text{K}}{}^0_\text{S} \to \pi^+ + \pi^- \left(\text{or } \pi^0 + \pi^0\right)$$

$$\overline{\Lambda}{}^0 \to \overline{\text{p}} + \pi^+ \qquad \overline{\text{n}} \to \overline{\text{p}} + \text{e}^+ + \nu_e$$ ◊

**P31.17** The following reactions or decays involve one or more neutrinos. In each case supply the missing neutrino ($\nu_e$, $\nu_\mu$, or $\nu_\tau$) or antineutrino.

(a) $\pi^- \to \mu^- + ?$

(b) $\text{K}^+ \to \mu^+ + ?$

(c) $? + \text{p}^+ \to \text{n} + \text{e}^+$

(d) $? + \text{n} \to \text{p}^+ + \text{e}^-$

(e) $? + \text{n} \to \text{p}^+ + \mu^-$

(f) $\mu^- \to \text{e}^- + ? + ?$

**Solution** (a) $\pi^- \to \mu^- + \overline{\nu}_\mu$ $\quad L_\mu$: $0 \to 1 - 1$ ◊

(b) $\text{K}^+ \to \mu^+ + \nu_\mu$ $\quad L_\mu$: $0 \to -1 + 1$ ◊

(c) $\overline{\nu}_e + \text{p}^+ \to \text{n} + \text{e}^+$ $\quad L_e$: $-1 + 0 \to 0 - 1$ ◊

(d) $\nu_e + \text{n} \to \text{p}^+ + \text{e}^-$ $\quad L_e$: $1 + 0 \to 0 + 1$ ◊

(e) $\nu_\mu + \text{n} \to \text{p}^+ + \mu^-$ $\quad L_\mu$: $1 + 0 \to 0 + 1$ ◊

(f) $\mu^- \to \text{e}^- + \overline{\nu}_e + \nu_\mu$ $\quad L_\mu$: $1 \to 0 + 0 + 1$ and $L_e$: $0 \to 1 - 1 + 0$ ◊

**P31.19** Determine which of the following reactions can occur. For those that cannot occur, determine the conservation law (or laws) violated:

(a) $p \to \pi^+ + \pi^0$

(b) $p + p \to p + p + \pi^0$

(c) $p + p \to p + \pi^+$

(d) $\pi^+ \to \mu^+ + \nu_\mu$

(e) $n \to p + e^- + \overline{\nu}_e$

(f) $\pi^+ \to \mu^+ + n$

**Solution**

(a) $p \to \pi^+ + \pi^0$ Baryon number is violated: $1 \to 0 + 0$ ◊

(b) $p + p \to p + p + \pi^0$ This reaction can occur. ◊

(c) $p + p \to p + \pi^+$ Baryon number is violated: $1 + 1 \to 1 + 0$ ◊

(d) $\pi^+ \to \mu^+ + \nu_\mu$ This reaction can occur. ◊

(e) $n \to p + e^- + \overline{\nu}_e$ This reaction can occur. ◊

(f) $\pi^+ \to \mu^+ + n$ Violates baryon number: $0 \to 0 + 1$ ◊
and violates muon-lepton number: $0 \to -1 + 0$ ◊

**P31.23** Determine whether or not strangeness is conserved in the following decays and reactions.

(a) $\Lambda^0 \to p + \pi^-$

(b) $\pi^- + p \to \Lambda^0 + K^0$

(c) $\overline{p} + p \to \overline{\Lambda}^0 + \Lambda^0$

(d) $\pi^- + p \to \pi^- + \Sigma^+$

*continued on next page*

(e) $\Xi^- \to \Lambda^0 + \pi^-$

(f) $\Xi^0 \to p + \pi^-$

**Solution** We look up the strangeness quantum numbers in Table 31.2.

(a) $\Lambda^0 \to p + \pi^-$ Strangeness: $-1 \to 0 + 0$
(–1 does not equal 0: strangeness is not conserved) ◊

(b) $\pi^- + p \to \Lambda^0 + K^0$ Strangeness: $0 + 0 \to -1 + 1$
($0 = 0$: strangeness is conserved) ◊

(c) $\overline{p} + p \to \overline{\Lambda}^0 + \Lambda^0$ Strangeness: $0 + 0 \to +1 - 1$
($0 = 0$: strangeness is conserved) ◊

(d) $\pi^- + p \to \pi^- + \Sigma^+$ Strangeness: $0 + 0 \to 0 - 1$
(0 does not equal –1: strangeness is not conserved) ◊

(e) $\Xi^- \to \Lambda^0 + \pi^-$ Strangeness: $-2 \to -1 + 0$
(–2 does not equal –1: strangeness is not conserved) ◊

(f) $\Xi^0 \to p + \pi^-$ Strangeness: $-2 \to 0 + 0$
(–2 does not equal 0: strangeness is not conserved) ◊

**P31.29** If a $K_S^0$ meson at rest decays in $0.900 \times 10^{-10}$ s, how far will a $K_S^0$ meson travel if it is moving at 0.960*c*?

**Solution** The motion of the $K_S^0$ particle is relativistic. Just like the astronaut who leaves for a distant star, and returns to find his family long gone, the kaon appears to us to have a longer lifetime.

That time-dilated lifetime is:

$$t = \gamma t_0 = \frac{0.900 \times 10^{-10}\ \text{s}}{\sqrt{1 - v^2/c^2}} = \frac{0.900 \times 10^{-10}\ \text{s}}{\sqrt{1 - (0.960)^2}} = 3.21 \times 10^{-10}\ \text{s}.$$

During this time, we see the kaon travel at 0.960*c*. It travels for a distance of

$$d = vt = \left[0.960\left(3.00 \times 10^8\ \text{m/s}\right)\right]\left(3.21 \times 10^{-10}\ \text{s}\right) = 0.092\,6\ \text{m} = 9.26\ \text{cm}.$$ ◊

**P31.35** Analyze each reaction in terms of constituent quarks.

(a) $\pi^- + p \rightarrow K^0 + \Lambda^0$

(b) $\pi^+ + p \rightarrow K^+ + \Sigma^+$

(c) $K^- + p \rightarrow K^+ + K^0 + \Omega^-$

(d) $p + p \rightarrow K^0 + p + \pi^+ + ?$

In the last reaction, identify the mystery particle.

**Solution** We look up the quark constituents of the particles in Tables 31.4 and 31.5.

(a) $d\bar{u} + uud \rightarrow d\bar{s} + uds$ nets 1u, 2d, 0s before and after ◊

(b) $\bar{d}u + uud \rightarrow u\bar{s} + uus$ nets 3u, 0d, 0s before and after ◊

(c) $\bar{u}s + uud \rightarrow u\bar{s} + d\bar{s} + sss$ nets 1u, 1d, 1s before and after ◊

(d) $uud + uud \rightarrow d\bar{s} + uud + u\bar{d} + uds$ nets 4u, 2d, 0s before and after ◊

A uds is either a $\Lambda^0$ or a $\Sigma^0$. ◊

**P31.51** The energy flux carried by neutrinos from the Sun is estimated to be on the order of 0.4 $\text{W/m}^2$ at Earth's surface. Estimate the fractional mass loss of the Sun over $10^9$ years due to the emission of neutrinos. (The mass of the Sun is $2 \times 10^{30}$ kg. The Earth-Sun distance is $1.5 \times 10^{11}$ m.)

**Solution** **Conceptualize:** Our Sun is estimated to have a life span of about 10 billion years, so in this problem, we are examining the radiation of neutrinos over a considerable fraction of the Sun's life. However, the mass carried away by the neutrinos is a very small fraction of the total mass involved in the Sun's nuclear fusion process, so even over this long time, the mass of the Sun may not change significantly (probably less than 1%).

**Categorize:** The change in mass of the Sun can be found from the energy flux received by the Earth and Einstein's famous equation, $E = mc^2$.

*continued on next page*

**Analyze:** Since the neutrino flux from the Sun reaching the Earth is 0.4 $\text{W/m}^2$, the total energy emitted per second by the Sun in neutrinos in all directions is

$$\left(0.4\ \text{W/m}^2\right)\left(4\pi r^2\right)=\left(0.4\ \text{W/m}^2\right)\left[4\pi\left(1.5\times10^{11}\ \text{m}\right)^2\right]=1.13\times10^{23}\ \text{W}$$

In a period of $10^9$ yr, the Sun emits a total energy of

$$\left(1.13\times10^{23}\ \text{J/s}\right)\left(10^9\ \text{yr}\right)\left(3.156\times10^7\ \text{s/yr}\right)=3.57\times10^{39}\ \text{J}$$

in the form of neutrinos. This energy corresponds to an annihilated mass of

$$E=m_\nu c^2=3.57\times10^{39}\ \text{J} \qquad \text{so} \qquad m_\nu=3.97\times10^{22}\ \text{kg}$$

Since the Sun has a mass of about $2\times10^{30}$ kg, this corresponds to a loss of only about 1 part in 50 000 000 of the Sun's mass over $10^9$ yr in the form of neutrinos. ◊

**Finalize:** It appears that the neutrino flux changes the mass of the Sun by so little that it would be difficult to measure the difference in mass, even over its lifetime.

**P31.53** Assume that the half-life of free neutrons is 614 s. What fraction of a group of free thermal neutrons with kinetic energy 0.040 0 eV will decay before traveling a distance of 10.0 km?

**Solution** The fraction that will remain is given by the ratio $\dfrac{N}{N_0}$,

where $$\frac{N}{N_0}=e^{-\lambda\Delta t}$$

and $\Delta t$ is the time it takes the neutron to travel a distance of $d=10.0$ km.

We use the particle under constant speed model. The time of flight is given by $\Delta t=\dfrac{d}{v}$.

Since $K=\dfrac{1}{2}mv^2$,
$$\Delta t=\frac{d}{\sqrt{\frac{2K}{m}}}=\frac{10.0\times10^3\ \text{m}}{\sqrt{\dfrac{2(0.040\,0\ \text{eV})\left(1.60\times10^{-19}\ \text{J/eV}\right)}{1.67\times10^{-27}\ \text{kg}}}}=3.61\ \text{s}.$$

*continued on next page*

The decay constant is then $\lambda = \dfrac{0.693}{T_{1/2}}$

$$\lambda = \frac{0.693}{614 \text{ s}} = 1.13 \times 10^{-3} \text{ s}^{-1}.$$

Therefore, $\lambda \Delta t = \left(1.13 \times 10^{-3} \text{ s}^{-1}\right)(3.61 \text{ s})$

$$\lambda \Delta t = 4.08 \times 10^{-3} = 0.004\,08$$

and $\dfrac{N}{N_0} = e^{-\lambda \Delta t} = e^{-0.004\,08} = 0.995\,9.$

Hence, the fraction that has decayed in this time is

$$1 - \frac{N}{N_0} = 0.004\,07 \qquad \text{or} \qquad 0.407\%. \qquad \Diamond$$

**P31.55** Determine the kinetic energies of the proton and pion resulting from the decay of a $\Lambda^0$ at rest: $\Lambda^0 \rightarrow \text{p} + \pi^-$.

**Solution** We first look up the rest energy of each particle:

$$m_\Lambda c^2 = 1\,115.6 \text{ MeV}; \quad m_p c^2 = 938.3 \text{ MeV}; \quad m_\pi c^2 = 139.6 \text{ MeV}.$$

The difference between starting rest energy and final rest energy is the kinetic energy of the products:

$$K_p + K_\pi = (1\,115.6 - 938.3 - 139.6) \text{ MeV} = 37.7 \text{ MeV}.$$

In addition, since momentum is conserved in the decay, $|p_p| = |p_\pi| = p$. We use one symbol for the magnitude of the momentum of each product particle. Applying conservation of relativistic energy:

$$\left[\sqrt{(938.3)^2 + p^2c^2} - 938.3\right] + \left[\sqrt{(139.6)^2 + p^2c^2} - 139.6\right] = 37.7 \text{ MeV}.$$

*continued on next page*

Solving the algebra yields $p_\pi c = p_p c = 100.4\ \text{MeV}$.

Thus, $$K_p = \sqrt{\left(m_p c^2\right)^2 + (100.4)^2} - m_p c^2 = 5.35\ \text{MeV}$$ ◊

and $$K_\pi = \sqrt{(139.6)^2 + (100.4)^2} - 139.6 = 32.3\ \text{MeV}.$$ ◊